Ajith Abraham, Aboul-Ella Hassanien, and
André Ponce de Leon F. de Carvalho (Eds.)

Foundations of Computational Intelligence Volume 4

# Studies in Computational Intelligence, Volume 204

**Editor-in-Chief**

Prof. Janusz Kacprzyk
Systems Research Institute
Polish Academy of Sciences
ul. Newelska 6
01-447 Warsaw
Poland
*E-mail:* kacprzyk@ibspan.waw.pl

Further volumes of this series can be found on our homepage: springer.com

Vol. 181. Georgios Miaoulis and Dimitri Plemenos (Eds.)
*Intelligent Scene Modelling Information Systems,* 2009
ISBN 978-3-540-92901-7

Vol. 182. Andrzej Bargiela and Witold Pedrycz (Eds.)
*Human-Centric Information Processing Through Granular Modelling,* 2009
ISBN 978-3-540-92915-4

Vol. 183. Marco A.C. Pacheco and Marley M.B.R. Vellasco (Eds.)
*Intelligent Systems in Oil Field Development under Uncertainty,* 2009
ISBN 978-3-540-92999-4

Vol. 184. Ljupco Kocarev, Zbigniew Galias and Shiguo Lian (Eds.)
*Intelligent Computing Based on Chaos,* 2009
ISBN 978-3-540-95971-7

Vol. 185. Anthony Brabazon and Michael O'Neill (Eds.)
*Natural Computing in Computational Finance,* 2009
ISBN 978-3-540-95973-1

Vol. 186. Chi-Keong Goh and Kay Chen Tan
*Evolutionary Multi-objective Optimization in Uncertain Environments,* 2009
ISBN 978-3-540-95975-5

Vol. 187. Mitsuo Gen, David Green, Osamu Katai, Bob McKay, Akira Namatame, Ruhul A. Sarker and Byoung-Tak Zhang (Eds.)
*Intelligent and Evolutionary Systems,* 2009
ISBN 978-3-540-95977-9

Vol. 188. Agustín Gutiérrez and Santiago Marco (Eds.)
*Biologically Inspired Signal Processing for Chemical Sensing,* 2009
ISBN 978-3-642-00175-8

Vol. 189. Sally McClean, Peter Millard, Elia El-Darzi and Chris Nugent (Eds.)
*Intelligent Patient Management,* 2009
ISBN 978-3-642-00178-9

Vol. 190. K.R. Venugopal, K.G. Srinivasa and L.M. Patnaik
*Soft Computing for Data Mining Applications,* 2009
ISBN 978-3-642-00192-5

Vol. 191. Zong Woo Geem (Ed.)
*Music-Inspired Harmony Search Algorithm,* 2009
ISBN 978-3-642-00184-0

Vol. 192. Agus Budiyono, Bambang Riyanto and Endra Joelianto (Eds.)
*Intelligent Unmanned Systems: Theory and Applications,* 2009
ISBN 978-3-642-00263-2

Vol. 193. Raymond Chiong (Ed.)
*Nature-Inspired Algorithms for Optimisation,* 2009
ISBN 978-3-642-00266-3

Vol. 194. Ian Dempsey, Michael O'Neill and Anthony Brabazon (Eds.)
*Foundations in Grammatical Evolution for Dynamic Environments,* 2009
ISBN 978-3-642-00313-4

Vol. 195. Vivek Bannore and Leszek Swierkowski
*Iterative-Interpolation Super-Resolution Image Reconstruction:*
*A Computationally Efficient Technique,* 2009
ISBN 978-3-642-00384-4

Vol. 196. Valentina Emilia Balas, János Fodor and Annamária R. Várkonyi-Kóczy (Eds.)
*Soft Computing Based Modeling in Intelligent Systems,* 2009
ISBN 978-3-642-00447-6

Vol. 197. Mauro Birattari
*Tuning Metaheuristics,* 2009
ISBN 978-3-642-00482-7

Vol. 198. Efrén Mezura-Montes (Ed.)
*Constraint-Handling in Evolutionary Optimization,* 2009
ISBN 978-3-642-00618-0

Vol. 199. Kazumi Nakamatsu, Gloria Phillips-Wren, Lakhmi C. Jain, and Robert J. Howlett (Eds.)
*New Advances in Intelligent Decision Technologies,* 2009
ISBN 978-3-642-00908-2

Vol. 200. Dimitri Plemenos and Georgios Miaoulis
*Visual Complexity and Intelligent Computer Graphics Techniques Enhancements,* 2009
ISBN 978-3-642-01258-7

Vol. 201. Aboul-Ella Hassanien, Ajith Abraham, Athanasios V. Vasilakos, and Witold Pedrycz (Eds.)
*Foundations of Computational Intelligence Volume 1,* 2009
ISBN 978-3-642-01081-1

Vol. 202. Aboul-Ella Hassanien, Ajith Abraham, and Francisco Herrera (Eds.)
*Foundations of Computational Intelligence Volume 2,* 2009
ISBN 978-3-642-01532-8

Vol. 203. Ajith Abraham, Aboul-Ella Hassanien, Patrick Siarry, and Andries Engelbrecht (Eds.)
*Foundations of Computational Intelligence Volume 3,* 2009
ISBN 978-3-642-01084-2

Vol. 204. Ajith Abraham, Aboul-Ella Hassanien, and André Ponce de Leon F. de Carvalho (Eds.)
*Foundations of Computational Intelligence Volume 4,* 2009
ISBN 978-3-642-01087-3

Ajith Abraham, Aboul-Ella Hassanien, and
André Ponce de Leon F. de Carvalho (Eds.)

# Foundations of Computational Intelligence Volume 4

Bio-Inspired Data Mining

 Springer

Dr. Ajith Abraham
Machine Intelligence Research Labs
(MIR Labs)
Scientific Network for Innovation
and Research Excellence
P.O. Box 2259 Auburn,
Washington 98071-2259
USA
E-mail: ajith.abraham@ieee.org
http://www.mirlabs.org
http://www.softcomputing.net

Prof. André Ponce de Leon F. de
Carvalho
Department of Computer Science
University of São Paulo
SCE - ICMSC - USP
Caixa Postal 668
13560-970 Sao Carlos, SP
Brazil
E-mail: andre@icmc.usp.br

Prof. Aboul-Ella Hassanien
College of Business Administration
Quantitative and Information System
Department
Kuwait University
P.O. Box 5486
Safat, 13055
Kuwait
E-mail: abo@cba.edu.kw

ISBN 978-3-642-10166-3          e-ISBN 978-3-642-01088-0

DOI 10.1007/978-3-642-01088-0

Studies in Computational Intelligence          ISSN 1860949X

*Typeset & Cover Design:* Scientific Publishing Services Pvt. Ltd., Chennai, India.

Printed in acid-free paper

9 8 7 6 5 4 3 2 1

springer.com

# Preface

## Foundations of Computational Intelligence

### Volume 4: Bio-Inspired Data Mining Theoretical Foundations and Applications

Recent advances in the computing and electronics technology, particularly in sensor devices, databases and distributed systems, are leading to an exponential growth in the amount of data stored in databases. It has been estimated that this amount doubles every 20 years. For some applications, this increase is even steeper. Databases storing DNA sequence, for example, are doubling their size every 10 months. This growth is occurring in several applications areas besides bioinformatics, like financial transactions, government data, environmental monitoring, satellite and medical images, security data and web. As large organizations recognize the high value of data stored in their databases and the importance of their data collection to support decision-making, there is a clear demand for sophisticated Data Mining tools. Data mining tools play a key role in the extraction of useful knowledge from databases. They can be used either to confirm a particular hypothesis or to automatically find patterns. In the second case, which is related to this book, the goal may be either to describe the main patterns present in dataset, what is known as descriptive Data Mining or to find patterns able to predict behaviour of specific attributes or features, known as predictive Data Mining. While the first goal is associated with tasks like clustering, summarization and association, the second is found in classification and regression problems.

Computational tools or solutions based on intelligent systems are being used with great success in Data Mining applications. Nature has been very successful in providing clever and efficient solutions to different sorts of challenges and problems posed to organisms by ever-changing and unpredictable environments. It is easy to observe that strong scientific advances have been made when issues from different research areas are integrated. A particularly fertile integration combines biology and computing. Computational tools inspired on biological process can be found in a large number of applications. One of these applications is Data Mining, where computing techniques inspired on nervous systems; swarms, genetics, natural selection, immune systems and molecular biology have provided new efficient alternatives to obtain new, valid, meaningful and useful patterns in large datasets.

This Volume comprises of 16 chapters, including an overview chapter, providing an up-to-date and state-of-the research on the application of Bio-inspired techniques for Data Mining.

The book is divided into 5 parts:

Part-I:   Bio-inspired approaches in sequence and data streams
Part-II:  Bio-inspired approaches in classification problem
Part-III: Evolutionary Fuzzy and Swarm in Clustering Problems
Part-IV: Genetic and evolutionary algorithms in Bioinformatics
Part-V:  Bio-inspired approaches in information retrieval and visualization

Part I on **Bio-inspired approaches in sequence and data streams** contains four chapters that describe several approaches bio-inspired approaches in sequence and data streams.

**In Chapter 1**, "Adaptive and Self-adaptive Techniques for Evolutionary Forecasting Applications Set in Dynamic and Uncertain Environments," "Adaptive and Self-adaptive Techniques for Evolutionary Forecasting Applications Set in Dynamic and Uncertain Environments," *Wagner and Michalewicz*, present recent studies on evolutionary forecasting, showing how adaptive and self-adaptive algorithms can be efficiently used for the analysis and prediction of dynamic time series. In these time series, the data-generating process can change with time, which is the case in real world time series. Authors point out that previous works usually do not consider this dynamic behaviour and they propose a self-adaptive windowing technique based on Genetic Programming.

**Chapter 2**, "Sequence Pattern Mining," by *Zhou, Shimada, Mabu and Hirasawa*, presents the main aspects of mining datasets where the data assume the format of sequences. Sequence datasets are found in several application areas, like bioinformatics, web and system use logs. The authors analyse the different nature of sequences and models found in the literature for sequence mining. They also propose a new model for sequence mining based on Evolutionary Algorithms.

**In Chapter 3**, "Growing Self-Organizing Map for Online Continuous Clustering," written by *Smith and Alahakoon*, the authors propose a hybrid intelligent learning algorithm based on Self-Organising Maps. The proposed algorithm combines a Growing Self-Organising Map with a Cellular Probabilistic Self-Organising Map. The authors illustrate the advantages of using their algorithm for dynamic clustering in data stream applications. For such, they show the results obtained in experiments using artificial and real world data.

**Chapter 4**, "Synthesis of Spatio-Temporal Models by the Evolution of Non-Uniform Cellular Automata," by *Romano, Villanueva, Zanetti and Von Zuben*, deals with the definition of transition rules for each cell in cellular automata. Motivated by the fact that the search space is very large, the authors employ Evolutionary Algorithms to optimize the definition of the set of transition rules in cellular automata. In the experiments performed, the authors considered one and two-dimensional regular lattices.

Part II on **Bio-inspired approaches in classification problems** contains three chapters discussing many approaches in classification problem.

**Chapter 5**, "Genetic Selection Algorithm and Cloning for Data Mining with GMDH Method," by *Jirina and Jirina, Jr*, is related to Artificial Immune Systems. In this chapter, the authors modify the well known GMDH MIA (Group Method Data Handling Multilayer Iterative Algorithm) neural networks model by employing a selection operation to select the parents of a new neuron. The cloning takes place by small modifications in the parameters present in the copies of the best neuron. The classification accuracy of the new model is compared with the previous models and four other classification techniques using several datasets.

**In Chapter 6**, "Inducing Relational Fuzzy Classification Rules by means of Cooperative Co-evolution," by *Akbarzadeh, Sadeghian and dos Santos*, the induction of fuzzy classification rules using Evolutionary Algorithms is investigated. The Evolutionary Algorithm employs two separate populations. The first population has fuzzy classification rules and is evolved by Genetic Programming. The second population is composed by definitions for the membership function and is evolved by a mutation-based Evolutionary Algorithm. Relational operators are fuzzified by evolutionary methods. The proposed approach is experimentally evaluated and compared with some other evolutionary approaches.

A new Evolutionary Algorithm able to evolve decision trees is presented in **Chapter 7**, "Post-processing Evolved Decision Trees," authored by Johansson, *König, Löfström, Sönströd and Niklasson*. The proposed algorithm iteratively builds a Decision Tree by progressively including new nodes in order to improve the tree accuracy for the training set. In the experiments performed using 22 datasets, Decision Trees have been induced either directly from a training set or from the Neural Networks ensembles.

**Fuzzy and swarm in clustering problems** is the third Part of the book. It contains two chapters discussing the issues of clustering using Evolutionary Fuzzy and Swarm bio-inspired approaches.

**In Chapter 8**, "Evolutionary Fuzzy Clustering: An Overview and Efficiency Issues," by *Horta, Naldi, Campello, Hruschka and de Carvalho*, the authors, after discussing the importance of clustering techniques for data Mining, and presenting a brief description of hybrid clustering techniques, describe the Evolutionary Algorithm for Clustering (EAC) algorithm. They show that EAC provides a efficient combination of Fuzzy clustering and Evolutionary algorithms. After presenting the main features of EAC, the algorithm is experimentally evaluated.

**In Chapter 9**, "Stability-based Model Order Selection for Clustering Using Multiple Cooperative Swarms," by *Ahmadi, Karray and Kamel*, the authors propose a clustering algorithm based on the cooperative work of a multiple swarms. They also investigate a stability analysis model able to define the number of clusters, also known as model order selection, in a given dataset. The Multiple Cooperative Swarm clustering has its performance compared with other clustering algorithms in four datasets. Different clustering validation indexes are used in these comparisons.

**Genetic and evolutionary algorithms in Bioinformatics** are the fourth part in this book. It contains three chapters discussing some Bio-inspired approach in bioinformatics applications.

Another interesting application of Data Mining in Bioinformatics is described in **Chapter 10,** "Data-mining Protein Structure by Clustering, Segmentation and Evolutionary Algorithms," by *Lexa, Snásel and Zelinka*. After a brief introduction to Bioinformatics, the authors discuss how Evolutionary Algorithms can be used to solve problems from Bioinformatics. Later, the authors describe how clustering techniques can group protein fragments and how short fragments can be combined to obtain a larger segment and therefore be able to infer higher level functions for a protein.

**Chapter 11,** "A Clustering Genetic Algorithm for Genomic Data Mining," by *Tapia, Morett and Vallejo*, proposes a new framework based on clustering for the reconstruction of functional modules of proteins. Authors formulate the problem of protein-protein interactions as a multi-objective optimization problem. The framework is evaluated for the analysis of phylogenetic profiles. After presenting the main features of the evolutionary-based clustering algorithm investigated in this chapter, the authors provide a set of experimental evaluations on the prediction of protein-protein functional interactions from different sources of genomic data.

**Chapter 12,** "Detection of Remote Protein Homologs Using Social Programming," by *Ramstein, Beaume and Jacques*, covers an important issue in Bioinformatics, the identification of the function of new, unknown, proteins by looking for the function of homologous proteins. For this application, the authors propose the use of Social Programming, particularly, Grammatical Swarms. In the proposed approach, Support Vector machines are use to identify remote homologs. Experiments are carried out using protein sequences extracted from the SCOP database.

The final Part of the book deals with the **Bio-inspired approaches in information retrieval and visualization.** It contains four chapters, which discusses the Information Retrieval using bio-inspired approaches including the optimizing and clustering information retrieval as well as mining network traffic data.

**Chapter 13,** "Optimizing Information Retrieval Using Evolutionary Algorithms and Fuzzy Inference System," by *Snásel, Abraham, Owais, Platos and Krömer*, investigates the use of two models for information retrieval, one based on a evolutionary algorithms and crisp membership function for document terms, crisp Information Retrieval (IR) framework, named BRIM (Boolean Information Retrieval Model), and another version that combines evolutionary algorithms and fuzzy systems, using fuzzy membership functions, fuzzy IR framework, named EBIRM (Extended BIRM). Experiments compare these two frameworks using different scenarios.

**In Chapter 14,** "Web Data Clustering," by *Húsek, Pokorný, Řezanková and Snášel*, the authors show the benefits of using clustering algorithms for information retrieval, particularly for analysing information from the web. After presenting the fundamentals of cluster analysis, with emphasis on connectionist-based algorithms, they present several applications of clustering in the Web environment.

**In Chapter 15,** "Efficient Construction of Image Feature Extraction Programs by Using Linear Genetic Programming with Fitness Retrieval and Intermediate-result Caching," by *Watchareeruetai, Matsumoto, Takeuchi, Kudo and Ohnishi*, the authors illustrate how bio-inspired algorithms can be used to evolve programs for feature extraction. In their approach, a variation of Linear Genetic Programming uses a population of feature extraction programs, derived from basic image processing operations. The authors show that the computational efficiency is improved by storing intermediate results. The computational efficiency of this approach is assessed by several experiments.

**Chapter 16,** "Mining Network Traffic Data for Attacks through MOVICAB-IDS," by *Herrero and Corchado* describe an Intrusion Detection System (IDS) called MOVICAB-IDS (MObile VIsualization Connectionist Agent-Based IDS). This system is based on a dynamic multiagent architecture combining case-base reasoning and an unsupervised neural projection model to visualize and analyze the flow of network traffic data. To illustrate the performance of the described IDS, it has been tested in different domains containing several interesting attacks and anomalous situations.

We are very much grateful to the authors of this volume and to the reviewers for their great efforts by reviewing and providing interesting feedback to authors of the chapter. The editors would like to thank Dr. Thomas Ditzinger (Springer Engineering Inhouse Editor, Studies in Computational Intelligence Series), Professor Janusz Kacprzyk (Editor-in-Chief, Springer Studies in Computational Intelligence Series) and Ms. Heather King (Editorial Assistant, Springer Verlag, Heidelberg) for the editorial assistance and excellent cooperative collaboration to produce this important scientific work. We hope that the reader will share our joy and will find it useful!

December 2008

Ajith Abraham Trondheim, Norway
Aboul Ella Hassanien, Cairo University
André Ponce de Leon F. de Carvalho,
Sao Carlos, SP, Brazil

# Contents

Part III: Evolutionary Fuzzy and Swarm in Clustering Problems

Part IV: Genetic and Evolutionary Algorithms in Bioinformatics

Part V: Bio-Inspired Approaches in Information Retrieval and
        Visualization

# Part I
# Bio-Inspired Approaches in Sequence and Data Streams

# Adaptive and Self-adaptive Techniques for Evolutionary Forecasting Applications Set in Dynamic and Uncertain Environments

Neal Wagner and Zbigniew Michalewicz

**Abstract.** Evolutionary Computation techniques have proven their applicability for time series forecasting in a number of studies. However these studies, like those applying other techniques, have assumed a static environment, making them unsuitable for many real-world forecasting concerns which are characterized by uncertain environments and constantly-shifting conditions. This chapter summarizes the results of recent studies that investigate adaptive evolutionary techniques for time series forecasting in non-static environments and proposes a new, self-adaptive technique that addresses shortcomings seen from these studies. A theoretical analysis of the proposed technique's efficacy in the presence of shifting conditions and noise is given.

## 1 Introduction

Forecasting is an important activity in any business or organization. A successful forecaster typically finds a model that accurately represents the current environment and uses this model to forecast the future. Many real-world forecasting tasks are set in dynamic and uncertain environments and, thus,

Neal Wagner
Department of Mathematics and Computer Science,
Augusta State University, Augusta, GA 30904, USA
e-mail: nwagner@aug.edu

Zbigniew Michalewicz
School of Computer Science, University of Adelaide, Adelaide, SA 5005, Australia

Institute of Computer Science, Polish Academy of Sciences, ul. Ordona 21, 01-237 Warsaw, Poland, and Polish-Japanese Institute of Information Technology, ul. Koszykowa 86, 02-008 Warsaw, Poland
e-mail: zbyszek@cs.adelaide.edu.au

A. Abraham et al. (Eds.): Foundations of Comput. Intel. Vol. 4, SCI 204, pp. 3–21.
springerlink.com                                      © Springer-Verlag Berlin Heidelberg 2009

it may be difficult for a forecaster to find such a model either because the environment is not well understood or because it is constantly changing.

Evolutionary and other biologically inspired computational methods have often been applied to time series forecasting tasks with favorable results.[1] However, these studies have assumed a static environment which is not often present in real-world applications. This chapter summarizes the results of recent studies that investigate adaptive data-windowing techniques for time series forecasting in non-static environments and proposes a new, self-adaptive data-windowing technique that addresses shortcomings seen from these studies. A theoretical analysis of the proposed technique's efficacy in the presence of shifting conditions and noise is given. Here, an adaptive technique is defined as using feedback from the evolutionary search process to adjust itself while a self-adaptive technique is defined as one that encodes the adjustment into the chromosonal representation of an individual [12].

The rest of this chapter is organized as follows: Sect. 2 gives a brief review of evolutionary time series forecasting studies which have assumed a static environment, Sect. 3 reviews recent evolutionary forecasting studies that use an adaptive data-windowing technique to handle non-static environments, Sect. 4 describes a new, self-adaptive data-windowing technique that seeks to improve upon the adaptive technique and provides a theoretical analysis of its efficacy, and Sect. 5 concludes.

## 2   Review of Evolutionary Forecasting

Most biologically inspired computational methods for time series forecasting fall into two major categories:

1. methods based on neural networks (NN), and
2. methods based on evolutionary computation.

We can refine the latter category by dividing it further into methods based on genetic algorithms (GA), evolutionary programming (EP), and genetic programming (GP).

NN attempt to solve problems by imitating the human brain. A NN is a graph-like structure that contains an input layer, zero or more hidden layers, and an output layer. Each layer contains several "neurons" which have weighted connections to neurons of the following layer. A neuron from the input layer holds an input variable. For forecasting models, this input is a previous time series observation or an explanatory variable. A neuron from the hidden or output layer consists of an "activation" function (usually the logistic function: $g(u) = \frac{1}{1+e^{-u}}$). A three-layer feed-forward NN (one hidden layer between an input and output layer) is commonly used for forecasting applications due to its ability to approximate virtually any non-linear model (if given a sufficient number of neurons at the hidden layer) [52]. Several

---

[1] These studies will be cited below.

applications of NN to forecasting are proffered in [16, 36, 45, 46]. General descriptions of NN can be found in [18] and [52].

For methods based on evolutionary computation, the process of biological evolution is mimicked. When GA is applied to forecasting, first an appropriate model (either linear or non-linear) is selected and an initial population of candidate solutions is created. A candidate solution is produced by randomly choosing a set of parameter values for the selected forecasting model. Each solution is then ranked based on its prediction error over a set of training data. A new population of solutions is generated by selecting fitter solutions and applying a crossover or mutation operation. Crossover is performed by swapping a subset of parameter values from two parent solutions. Mutation causes one (random) parameter from a solution to change. New populations are created until the fittest solution has a sufficiently small prediction error or repeated generations produce no reduction of error.

GA has been used successfully for a wide variety of difficult optimization problems including the forecasting of real-world time series. [4, 37, 38] give detailed descriptions of GA while [5, 10, 11, 17, 22, 24, 30, 47] provide additional examples of GA applied to forecasting.

For EP each candidate solution is represented as a finite state machine (FSM) rather than a numeric vector. FSM inputs/outputs correspond to appropriate inputs/outputs of the forecasting task. An initial population of FSMs is created and each is ranked according to its prediction error. New populations are generated by selecting fitter FSMs and randomly mutating them to produce offspring FSMs. EP was devised by Fogel [15] and has applications in many areas. Some examples of successful EP forecasting experiments include [13, 14, 15, 43].

In GP solutions are represented as tree structures instead of numeric vectors or finite state machines. Internal nodes of solution trees represent appropriate operators and leaf nodes represent input variables or constants. For forecasting applications, the operators are mathematical functions and the inputs are lagged time series values and/or explanatory variables. Figure 1 gives an example solution tree for time series forecasting. Variables $x_{t1}$ and $x_{t2}$ represent time series values one and two periods in the past, respectively. Crossover in GP is performed by (randomly) selecting a single subtree from each of two parent trees and then swapping them to produce two offspring

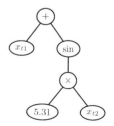

**Fig. 1** GP representation of forecasting solution $x_{t1} + \sin(5.31x_{t2})$

trees. Mutation is performed by (randomly) selecting a single subtree from a single parent tree and replacing it with a randomly generated tree.

GP was developed by Koza [31] as a problem-solving tool with applications in many areas. He was the first to use GP to search for model specifications that can replicate patterns of observed time series.[2] Numerous studies have applied GP to time series forecasting with favorable results. Some examples of these include [2, 6, 7, 19, 21, 20, 23, 29, 28, 25, 27, 26, 32, 39, 41, 48, 49, 51].

Also prevalent in the literature are forecasting studies which employ a hybrid technique. A common hybrid method combines NN and GA. In these applications a GA is used to optimize several aspects of a NN architecture [1, 3, 9, 8, 33, 35, 40, 42, 44, 53]. The optimized NN is then used to produce the desired forecasts. Another hybrid method utilizes EP to evolve both the weights and the topology (i.e., the connectivity) of a NN simultaneously [34, 54].

The following section describes an adaptive evolutionary forecasting technique.

## 3    Adaptive Evolutionary Forecasting

All of the methods described in the previous section assume a static environment. If the underlying data generating process of a time series shifts, the methods must be reevaluated in order to accomodate the new process. Additionally, these methods require that the number of historical time series data used for analysis be designated *a priori*. This presents a problem in non-static environments because different segments of the time series may have different underlying data generating processes. For example, a time series representing the daily stock value of a major U.S. airline is likely to have a different underlying process before September 11, 2001 than it does afterwards. If analyzed time series data span more than one underlying process, forecasts based on that analysis may be skewed.

$$22, 33, 30, 27, 24, 20, 21, 20, 20, \underbrace{23, 26, 29, 30, 28, 29, 32, 30, 31}_{segment2} \mid \underbrace{\ldots}_{future}$$
$$\underbrace{\phantom{22, 33, 30, 27, 24, 20, 21, 20, 20,}}_{segment1}$$

**Fig. 2** Time series containing segments with differing underlying processes

Consider the subset of time series data shown in Fig. 2. Suppose this represents the most recent historical data and has been chosen for analysis. Suppose further that the subset consists of two segments each with a different underlying process. The second segment's underlying process represents the current environment and is valid for forecasting future data. The first segment's process represents an older environment that no longer exists.

---

[2] In [31] Koza refers to this as "symbolic regression."

Because both segments are analyzed, the forecasting model is flawed and forecasts based on it may be flawed as well.

Recent studies conducted by Wagner and Michalewicz [50, 51] investigate an adaptive algorithm based on GP that automatically adjusts "windows" of analyzed data in order to hone in on the currently active process in an environment where the underlying process varies with time. This algorithm (called "Dynamic Forecasting GP" or "DyFor GP") uses feedback from the evolutionary search to control the adjustment. In DyFor GP, analysis starts at the beginning of available historical time series data. Some initial windowsize is set and several generations are run (in the manner described in the previous section) to evolve a population of solutions. Then the data window slides to include the next time series observation. Several generations are run with the new data window and then the window slides again. This process is repeated until all available data have been analyzed up to and including the most recent historical data. Figure 3 illustrates this process. In the figure, | marks the end of available historical data. The set of several generations run on a single data window is referred to as a "dynamic generation." Thus, a single run of DyFor GP includes several dynamic generations (one for each window slide) on several different consecutive data windows.

DyFor GP adjusts the data windowsize using feedback from the GP search process. This is accomplished in the following way.

1. Select two initial windowsizes, one of size $n$ and one of size $n+i$ where $n$ and $i$ are positive integers.
2. Run a dynamic generation at the beginning of the historical data with windowsize $n$.
3. At the end of the dynamic generation, use the best evolved solution (forecasting model) to predict a number of future data and then measure the prediction's accuracy.
4. Run another dynamic generation also at the beginning of the historical data with windowsize $n+i$.

$$\underbrace{22, 33, 30, 27, 24, 20, 21, 20, 20}_{\text{window1}}, 23, 26, 29, 30, 28, 29, 32, 30, 31| \underbrace{\ldots}_{\text{future}}$$

$$22, \underbrace{33, 30, 27, 24, 20, 21, 20, 20, 23}_{\text{window2}}, 26, 29, 30, 28, 29, 32, 30, 31| \underbrace{\ldots}_{\text{future}}$$

$$\vdots$$

$$22, 33, 30, 27, 24, 20, 21, 20, 20, \underbrace{23, 26, 29, 30, 28, 29, 32, 30, 31}_{\text{windowi}}| \underbrace{\ldots}_{\text{future}}$$

**Fig. 3** A sliding data window

5. At the end of the dynamic generation, predict future data and measure the prediction's accuracy. Note which windowsize generated the better prediction.
6. Select another two windowsizes based on which windowsize had better accuracy. For example if the smaller of the 2 windowsizes (size $n$) predicted more accurately, then choose 2 new windowsizes, one of size $n$ and one of size $n - i$. If the larger of the 2 windowsizes (size $n + i$) predicted more accurately, then choose windowsizes $n + i$ and $n + 2i$.
7. Slide the data windows to include the next time series observation. Use the two selected windowsizes to run another two dynamic generations, predict future data, and measure their prediction accuracy.
8. Repeat the previous two steps until the the data windows reach the end of historical data.

Thus, predictive accuracy is used to determine the direction in which to adjust the windowsize toward the optimal. Successive window slides bring the windowsize closer and closer to this destination.

Consider the following example. Suppose the time series given in Fig. 2 is to be analyzed and forecasted. As depicted in the figure, this time series consists of two segments each with a different underlying data generating process. The second segment's underlying process represents the current environment and is valid for forecasting future data. The first segment's process represents an older environment that no longer exists. If there is no knowledge available concerning these segments, automatic techniques are required to discover the correct windowsize needed to forecast the current setting. DyFor GP starts by selecting two initial windowsizes, one larger than the other. Then, two separate dynamic generations are run at the beginning of the historical data, each with its own windowsize. After each dynamic generation, the best solution is used to predict some number of future data and the accuracy of this prediction is measured. Figure 4 illustrates these steps. In the figure **win1** and **win2** represent data windows of size 3 and 4, respectively, and **pred** represents the future data predicted.

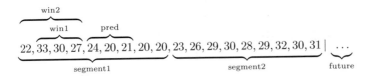

**Fig. 4** Initial steps of window adaptation

The data predicted in these initial steps lies inside the first segment's process and, because the dynamic generation involving data window **win2** makes use of a greater number of appropriate data than that of **win1**, it is likely that **win2**'s prediction accuracy is better. If this is true, two new windowsizes for **win1** and **win2** are selected with sizes of 4 and 5, respectively. The data

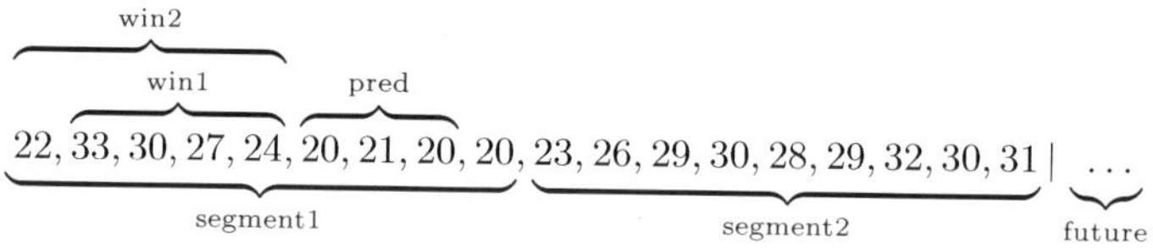

**Fig. 5** Window adaptation after the first window slide. Note: **win1** and **win2** have size 4 and 5, respectively

windows then slide to include the next time series value, two new dynamic generations are run, and the best solutions for each used to predict future data. Figure 5 depicts these steps. In the figure data windows **win1** and **win2** now include the next time series value, 24, and **pred** has shifted one value to the right.

This process of selecting two new windowsizes, sliding the data windows, running two new dynamic generations, and predicting future data is repeated until the data windows reach the end of historical data. It may be noted that while the prediction data, **pred**, lies entirely inside the first segment, the data windows, **win1** and **win2**, expand to encompass a greater number of appropriate data. However, after several window slides, when the data windows span data from both the first and second segments, it is likely that the window adjustment reverses direction. Figures 6 and 7 show this phenomenon. In Fig. 6 **win1** and **win2** have sizes of 4 and 5, respectively. As the prediction data, **pred**, lies inside the second segment, it is likely that the dynamic generation involving data window **win1** has better prediction accuracy than that involving **win2** because **win1** includes less erroneous data. If this is so, the two new windowsizes selected for **win1** and **win2** are sizes 3 and 4, respectively. Thus, as the data windows slide to incorporate the next

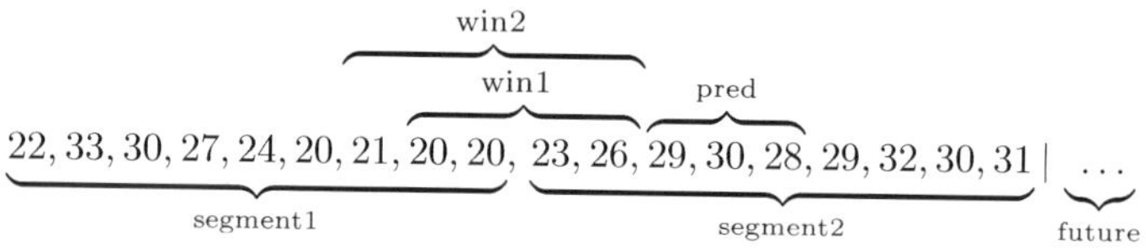

**Fig. 6** Window adaptation when analysis spans both segments. Note: the smaller window, **win1**, is likely to have better prediction accuracy because it includes less erroneous data

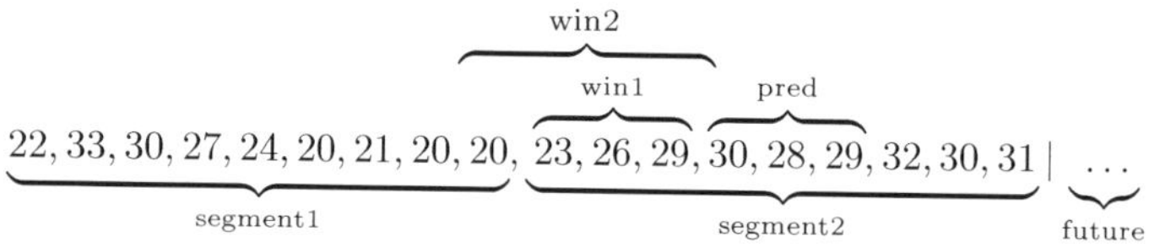

**Fig. 7** Window adaptation when analysis spans both segments. Note: **win1** and **win2** have contracted to include less erroneous data

time series value, they also contract to include a smaller number of spurious data. In Fig. 7 this contraction is shown.

As illustrated in the above example, DyFor GP uses predictive accuracy to adapt the size of its data window automatically towards the optimal. Automatic determination of an appropriate windowsize is beneficial for forecasting concerns in which the number of recent historical data whose underlying data generating process corresponds to the current environment is not known. Furthermore, this adaptation takes place *dynamically*. This means that as DyFor GP moves through (analyzes) historical time series data, the data window expands or contracts depending on the environment encountered. In the studies conducted by Wagner and Michalewicz [50, 51], DyFor GP with adaptive data windowing is compared to conventional GP for forecasting performance. Experimental results show that DyFor GP yields the more efficient forecasts and that this performance difference is due to DyFor GP's ability to hone in on the currently active process in an environment where the underlying process is non-static.

The following section describes a self-adaptive evolutionary forecasting technique.

## 4   Self-adaptive Evolutionary Forecasting

The data-windowing technique described in the previous section is categorized as "adaptive" because it uses feedback from the evolutionary search to adjust itself [12]. In this section, we discuss shortcomings of this approach and detail a "self-adaptive" data-windowing technique that seeks to address these shortcomings. A self-adaptive evolutionary algorithm is one that encodes one or more of its parameters into the chromosonal representation of an individual and can be thought of as "evolution of evolution" [12].

While the studies discussed in the previous section show DyFor GP (with adaptive data windowing) to be a viable model for non-static environments, they also give insight into the shortcomings of the adaptive windowing technique and point to further enhancements that could be made. One shortcoming made apparent by these studies is that once the model notices that the underlying data-generating process has shifted, it takes a long time (i.e., many successive contractions over many slides of the data window) to correctly adjust to the new process. If, after a relatively short period, the process shifts again, DyFor GP is now in "catch up" mode where its window adjustment is always lagging behind the actual process movements. This is because the adaptive windowing technique compares only two different windowsizes and can only adjust itself by small increments at each dynamic generation (slide of the data window). The windowing technique could be improved by comparing more windowsizes at each dynamic generation or by allowing for greater adjustment increments when conditions call for them. If computational resources were not an issue, it could easily be improved by comparing several different windowsizes (rather than just two) at each dynamic generation. This would give a more accurate estimate of the optimal windowsize and would allow the model to "jump" to a good windowsize without having to go through several small adjustments.

While the improvement to DyFor GP's adaptive windowing technique suggested above may not be feasible for computational resources available today, the following describes a self-adaptive data-windowing technique that might be. Instead of running a single dynamic generation for each compared windowsize as is done for the adaptive windowing technique, the windowsize itself could be included in the GP chromosome and evolved along with the functional form of the forecasting model. This means that each individual in the GP population would contain two parts: a solution tree (representing the functional form of the forecasting model) and a corresponding windowsize for that forecasting model to train on. Figure 9 depicts two example GP individuals using this scheme for the time series given in Fig. 8.

$$22, 33, 30, 27, 24, 20, 21, 20, 20, 23, 26, 29, 30, 28, 29, 32, 30, 31 | \ldots$$

future

**Fig. 8** Example time series

In Fig. 9 each individual consists of a solution tree (depicted in the figure as a freehand circular shape) and a windowsize specifying a window of training data. Individual #1 in the figure corresponds to Solution Tree 1 and windowsize = 13 while Individual #2 corresponds to Solution Tree 2 and windowsize = 5. Note that although the two individuals have different windows of training data, their windows are synchronized such that both end at the most recent historical data of the time series of Fig. 8.

Thus, each GP individual would contain a windowsize as part of their chromosone and this windowsize would be subject to the crossover and mutation

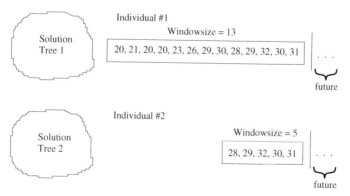

**Fig. 9** Example of GP individuals for self-adaptive data windowing

genetic operators. Windowsize crossover could be performed in the following way.

1. Define a range of windowsizes based on the windowsizes of two selected parent individuals where the parent with the larger windowsize sets the upper bound and the parent with the smaller windowsize sets the lower bound.
2. Randomly choose a new windowsize from this range and assign to the offspring individual.

Solution tree crossover and mutation would be performed in the usual way prescribed by GP while windowsize mutation would be to simply assign a new (random) windowsize. An individual's fitness is calculated by measuring how well its solution tree fits its training window.

In this way a large number of windowsizes (one for each individual) are compared during only one dynamic generation and DyFor GP could, potentially, find the optimal (or near-optimal) windowsize with much less computation. For example, when a process shift is first noticed, the model could potentially find the correct (smaller) windowsize after only one slide of the data window because many windowsizes are compared. Thus, it would spend less time adjusting to the latest process movement and more time training on the currently active process.

It is reasonable to believe that this self-adaptive approach may be biased toward evolving individuals with smaller windowsizes since intuitively it seems easier to evolve a solution tree to fit a smaller number of data than it is to evolve one to fit a larger number of data. If this bias exists, this would severely hamper the proposed method's efficacy since it would evolve only small windowsizes even when environmental conditions called for larger windowsizes (e.g., when the currently active process has remained stable for a number of time periods). In order to investigate this possibility and others, an analysis of the proposed self-adaptive method's behavior for cases which include both process shifts and stable processes is provided in the following sections.

## 4.1 Analysis of the Self-adaptive Algorithm in the Presence of a Process Shift

Suppose the time series given in Fig. 10 represents historical data that is to be used to build a forecasting model. Suppose further that this series contains two segments generated by two different underlying processes. Segment 1 is generated by an old process that no longer is in effect. Segment 2 is generated by the currently active process and, thus, is valid for forecasting future data. This example represents a series that has undergone a shift in the underlying data-generating process. As discussed in Sect. 3, if data from both segments are used for training, then the forecasting model is likely to be flawed.

$$22, 33, 30, 27, 24, 20, 21, 20, 20, \underbrace{23, 26, 29, 30, 28, 29, 32, 30, 31} \mid \; \cdots$$

$$\underbrace{\phantom{22, 33, 30, 27, 24, 20, 21, 20, 20}}_{\text{segment1}} \quad \underbrace{\phantom{23, 26, 29, 30, 28, 29, 32, 30, 31}}_{\text{segment2}} \quad \underbrace{\phantom{\cdots}}_{\text{future}}$$

**Fig. 10** Time series containing segments with differing underlying processes

Given the time series of Fig. 10, the desired behavior for an adaptive or self-adaptive forecasting model is to automatically hone in on the currently active process, that is to adjust its data window to include only data from segment 2. Recall that fitness of an individual in the proposed self-adaptive scheme is calculated by measuring how well the individual's solution tree fits its data window. Figure 11 depicts two example individuals from this scheme for the time series of Fig. 10. In the figure Individual #1 has Solution Tree 1 and windowsize = 18 and Individual #2 has Solution Tree 2 and windowsize = 9.

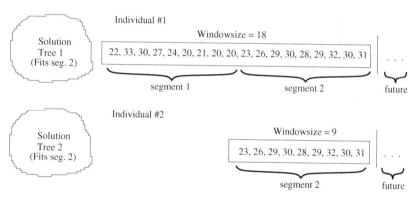

**Fig. 11** Example of GP individuals for self-adaptive data windowing during a process shift

Suppose that Individual #2 from Fig. 11 has a solution tree that fits segment 2, that is its solution tree represents a forecasting model that predicts data from segment 2 with little error. It is clear that this individual would have high fitness since its data window includes only data from segment 2. Suppose that Individual #1 from the figure also has a solution tree that fits segment 2. Because this individual's data window contains data from both segments 1 and 2, its fitness would be lower than the fitness of Individual #2. Note that any individual with a solution tree that does not fit segment 2 must have low fitness regardless of its windowsize since all data windows are sychronized to end at the most recent historical data (which includes segment 2 data).

For this example individuals whose solution trees fit segment 2 and have smaller windowsizes (size ≈ 9) will have higher fitness and be favored by the evolutionary search. This example shows that, for time series that have recently undergone a process shift, the self-adaptive data-windowing algorithm correctly selects individuals whose data windows are honed in on the currently active process. Because training is focused only on data from the currently active process, this algorithm will produce better forecasts than ones that train in part on data from older processes that are no longer in effect.

## 4.2   Analysis of the Self-adaptive Algorithm in the Presence of a Stable Process

Suppose the time series given in Fig. 12 represents historical data that is to be used to build a forecasting model. Suppose further that this series contains one segment generated by only one underlying processes. This example represents a series that is the product of a process that has remained stable (unshifted) for a number of time periods.

$$22, 33, 30, 27, 24, 20, 21, 20, 20, 23, 26, 29, 30, 28, 29, 32, 30, 31 \mid \ \cdots$$

segment1                                                                      future

**Fig. 12** Time series containing one segment with one underlying process

Given the time series of Fig. 12, the desired behavior for an adaptive or self-adaptive forecasting model would be to train on as much relevant data as possible in order to build an accurate model. Therefore, it should adjust its data window to include all (18) data from segment 1. Figure 13 depicts two example individuals from the self-adaptive scheme for the time series of Fig. 12. In the figure individuals are represented in the same way as they are in Figs. 9 and 11.

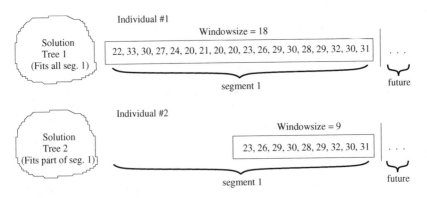

**Fig. 13** Example of GP individuals for self-adaptive data windowing during a stable process

In Fig. 13 Individual #1 has a data window that includes all data from segment 1 and a solution tree that fits all of this data while Individual #2 has a data window that includes only part of segment 1's data and has a solution tree that fits only the data that it has. Because Individual #1 has trained on a greater number of relevant data, its solution tree should represent the more accurate forecasting model. The question is: "will the fitness of the two individuals reflect this?" For this example both individuals would have equivalently high fitness because both of their solution trees fit their respective data windows. However, Individual #1 should be preferred since its solution tree fits all of segment 1 and, thus, captures segment 1's underlying process more accurately. In order to model this correctly, the self-adaptive algorithm includes an additional heuristic that favors individuals with larger windowsizes when fitness is equivalent (or nearly equivalent as specified by some threshold parameter). Using this heuristic, Individual #1 would be correctly selected over Individual #2.

This example shows that the self-adaptive data-windowing algorithm (with additional heuristic) can correctly select individuals that model a stable process in its entirety over those individuals that provide only a partial model. The following section considers examples in which the time series to be forecast contains noise.

## 4.3 Analysis of the Self-adaptive Algorithm in the Presence of Noise

It is reasonable to question whether the presence of noise in a time series will significantly affect the self-adaptive algorithm's ability to hone in on all relevant data corresponding to the currently active process. Suppose the time series of Fig. 14 is to be used to build a forecasting model. This time series is the same as the one depicted in Fig. 12 (i.e., it is generated by only one underlying process) except that now one of the time series datum is noise (represented by an $X$ in the figure).

$$22, 33, 30, 27, 24, 20, 21, 20, 20, 23, 26, 29, 30, 28, 29, X, 30, 31 \mid \; \cdots$$

$$\underbrace{\phantom{22, 33, 30, 27, 24, 20, 21, 20, 20, 23, 26, 29, 30, 28, 29, X, 30, 31}}_{\text{segment 1}} \quad \underbrace{\phantom{\cdots}}_{\text{future}}$$

**Fig. 14** Time series containing one segment with one underlying process and noise

If the same two individuals as depicted in Fig. 13 are considered, will the algorithm still correctly select Individual #1? Note that in Fig. 14 the noise occurs at a datum that would be included in the data windows of both individuals and would have some effect on both of their fitnesses, respectively. In this case Individual #1 will have higher fitness than Individual #2 because its larger window will mollify the effects of the noisy datum. Thus, Individual #1 will be correctly selected. A generalization can be made concerning any two individuals in which one individual's window is larger than the other's, both have solution trees that fit the data in their respective windows, and noise occurs in the windows of both. If the difference between the windowsizes of the two individuals is large, then the individual with the larger windowsize will have significantly higher fitness than the one with the smaller windowsize because the noise will have a greater effect on the smaller window. If the windowsize difference between the two individuals is small, then both individuals will have nearly equivalent fitness and the individual with the larger windowsize will still be selected as specified by the additional heuristic discussed in Sect. 4.2.

What if the noise occurs only in the data window of the individual with the larger windowsize? Suppose the same time series as shown in Fig. 12 is given, this time with noise occurring at an earlier time period. Figure 15 depicts such a series; the noise is represented by an $X$ in the figure. For this case, the

$$22, 33, X, 27, 24, 20, 21, 20, 20, 23, 26, 29, 30, 28, 29, 32, 30, 31 \mid \ \cdots$$

$$\underbrace{\phantom{22, 33, X, 27, 24, 20, 21, 20, 20, 23, 26, 29, 30, 28, 29, 32, 30, 31}}_{\text{segment1}} \quad \underbrace{\phantom{\cdots}}_{\text{future}}$$

**Fig. 15** Time series containing one segment with one underlying process and noise at an early time period

noise occurs in the data window of Individual #1 of Fig. 13 but not in the window of Individual #2 from that figure. Here, Individual #2 would have the higher fitness because the noise affects only Individual #1. Recall that the additional heuristic discussed in Sect. 4.2 specifies a threshold to determine if two individuals have "nearly equivalent" fitness values. Whether or not the effect of the noisy datum is great enough to overcome this threshold depends on the windowsize difference between the two individuals. If the windowsize difference is large (i.e., Individual #1 has a much larger windowsize relative to Individual #2), then the larger windowsize will significantly reduce the noisy datum's impact and the two individuals will likely have a fitness difference within the specified threshold. Thus, the two individuals will be considered as having "nearly equivalent" fitness and the one with the larger windowsize (Individual #1) will be (correctly) selected.

If the windowsize difference is small (i.e., Individual #1 has only a slightly bigger window than Individual #2), then the effect of the noisy datum on Individual #1's fitness will not be reduced and Individual #2 will be (incorrectly) selected. Here, the proposed algorithm fails to select the individual with the correct windowsize (i.e., the individual that has trained on a greater number of data relevant to the active process). However, for this situation to occur the windowsizes of the two compared individuals must be nearly the same. This scenario is unlikely as it is unlikely for noise to appear in only one of two data windows that have nearly the same size.

The above discussion focused on time series generated by an underlying process that has remained stable for a number of time periods. Now we turn

$$22, 33, 30, 27, 24, 20, 21, 20, 20, 23, 26, 29, 30, 28, 29, X, 30, 31 \mid \; \ldots$$

$$\underbrace{\phantom{22, 33, 30, 27, 24, 20, 21, 20, 20}}_{\text{segment1}} \quad \underbrace{\phantom{23, 26, 29, 30, 28, 29, X, 30, 31}}_{\text{segment2}} \quad \underbrace{\phantom{\ldots}}_{\text{future}}$$

**Fig. 16** Time series containing segments with differing underlying processes and noise near the end

$$22, 33, X, 27, 24, 20, 21, 20, 20, 23, 26, 29, 30, 28, 29, 32, 30, 31 \mid \; \ldots$$

$$\underbrace{\phantom{22, 33, X, 27, 24, 20, 21, 20, 20}}_{\text{segment1}} \quad \underbrace{\phantom{23, 26, 29, 30, 28, 29, 32, 30, 31}}_{\text{segment2}} \quad \underbrace{\phantom{\ldots}}_{\text{future}}$$

**Fig. 17** Time series containing segments with differing underlying processes and noise near the beginning

our attention to time series generated by an underlying process that has recently shifted. Suppose the time series of Figs. 16 and 17 are to be used to build forecasting models. Both of these series represent the same series as depicted in Fig. 10 (i.e., they both are generated by an underlying process that has shifted) except that now noise has been added. In Fig. 16 the noise occurs late in the time series while in Fig. 17 the noise occurs earlier.

Considering the two individuals depicted in Fig. 11, will the self-adaptive algorithm correctly select Individual #2? For the time series given in Fig. 17, the noise does not affect Individual #2 since it is not included in Individual #2's data window. Thus, Individual #2 would still have the higher fitness and would be correctly selected. For the time series of Fig. 16, the noise is included in the data windows of both individuals and thus affects them both to some degree. However, Individual #1 already has significantly lower fitness than Individual #2 because its solution tree fits segment 2 but its data window includes several data from segment 1. Therefore it is unlikely that a single noisy datum in Individual #2's data window will cause its fitness to drop to a value near that of Individual #1 given that the noisy datum also occurs in the data window of Individual #1 and thus adversely affects its fitness as well.

The above examples show that the self-adaptive data-windowing algorithm can correctly hone in on the currently active process in the presence of either a process shift or a long-running stable process. The effect of noise on the behavior of the self-adaptive data-windowing algorithm is considered and it is shown that the algorithm is robust in the presence of noise assuming that the noise occurs only once in any data window. The motivation for this self-adaptive algorithm was given by the shortcomings seen in experiments conducted using the adaptive data-windowing algorithm (described in Sect. 3). Specifically, it was seen that the adaptive algorithm takes a longer time (i.e., more computation) to adjust to a process shift and, thus would spend more time tracking process shifts than training on the currently active process in an environment that changes frequently.

# 5  Conclusion

This chapter discusses both adaptive and self-adaptive techniques for evolutionary forecasting applications set in dynamic and uncertain environments. Recent studies conducted using an adaptive data-windowing technique to automatically hone in on the currently active process in a non-static environment are summarized and lessons learned from those studies are elucidated. A new "self-adaptive" data-windowing technique is detailed that seeks to address the shortcomings seen in the adaptive windowing technique and an analysis is given concerning the self-adaptive algorithm's behavior in the presence of both an underlying data-generating process that has shifted and one that has been stable for several time periods.

The proposed self-adaptive data-windowing technique is currently being implemented and future work would include experiments comparing this technique to the adaptive data-windowing technique and other static (non-adaptive) techniques as well.

# References

1. Andreou, A., Georgopoulos, E., Likothanassis, S.: Exchange rates forecasting: a hybrid algorithm based on genetically optimized adaptive neural networks. Computational Economics 20, 191–202 (2002)
2. Andrew, M., Prager, R.: Genetic programming for the acquistion of double auction market strategies. In: Kinnear, K. (ed.) Advances in Genetic Programming, pp. 355–368. MIT Press, Cambridge (1994)
3. Back, B., Laitinen, T., Sere, K.: Neural networks and genetic algorithms for bankruptcy predictions. Expert Systems with Applications 11, 407–413 (1996)
4. Back, T.: Evolutionary Algorithms in Theory and Practice: Evolution Strategies, Evolutionary Programming, and Genetic Algorithms. Oxford University Press, Oxford (1996)
5. Chambers, L. (ed.): Practical Handbook of Genetic Algorithms: Applications. CRC Press, Boca Raton (1995)
6. Chen, S., Yeh, C.: Toward a computable approach to the efficient market hypothesis: an application of genetic programming. Journal of Economics Dynamics and Control 21, 1043–1063 (1996)
7. Chen, S., Yeh, C., Lee, W.: Option pricing with genetic programming. In: Genetic Programming 1998: Proceedings of the Third Annual Conference, pp. 32–37 (1998)
8. Chen, Y., Yang, B., Abraham, A.: Time series forecasting using flexible neural tree model. Information Sciences 174, 219–235 (2005)
9. Chen, Y., Yang, B., Abraham, A.: Flexible neural trees ensemble for stock index modeling. Neurocomputing Journal 70, 305–313 (2007)
10. Chiraphadhanakul, S., Dangprasert, P., Avatchanakorn, V.: Genetic algorithms in forecasting commercial banks deposit. In: Proceedings of the IEEE International Conference on Intelligent Processing Systems, pp. 557–565 (1997)
11. Deboeck, G. (ed.): Trading on the Edge: Neural, Genetic, and Fuzzy Systems for Chaotic and Financial Markets. John Wiley and Sons, Inc., Chichester (1994)

12. Eiben, A., Hinterding, R., Michalewicz, Z.: Parameter control in evolutionary algorithms. IEEE Transactions on Evolutionary Computation 3, 124–141 (1999)
13. Fogel, D., Chellapilla, K.: Revisiting evolutionary programming. In: SPIE Aerosense 1998, Applications and Science of Computational Intelligence, pp. 2–11 (1998)
14. Fogel, L., Angeline, P., Fogel, D.: An evolutionary programming approach to self-adaptation on finite state machines. In: Proceedings of the 4th Annual Conference on Evolutionary Programming, pp. 355–365 (1995)
15. Fogel, L., Owens, A., Walsh, M.: Artificial Intelligence through Simulated Evolution. Wiley, Inc., Chichester (1966)
16. Gately, E.: Neural Networks for Financial Forecasting. John Wiley and Sons, Inc., Chichester (1996)
17. Goto, Y., Yukita, K., Mizuno, K., Ichiyanagi, K.: Daily peak load forecasting by structured representation on genetic algorithms for function fitting. Transactions of the Institute of Electrical Engineers of Japan 119, 735–736 (1999)
18. Gurney, K.: An Introduction to Neural Networks. UCL Press (1997)
19. Hiden, H., McKay, B., Willis, M., Tham, M.: Non-linear partial least squares using gentic programming. In: Genetic Programming 1998: Proceedings of the Third Annual Conference, pp. 128–133 (1998)
20. Iba, H., Nikolaev, N.: Genetic programming polynomial models of financial data series. In: Proceedings of the 2000 Congress of Evolutionary Computation, pp. 1459–1466. IEEE, Los Alamitos (2000)
21. Iba, H., Sasaki, T.: Using genetic programming to predict financial data. In: Proceedings of the Congress of Evolutionary Computation, pp. 244–251 (1999)
22. Jeong, B., Jung, H., Park, N.: A computerized causal forecasting system using genetic algorithms in supply chain management. The Journal of Systems and Software 60, 223–237 (2002)
23. Jonsson, P., Barklund, J.: Characterizing signal behavior using genetic programming. In: Fogarty, T.C. (ed.) AISB-WS 1996. LNCS, vol. 1143, pp. 62–72. Springer, Heidelberg (1996)
24. Ju, Y., Kim, C., Shim, J.: Genetic based fuzzy models: interest rate forecasting problem. Computers and Industrial Engineering 33, 561–564 (1997)
25. Kaboudan, M.: Forecasting stock returns using genetic programming in c++. In: Proceedings of 11th Annual Florida Artificial Intelligence International Research Symposium, pp. 502–511 (1998)
26. Kaboudan, M.: Genetic evolution of regression models for business and economic forecasting. In: Proceedings of the 1999 Congress of Evolutionary Computation, pp. 1260–1268. IEEE, Los Alamitos (1999)
27. Kaboudan, M.: Genetic programming prediction of stock prices. Computational Economics 6, 207–236 (2000)
28. Kaboudan, M.: Genetically evolved models and normality of their residuals. Journal of Economics Dynamics and Control 25, 1719–1749 (2001)
29. Kaboudan, M.: Forecasting with computer-evolved model specifications: a genetic programming application. Computer and Operations Research 30, 1661–1681 (2003)
30. Kim, D., Kim, C.: Forecasting time series with genetic fuzzy predictor ensemble. IEEE Transactions on Fuzzy Systems 5, 523–535 (1997)
31. Koza, J.: Genetic Programming: On the Programming of Computers by Means of Natural Selection. MIT Press, Cambridge (1992)

32. Lee, D., Lee, B., Chang, S.: Genetic programming model for long-term forecasting of electric power demand. Electric Power Systems Research 40, 17–22 (1997)
33. Leigh, W., Purvis, R., Ragusa, J.: Forecasting the nyse composite index with technical analysis, pattern recognizer, neural network, and genetic algorithm: a case study in romantic decision support. Decision Support Systems 32, 361–377 (2002)
34. Liu, Y., Yao, X.: Evolving neural networks for hang seng stock index forecasting. In: CECCO 2001: Proceedings of the 2001 Congress on Evolutionary Computation, pp. 256–260 (2001)
35. Maniezzo, V.: Genetic evolution of the topology and weight distribution of neural networks. IEEE Transactions on Neural Networks 5, 39–53 (1994)
36. Maqsood, I., Abraham, A.: Weather analysis using and ensemble of connectionist learning paradigms. Applied Soft Computing Journal 7, 995–1004 (2007)
37. Michalewicz, Z.: Genetic Algorithms + Data Structures = Evolution Programs. Springer, Heidelberg (1992)
38. Mitchell, M.: An Introduction to Genetic Algorithms. MIT Press, Cambridge (1996)
39. Mulloy, B., Riolo, R., Savit, R.: Dynamics of genetic programming and chaotic time series prediction. In: Genetic Programming 1996: Proceedings of the First Annual Conference, pp. 166–174 (1996)
40. Nag, A., Mitra, A.: Forecasting daily foreign exhange rates using genetically optimized neural networks. Journal of Forecasting 21, 501–511 (2002)
41. Neely, C., Weller, P.: Predicting exchange rate volatility: genetic programming versus GARCH and RiskMetrics. Technical report, The Federal Reserve Bank of St. Louis (2002)
42. Phua, P., Ming, D., Lin, W.: Neural network with genetically evolved algorithms for stocks prediction. Asia-Pacific Journal of Operational Research 18, 103–107 (2001)
43. Sathyanarayan, R., Birru, S., Chellapilla, K.: Evolving nonlinear time series models using evolutionary programming. In: CECCO 1999: Proceedings of the 1999 Congress on Evolutionary Computation, pp. 243–253 (1999)
44. Sexton, R.: Identifying irrelevant input variables in chaotic time series problems: using genetic algorithms for training neural networks. Journal of Computational Intelligence in Finance 6, 34–41 (1998)
45. Smith, K., Gupta, J.: Neural Networks in Business: Techniques and Applications. Idea Group Pub. (2002)
46. Trippi, R., Turban, E. (eds.): Neural Networks in Finánce and Investing: Using Artificial Intelligence to Improve Real-World Performance. Irwin Professional Pub. (1996)
47. Venkatesan, R., Kumar, V.: A genetic algorithms approach to growth phase forecasting of wireless subscribers. International Journal of Forecasting 18, 625–646 (2002)
48. Wagner, N., Michalewicz, Z.: Genetic programming with efficient population control for financial times series prediction. In: 2001 Genetic and Evolutionary Computation Conference Late Breaking Papers, pp. 458–462 (2001)
49. Wagner, N., Michalewicz, Z.: Forecasting with a dynamic window of time: the dyfor genetic program model. In: Bolc, L., Michalewicz, Z., Nishida, T. (eds.) IMTCI 2004. LNCS, vol. 3490, pp. 205–215. Springer, Heidelberg (2005)

50. Wagner, N., Michalewicz, Z.: An analysis of adaptive windowing for time series forecasting in dynamic environments: Further tests of the DyFor GP model. In: Proceedings of the Genetic and Evolutionary Computation Conference (GECCO 2008), Atlanta, GA (July 2008)
51. Wagner, N., Michalewicz, Z., Khouja, M., McGregor, R.: Time series forecasting for dynamic environments: the DyFor genetic program model. IEEE Transactions on Evolutionary Computation 11(4), 433–452 (2007)
52. White, H.: Artificial neural networks: approximation and learning theory. Blackwell, Malden (1992)
53. White, J.: A genetic adaptive neural network approach to pricing option: a simulation analysis. Journal of Computational Intelligence in Finance 6, 13–23 (1998)
54. Yao, X., Liu, Y.: Epnet for chaotic time series prediction. In: First Asia-Pacific Complex Systems Conference, pp. 146–156 (1997)

# Sequence Pattern Mining

Huiyu Zhou, Kaoru Shimada, Shingo Mabu, and Kotaro Hirasawa

**Abstract.** In this chapter, we introduce a method of association rule mining using Genetic Network Programming (GNP) with time series processing mechanism and attributes accumulation mechanism in order to find time related sequence rules efficiently in association rule extraction systems. This process is called Sequence Pattern Mining. In order to deal with a large number of attributes, GNP individual accumulates fitter attributes gradually during rounds, and the rules of each round are stored in a Small Rule Pool using a hash method, then, the rules are finally stored in a Big Rule Pool after the check of the overlap at the end of each round. And, we also present experimental results using the traffic prediction problem. The aim of sequential pattern mining is to better handle association rule extraction of the databases in a variety of time-related applications, for example, in the traffic prediction problems. The algorithm which can find the important sequential association rules is described and several experimental results are presented considering a traffic prediction problem.

## 1 Introduction

Nowadays, data mining has become an important field since huge amounts of data have been collected in various applications. Mining these data sets

Huiyu Zhou
Graduate School of Information, Production and Systems, Waseda University
e-mail: zhy836@toki.waseda.jp

Kaoru Shimada
Information, Production and Systems Research Center, Waseda University
e-mail: k.shimada@aoni.waseda.jp

Shingo Mabu
Graduate School of Information, Production and Systems, Waseda University
e-mail: mabu@aoni.waseda.jp

Kotaro Hirasawa
Graduate School of Information, Production and Systems, Waseda University
e-mail: hirasawa@waseda.jp

A. Abraham et al. (Eds.): Foundations of Comput. Intel. Vol. 4, SCI 204, pp. 23–48.
springerlink.com                                          © Springer-Verlag Berlin Heidelberg 2009

efficiently and effectively are too difficult and intricate when using conventional methods, especially for the sequential database in dynamic systems.

"**What is sequence pattern mining?**" Sequence pattern mining is one of the main focuses of mining sequential databases, and it is the analytical process of digging through and exploring the enormous sets of sequence data for the search of consistent patterns and/or systematic relationships between attributes in sequence datasets. It aims at extracting implicit, previously unknown sequential information from data sets, which could be potentially useful in the future.

In many applications, such as bioinformatics, web access traces, system utilization logs, transportation systems, etc., the data is naturally in the form of sequences. For example, in traffic systems, the sequential information such as "The 5th road have high traffic volume, then, the 4th road will also have high traffic volume on the same day" has been of great interest for analyzing the time-related data to find its inherent characteristics.

To meet the different needs of various applications, and also, to analyze and understand the nature of various sequences, several models of sequential pattern mining have been proposed. This chapter not only briefly studies definitions and application domains of these models, but more deeply introduces an evolutionary algorithm on how to effectively and efficiently find these patterns.

## 1.1 Sequence Pattern Mining Environments

"What is time related database? What is sequence database?" **Time related database** is the database that consists of the tuples with data indexed by time, in other words, it is the database composed of sequential values or events which occur along the time. Time related database is the typical representation of the data gathered in many application fields, for example, stock markets, production systems, scientific experiments, medical applications, etc. Therefore, the time related data base can be considered as a kind of sequence database. On the other hand, **sequence database** is a kind of database which consist of sequential events, so, it could be, but not limited to, time related sequences with time tags. For example, Web pages searching sequences are a kind of sequential data, but not necessarily time related data.

**Table 1** Conventional Transactional Database

| T | $A_1$ | $A_2$ | $A_3$ |
|---|---|---|---|
| 1 | 8 | 3 | 6 |
| 2 | 5 | 4 | 6 |
| 3 | 8 | 1 | 10 |
| 4 | 2 | 9 | 9 |

**Table 2** Time Related Database after Discretization

| Time | $A_1$ | | | $A_2$ | | | $A_3$ | | |
|------|-----|--------|------|-----|--------|------|-----|--------|------|
|      | Low | Middle | High | Low | Middle | High | Low | Middle | High |
| 0001 | 0   | 0      | 1    | 1   | 0      | 0    | 0   | 1      | 0    |
| 0002 | 0   | 1      | 0    | 0   | 1      | 0    | 0   | 1      | 0    |
| 0003 | 0   | 0      | 1    | 1   | 0      | 0    | 0   | 0      | 1    |
| 0004 | 1   | 0      | 0    | 0   | 0      | 1    | 0   | 0      | 1    |

Middle threshold=4; High threshold=7;

While the databases which consist of transactions are called transactional databases. Most modern relational database management systems fall into this category as shown in Table 1, where each tuple in the database represents one transaction. However, when processing real-time information systems, the databases are to be designed and developed to emphasize the need to satisfy time related requirements, like the necessity to predict the occurrence of the critical events in the time domain. For example, traffic prediction is one of these problems. As the traffic flow in each section of the roads is changing consistently as time goes on, the database should represent at what time the events in the road section occur, in order to better predict the traffic flow of the roads in the future.

Traditional transaction-based rules without time series can tell what interesting events happen together, however, it can not represent at what time the event will happen or how long the event will last.

Thus, the database we handle here is no longer transaction-based, but time related-based, i.e., the tuples in the time series database represent the time unit instead of the transaction. The "Time" in Table 2 represents the time unit, it can be very small ones as one second, one minute, or long ones like one year or one episode of a process and so on, thus its concrete meaning is related to the concrete problem to solve.

In the traffic systems we deal with, the database have continuous attributes like the traffic volume of each section, so, we divided the continuous values to three categories, i.e., Low, Middle and High. Supposing that the Middle threshold is 4 and High threshold is 7, the attribute $A = 9$ is ranked as $A - High$, i.e., $(A - Low, A - Middle, A - High)=(0, 0, 1)$. Thus, Table 2 shows the binary values of 0s and 1s after discretizing Table 1 using the above thresholds.

In the traffic prediction problem, we assume there exist many cars on each section of the roads, thus rows of the database are consisted of time units and the number of cars on every section becomes the attribute columns of the database. "The section named A has the traffic volume of 10 cars on the time unit 2." is a typical case of the events in the traffic systems. This event information is recorded in row 2 and the column which represents A in the database. Since the value of the event is 10, we can classify it to $A - High$ using the thresholds mentioned above, and the event can be represented as

$(A - Low, \ A - Middle, \ A - High)$=$(0,0,1)$ in the discretized database, where, the $A - Low$, $A - Middle$ and $A - High$ with binary values are called **attribute** here. Now the problem is to find time related interesting sequential relationships among attributes in the discretized database.

## 1.2  Related Data Mining Methods

Among data mining methods with a wide range of applicability, there is an association rule mining, where its major task is to detect relationships or associations between attributes in large databases. That is, the association rule mining aims at discovering association relations or correlations among attributes encoded within a database[1]. The relationship between data sets can be represented as association rules. In this chapter, we focus on how to extract sequence association rules in data mining.

One of the most popular models in the association rule mining is Apriori-based algorithm, in which Agrawal et al proposed the support-confidence framework [2]. This algorithm measures the importance of association rules with two factors: support and confidence. Chi-squared value has also been applied to association rule mining. Brin et al. suggested to measure the significance of associations via the chi-squared test for correlation used in classical statistics [3].

However, Apriori algorithm may suffer from large computational complexity for rule extraction when extracting from dense databases. Many approaches have been proposed to extract association information in various directions, including efficient apriori-like mining methods [4], [5]. The variations of Apriori approach such as using the hash-based algorithm have also been studied for efficiency [6], [7].

Another kind of data mining uses Neural Networks [8], [9], which is a collection of neuron-like processing units with weighted connections between the units. However, they often produce incomprehensible models and require training data and long training times.

Genetic Algorithm (GA)[10],[11] imitating the natural selection process of evolution, proposed by J.H. Holland, has also been applied to data mining research in order to deal with dense databases. Holland proposed two kinds of approaches, Pittsburgh approach and Michigan approach. Pittsburgh approach represents an entire rule set as an individual, and evolve the population of candidate rule sets, while in Michigan approach, members of the population are individual rules and a rule set is represented by the entire population. Both of the approachs evolve the rules during generations and the individuals or population themselves represent the association relationships capable of acquiring inference rules. However, because a rule is represented as an individual or a part of an individual of GA, it is hard for them to give us a complete picture of the underlying relationships in problem domains, thus not easy to extract enough number of association rules.

Genetic Programming (GP) [12], [13] improved the expression ability of GA by evolving individuals as tree structures. Although GP enables a more explicit representation of reference rules, it may suffer from the problems of loose structure and bloating, especially in dynamic problems.

Recently, many data mining methods to deal with sequence databases have been proposed to satisfy the increasing demand for efficient information mining in time-related dynamic systems. A. K. H. Tung proposed the inter-transaction association rule mining [14], which not only extracts relationship within the transaction, but also uses an extended Apriori method to obtain inter-transaction associations by defining and mining frequent intertransaction itemsets(FITI). This method uses the sequential association between transactions to represent temporal relations between transaction items. Since the method is basically an Apriori-based mining method, it shares the same disadvantage of not able to deal with the databases with a very large number of attributes.

TPrefixSpan algorithm [15] is a kind of nonambiguous temporal pattern mining and it can deal with temporal databases using temporal relationships between two intervals which was proposed by Kam and Fu in 2000 [16]. This method deals with interval-based event data, however, it overcomes the ambiguous problem with interval-based temporal mining by correctly describing the temporal relationship between every pair of events. As a result, the method is capable of building an unique temporal sequence for every time sequence of interval events. However, since there are 14 kinds of relationships between every pair of attributes, it would suffer from efficiency problems when processing the database with a large number of attributes.

Recurrent neural networks and associative memory can also deal with the time related database. Recurrent neural networks are used for extracting rules [17] as a form of Deterministic Finite-state Automata. The recurrent connection and time delay between neurons allows the network to generate time-varying patterns. Associative memory is mainly a content-addressable structure that maps specific input representations to specific output representations, and the time delay between neurons also enables to mimic the time related relationship patterns in real-time systems.

## 1.3  GNP-Based Data Mining Method

Genetic Network Programming (GNP) [18],[19],[20] we introduce in section 2 is an further extension of Genetic Algorithm (GA). GNP uses directed graph structures as solutions. The directed graph of GNP contributes to creating quite compact programs and implicitly memorizing past action sequences in the network flows. Also, GNP can find solutions without the bloating problem compared with GP, because of the fixed number of nodes in GNP.

Unlike the other methods mentioned in the above, GNP-based data mining allows the GNP individuals to evolve and extract association rules in the rule pool as many as possible. What's more, it uses evolved individuals(directed graphs of GNP) just as a tool to extract candidate association rules. Thus, the structure of GNP individuals does not necessarily represent the association relations of the database. Instead, the extracted association rules are stored together in the rule pool separatively from the individuals. As a result, the structures of GNP individuals are less restricted than the structures of GA and GP, and GNP-based data mining becomes capable of producing a great number of association rules. The major points of GNP-based data mining method with time related sequence processing are as follows:

- The database is time-related considering the real-time factor into consideration. Then, the same attribute now has different meanings at different time points.
- The extracted association rule has the following form: "$X \Rightarrow Y$", where each attribute of $X$ and $Y$ should have its time tag, and the rules obtained are actually show the relationships between data sets with time sequence.
- Time delay tags are used for the connections of the judgment nodes in GNP-based data mining, for example "$A(t = p)$" means the judgment of the attribute $A$ at time $p$. Therefore, the searching for the database for calculating the confidence, support and chi-squared value becomes the two dimensional search.
- The concept of an Attribute Accumulation Mechanism is introduced in order to systematically and efficiently explore the search space. Each round has its own attribute sub set and accumulate fitter attributes in the sub set gradually.

This chapter is organized as follows: In Section 2, the basic structure of Genetic Network Programming(GNP) are introduced. In Section 3, the algorithm of GNP-based time related sequence mining method is described. Section 4 shows examples of applying the proposed method to traffic prediction using a simple simulator. Section 5 is devoted to conclusions.

## 2   Genetic Network Programming (GNP)

Evolutionary Algorithm (EA) is a stochastic heuristic method for opimization. Since the 1960s, there has been increasing interests in imitating living things to develop powerful algorithms for difficult optimization problems.

EA is now the general term for several computational techniques which are based on the evolution of biological life in the natural world. Individuals in EA are solutions of the problems, where the fitness function is defined in order to evolve the individuals. Only the fitter ones in the generation can survive to the next generation and produce offspring by mutation and crossover imitating the evolution process in the nature. Using this kind of selection and evolving process, the fitter individual(solution) for the problem can be obtained and applied to the problem to solve.

In this section, Genetic Network Programming(GNP) is briefly introduced [18],[19],[20]. GNP is a sort of evolutionary optimization techniques(EA), which evolves arbitrary directed graph programs as solutions(individuals). Because of the strong expression ability of the directed graph structures, GNP has an implicit memory function, which can express the temporal information and past information implicitly.

The basic individual structure of GNP is showns in Fig.1. The graph structure(penotype) of GNP includes three kinds of nodes: start nodes, judgment nodes and processing nodes. Judgment nodes are the set of $J_1$, $J_2$,..., which work as if-then type logical decision making functions. On the other hand, processing nodes are the set of $P_1$, $P_2$,..., which work as certain kinds of action/processsing functions. The concrete roles of these nodes are predefined and stored in the library.

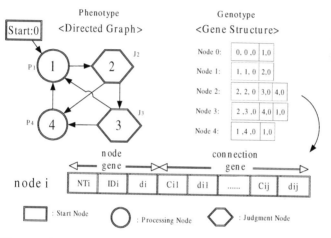

**Fig. 1** The basic structure of GNP individual

The genotype expression of GNP nodes is also shown in Fig.1. This describes the gene of node $i$, then the set of these genes represents the genotype of GNP individuals. $NT_i$ describes the node type, $NT_i = 0$ means node $i$ is start node, $NT_i = 1$ represents node $i$ is judgment node and $NT_i = 2$ represents node $i$ is processing node. $ID_i$ is an identification number, for example, $NT_i = 1$ and $ID_i = 1$ means node $i$ is $J_1$(Judgment node with $ID$ 1). $C_{i1}$, $C_{i2}$,..., denote the connections from node $i$. $d_i$ and $d_{ij}$ are the delay time required to execute the processing of the nodes and transition between nodes.

Each individual of GNP represents a solution of the problem, and judgement node is in charge of judging situation, while processing node do the real processing. Once GNP is started up, firstly the execution starts from the start node; consequently, the next node to execute is determined according to the connection and if-then judgment results of the current node. After the execution of the task, each individual gets the fitness value of itself depending on the completeness of the task.

## Evolution Process

The flow chart of the GNP evolution process is shown in Fig. 2. The first step is to initialize the population by randomly generating Judgement and Processing nodes of each individual in the initial generation. Then, execute each individual(solution) in the problem domain(environment) to obtain the fitness value. This process is the **evaluation** of the current generation.

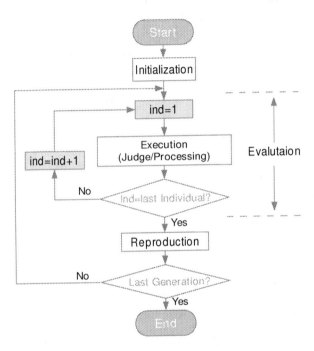

**Fig. 2** The flow chart of GNP evolution

After the evaluation, we select the individuals to reproduce new individuals in the next generation. Generally, the individuals with higher fitness values have the higher chance to be selected in the reproduction. Most EA systems avoid selecting only the fittest individual in reproduction, but rather a random (or semi-random) selection among the fitter individuals is carried out, thus, the diversity of the whole evolution can be maintained.

There are basically two methods of reproduction: **mutation** and **crossover**. In evolutionary algorithms, **mutation** is a genetic operator used to maintain genetic diversity in the population from one generation to the next generation. It is analogous to biological mutation which mutate the gene structure of individuals. The purpose of mutation in EAs is to help the algorithm to avoid local minima by preventing individuals from becoming too similar to each other.

Generally, mutation is to change some of the nodes in GNP individual, randomly, and connections and delays are also changed by mutation operator in GNP.

**Crossover** is also a genetic operator used to change individuals from one generation to the next generation. It is analogous biological crossover, upon which genetic algorithms are based. Two parents exchange their gene and produce two offspring. The crossover operation is as follows:

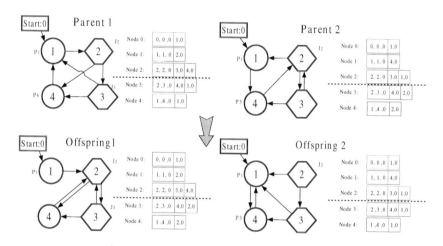

**Fig. 3** The basic procedure of GNP crossover

- Use any selection method to choose two parents for crossover
- In one point crossover as shown in Fig. 3, randomly choose a crossing point by probability $P_c$.
- Exchange the parts divided by the crossing point as shown in Fig. 3 to generate new offspring.

Using the above reproduction method, GNP generates the new population for the next generation. Thus, GNP evolves individuals using the above evaluation and reproduction sequence generation by generation.

## 3   Association Rules

An association rule has the form of $(X \Rightarrow Y)$, where $X$ represents antecedent and $Y$ represents consequent. The association rule "$X \Rightarrow Y$" can be interpreted as: the set of attributes satisfying $X$ is likely to satisfy $Y$.

**Table 3** The contingency of $X$ and $Y$

| | $Y$ | $\neg Y$ | $\sum_{row}$ |
|---|---|---|---|
| $X$ | $Nxy$ $Nz$ | $N(x - xy)$ $N(x - z)$ | $Nx$ |
| $\neg X$ | $N(y - xy)$ $N(y - z)$ | $N(1 - x - y + xy)$ $N(1 - x - y + z)$ | $N(1 - x)$ |
| $\sum_{col}$ | $Ny$ | $N(1 - y)$ | $N$ |

( $N$: the number of tuples ($= |TID|$) )

The following is a formal statement of the problem of mining association rules. Let $A = \{A_1, A_2, \ldots, A_l\}$ be a set of events, called items or attributes. Let $G$ be a large set of transactions, where each transaction $T$ is a set of items such that $T \subseteq A$. Each transaction is associated with a unique identifier whose set is called $TID$. We define that a transaction $T$ contains $X$, which is a set of some items in $A$, if $X \subseteq T$. An association rule is an implication of the "$X \Rightarrow Y$" where $X \subseteq A$, $Y \subseteq A$, and $X \cap Y = \emptyset$. As a result, $X$ is called antecedent and $Y$ is called consequent of the association rule. In general, a set of items is called an itemset. Each itemset has its own associated measure of statistical significance called support. If the number of transactions containing $X$ in $G$ equals $t$, and the total number of transactions in $G$ is $N$, then we say that $support(X) = t/N$. The rule $X \Rightarrow Y$ has a measure of its strength called confidence defined as the ratio of $support(X \cup Y)/support(X)$. Calculation of the chi-squared value of the rule $X \Rightarrow Y$ is described as follows. Let $support(X) = x$, $support(Y) = y$, $support(X \cup Y) = z$ and the number of database tuples equals $N$. If the events $X$ and $Y$ are independent, we can get $support(X \cup Y) = xy$. Table 3 is the contingency of $X$ and $Y$ ; the upper parts are the expectation values under the assumption of independence, and the lower parts are observational values.

Now, let $E$ denote the value of the expectation under the assumption of independence. $O$ is the value of the observation. Then the chi-squared value is defined as follows:

$$\chi^2 = \sum_{AllCells} \frac{(O - E)^2}{E}, \tag{1}$$

We can calculate the chi-squared value using $x$, $y$, $z$ and $N$ of Table 3 as follows:

$$\chi^2 = \frac{N(z - xy)^2}{xy(1 - x)(1 - y)}. \tag{2}$$

This has 1 degree of freedom. If it is higher than a threshold value (3.84 at the 95% significance level, or 6.63 at the 99% significance level), we should reject the independence assumption.

The time related association rule is used to represent the sequence pattern between attributes in the database in the GNP-based data mining method. Let $A_i(*)(t = p)$ be an attribute in a database at time $p$ and its value is binary values of 1 or 0(after discretization). Here, $A_i(*)$ represents $A_i(Low)/A_i(Middle)/A_i(High)$. The proposed method extracts the following association rules:

$$(A_j(*)(t = p) = 1) \wedge \ldots \wedge (A_k(*)(t = q) = 1) \Rightarrow$$
$$(A_m(*)(t = r) = 1) \wedge \ldots \wedge (A_n(*)(t = s) = 1)$$
$$\text{(briefly, } A_j(*)(t = p) \wedge \ldots \wedge A_k(*)(t = q) \Rightarrow$$
$$A_m(*)(t = r) \wedge \ldots \wedge A_n(*)(t = s))$$

Here, $p \leq q \leq r \leq s$, and the first $t$ always equals 0, other time points are the relative time shifts from the first attribute. For example: $A_1(Low)(t = 0) \wedge A_2(Low)(t = 6) \Rightarrow A_3(High)(t = 22)$ means that $A_1$ is Low at time 0 and $A_2$ is Low at time 6, then $A_3$ becomes High at time 22. These kinds of rules could find time related sequential relations between attributes and would be used, for example, in prediction problems.

## 4 Time Related Association Rule Mining Using GNP

### 4.1 GNP for Association Rule Mining with Time Series

Time related Association rule mining is an extension of the GNP-based data mining [21] in terms of treating time related sequence rules. The general flow chart of GNP-based data mining method is shown in Fig. 4, For every individual there are generally 3 steps:

- Firstly, generate candidate rules with possible connections and time delays using GNP individual structure.
- Secondly, we need to examine the frequency of the appearance of the attributes in the node connections of GNP, in order to testify whether the candidate rule is interesting or not.
- Finally, calculate the criteria of the rule, i.e., support and chi-squared value using the frequency counts. Only the rules with higher criteria can be considered as the association rules and stored in the rule pool.

GNP individual examines the attribute values of database tuples using judgment nodes and calculates the measurements of association rules using processing nodes. Attributes of the database and their values correspond to judgment nodes and their judgment in GNP, respectively. Therefore, the connections of nodes are represented as the candidates of association rules. The measurements include support and Chi-squared values. Judgment node determines the next node by a judgment result of (Yes/No). Fig.5 shows a basic structure of GNP for time-related association rule mining. $P_1$ here is a processing node and is a starting point of association rule mining. Each Processing node has an inherent numeric order $(P_1, P_2, ...)$ and is connected to a judgment node. Yes-side of the judgment node is connected to another

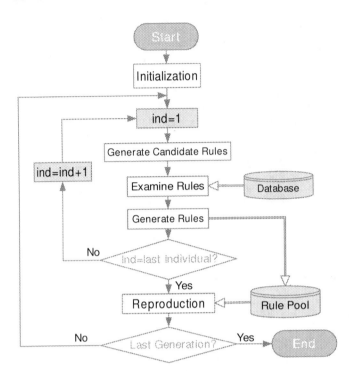

**Fig. 4** The flow chart of GNP-based data mining

judgment node. No-side of the judgment node is connected to the next num-
bered processing node. The total number of tuples in the database moving to
the Yes-side at each judgment node is calculated for every processing node.
These numbers are the fundamental values for calculating criterions(support
and chi-squared values) of the association rules.

In the proposed method, the examination should consider both the attribute dimension and the time dimension concurrrently, thus the method is
basicly two dimensional. That is, not only the attributes but also the corresponding time delays should be considered: For example, as described in
Fig.6: the judgment is not merely executed row by row, but the procedure
is like the following: firstly judge the tuple at time 0000, and according to
GNP individual structure of Fig.6, first $A_1(Low)$ is judged and if the value of
$A_1(Low)$ at time 0000 is '1', then move to the next judgement node named
$A_2(Mid)$. Then, due to the time delay $T = 2$ from $A_1(Low)$ to $A_2(Mid)$, we
check the value of $A_2(Mid)$ at time 0000+2=0002. If the value of $A_2(Mid)$
at time 0002 is '1', continue the judgment likewise, if not, execute another
turn of the judgment which begins from time 0001, 0002, 0003, ..., until the
end of the tuple.

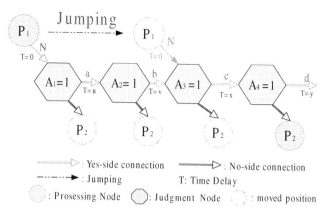

Fig. 5 The basic structure of individual

## 4.2 Extraction of Association Rule

The total number of moving to Yes-side from the processing node at each judgment node is calculated for every processing node, which is a starting point for calculating association rules. In Fig.6, $N$ is the number of the total search, and $a$, $b$, $c$ and $d$ are the number of the search moving to the Yes-side for each judgment node. The measurements are calculated by these numbers.

Table 4 shows the measurements of the support and confidence of association rules. After the search described in Fig.6, we calculate the support and confidence of the rules like Table 4, but in order to calculate the chi-squared value of the rules, we need the support of the consequent part of the rules. Therefore in the next step, we employ a jump mechanism, which make the

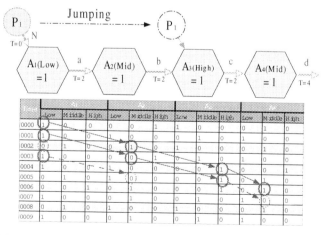

Fig. 6 Search method of time related data mining

**Table 4** Measure of association rules

| Association rules | support | confidence |
|---|---|---|
| $A_1(Low) \Rightarrow A_2(Mid)$ | b/N | b/a |
| $A_1(Low) \Rightarrow A_2(Mid) \wedge A_3(High)$ | c/N | c/a |
| $A_1(Low) \Rightarrow A_2(Mid) \wedge A_3(High) \wedge A_4(Mid)$ | d/N | d/a |
| $A_1(Low) \wedge A_2(Mid) \Rightarrow A_3(High)$ | c/N | c/b |
| $A_1(Low) \wedge A_2(Mid) \Rightarrow A_3(High) \wedge A_4(Mid)$ | d/N | d/b |
| $A_1(Low) \wedge A_2(Mid) \wedge A_3(High) \Rightarrow A_4(Mid)$ | d/N | d/c |

processing node randomly jump to another judgement node following the connection between nodes, where the judgment nodes after the jumped processing node correspond to the consequent part of the rules. Then, by counting the number of the moving to the Yes-side for each judgment node from the jumped processing node, we can calculate the support of the consequent part of the rules for obtaining the chi-squared values. These kind of jump mechanism is taken several times. For example, as represented in Fig.6, the processing node $P_1$ jumps from $A_1(Low)$ to $A_3(High)$, then, the antecedent part of the candidate rules becomes "$A_1(Low) \wedge A_2(Mid)$" and the judgement nodes after the jumped $P_1$ will be the consequent part of the candidate rule like $A_3(High) \wedge A_4(Mid) \ldots$ As another example, if $P_1$ node jumps from "$A_1(Low) = 1$" to "$A_2(Mid) = 1$", we are able to calculate the support of $A_2(Mid)$, $A_2(Mid) \wedge A_3(High)$ and so on. As a result, $chi - squared$ value can be calculated considering both the antecedent and consequent part of the candidate rules. We can repeat this like a chain operation in each generation. Thus, we can obtain the values for calculating the importance of the rules. Now, we define the important association rules as the ones which satisfy the following:

$$\chi^2 > \chi^2_{min}, \tag{3}$$

$$support \geq sup_{min}, \tag{4}$$

where, $\chi^2_{min}$ and $sup_{min}$ are the thresholds of the minimum chi-squared and support value given by supervisors. In this definition, if the rule "$X \Rightarrow Y$' is important, then, $X \Rightarrow \neg Y, \neg X \Rightarrow Y, \neg X \Rightarrow \neg Y, Y \Rightarrow X, Y \Rightarrow \neg X, \neg Y \Rightarrow X$ and $\neg Y \Rightarrow \neg X$" are also important rules. If necessary, we can also use the confidence value in the definition. The extracted important association rules are stored in a pool all together through generations in order to find new important rules.

## 4.3 Attribute Accumulation Mechanism

GNP individuals with different initial population may have different solutions, which is especially important in the databases with a large number of attributes. In order to overcome this problem and improve the robustness and stability of the algorithm, an Attribute Accumulation Mechanism is introduced to deal with a large number of attributes, where GNP individual accumulates better attributes in it gradually round by round of a sequence of generations.

The attribute accumulation mechanism proposed here firstly selects a small attribute set of size $s$ from the whole attribute set of size $S(S \geq s)$ at random, then uses GNP-based data mining method using GNP individuals generated exclusively from the selected attribute set and finally stores the extracted association rules in the corresponding Small Rule Pool(SRP) using hash functions. This whole procedure is called Round 0.

After the processing of Round 0, we get the corresponding SRP(0). For each of the rules stored in SRP(0), we check its overlap and sum up the count of the appearance of the attributes in the chosen attribute set, and finally sort the attributes from the most frequently used one to the least one. Then, the top $v\%$ $(0 \leq v \leq 100)$ of the attributes will be remained in the chosen attribute set. The attribute set of the next Round 1 is then composed of

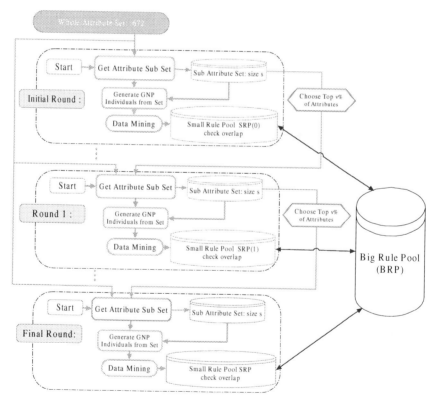

**Fig. 7** Attribute accumulation mechanism

the top $v\%$ of the attributes and attribute set randomly selected from the original whole attribute set. Using the newly generated set, Round 1 searches the important association rules and stores the newly generated rules in the corresponding SRP, likewise.

The similar procedure is repeated Round by Round until the final condition is satisfied. The procedure is shown in Fig.7.

Now, the following should be stressed, firstly, when an important rule is extracted by GNP, it is checked whether an important rule is new, i.e., whether it is already in the rule pool or not in each generation of the rounds. If the rule is new, it is stored in the rule pool, and only the rule never exist in the rule pool before can be considered as a real new rule. The appearance frequency of the attributes are also calculated using the rules in the rule pool.

Secondly, frequently extracted attributes in the rule pool have the important associations between each other potentially, thus these attributes are remained in the selected set of the next round.

Finally, we can see that the GNP individuals are just used as tools to pick up possible combinations of time related attribute sequences and to represent these sequences as time related association rules, and finally to store these rules in the big rule pool. The structure of the GNP individuals only represents some association sequences partially in the database, but does not represent all the complex association sequences in the whole datasets. Thus, the structure of the GNP individual would be less confined by the association rules already extracted in the past generations.

## 4.4   Fitness and Genetic Operator

When a new rule is generated, the rule like $"A \wedge A \Rightarrow A"$ is considered useless in the general GNP-based data mining method. However, taking account of the time related algorithm, even the same attribute has different meanings at different time points, so the rule like :

$$A_i(*)(t = p) \wedge \ldots \wedge A_i(*)(t = q) \Rightarrow$$
$$A_i(*)(t = r) \wedge \ldots \wedge A_i(*)(t = s),$$

where $p \leq q \leq r \leq s$, has the meaning. However, the number of this kind of rules should be controlled, since too many rules of this kind would harm the evolving process and produce a large number of GNP individuals with the same attribute. Now, we define the concept of multiplicity in order to overcome the problem. **Multiple** rules here mean the rules which contain many different kinds of attributes. Therefore, the fitness function of GNP is defined as:

$$F = \sum_{r \in R} \{\chi^2(r) + 10(n_{ante}(r) - 1) + 10(n_{con}(r) - 1)$$

$$+ \alpha_{new}(r) + \alpha_{mult}(r)\}.$$

The symbols are as follows:

$R$ : set of suffixes of important association rules which satisfy Eq.(3) and Eq.(4) in GNP individuals.

$\chi^2(r)$ : chi-squared value of rule $r$.

$n_{ante}(r)$ : the number of attributes in the antecedent of rule $r$.

$n_{con}(r)$ : the number of attributes in the consequent of rule $r$.

$\alpha_{new}(r)$ : constant defined as

$$\alpha_{new}(r) = \begin{cases} \alpha_{new}, & \text{if rule } r \text{ is new} \\ 0, & \text{otherwise} \end{cases}$$

$\alpha_{mult}(r)$: constant defined as

$$\alpha_{mult}(r) = \begin{cases} \alpha_{mult}, & \text{if rule } r \text{ has many kinds of} \\ & \quad \text{different attributes} \\ 0, & \text{otherwise} \end{cases}$$

$\chi^2(r)$, $n_{ante}(r)$, $n_{con}(r)$, $\alpha_{new}(r)$ and $\alpha_{mult}(r)$ are concerned with the importance, complexity, novelty and diversity of rule $r$, respectively. This fitness function is essential to evaluate GNP individuals in each generation, which means the GNP individuals which generate new rules with long antecedent and consequent attributes and with various kinds of attributes will have higher fitness value and consequently the higher chance to survive.

At each generation, GNP individuals are replaced with the new ones by the selection policy and other genetic operations. We use four kinds of genetic operators:

- Crossover: Judgment nodes are selected as the crossover nodes with the crossover rate. Two parents exchange the gene of the corresponding nodes, thus the potential useful building blocks(part of the gene structure) of the current individual might be maintained in the offspring of the next generation. This is a kind of exploitation in the evolutionary process.
- Mutation-1: The connection of the judgment nodes is changed randomly by mutation rate-1. The connection of the judgement node should be consistently changed in order to explore the various combinations of the attributes in the whole attribute set.
- Mutation-2: The function of the judgment nodes is changed randomly by mutation rate-2. The function of judgment nodes here means the attribute the judgment nodes judge.
- Mutation-3: The time delay between the judgment nodes is changed by mutation rate-3. The mutation range should be decided depending on the concrete problems to solve.

The individuals are ranked by their fitness and upper $1/4$ individuals are selected. After that, they are reproduced four times, then four kinds of genetic operators are executed to them. These operators are executed for the gene of judgment nodes of GNP individuals. All the connections of the processing nodes are changed randomly in order to extract rules efficiently.

## 4.5 Summary of the Proposed Mechanism

The whole procedure of the proposed algorithm is shown in Fig.8. Two kinds of iterations are included here, the outer iteration represents the attribute accumulation mechanism, which means better attributes are gradually accumulated in the attribute sub set of each round and GNP individuals are generated based on the attributes in the attribute sub set during each round. The inner iteration represents the basic procedure of the time related data mining method, which uses GNP individuals to generate the candidate rules, calculate the support and chi-squared values of the candidate rules based on the time related databases.

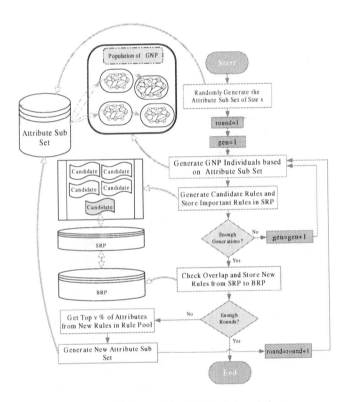

**Fig. 8** Flowchart of time related GNP data mining

# 5 Application Example

In this section, the effectiveness and efficiency of the proposed method are studied by simple traffic simulations.

## 5.1 Traffic Simulator

The main task of our simulator is to generate the databases to which we apply the proposed method.

Each section between two intersections in the road has two directions, and we assume each direction of the section represents different literals, i.e., items or attributes. The traffic simulator used in our simulations consists of the road model with $7 \times 7$ roads like Fig.9, i.e., each section has the same length, and the shape of the total road is like a grid network. Time shift in road setting of Fig.9 represents the time delay of the traffic lights between neighboring intersections.

Although the cars on the map share the same timer, they do not exactly have the same speed. Every time unit, all the cars can move forward by length 1 if and only if there exist spaces before them, thus the actual speed of all cars are influenced by the traffic lights or traffic jams. For example, if a car encounters the red light or traffic jam, it has to wait until the red light period passes or all the hindrances before it are moved. Therefore, the cars have different speeds depending on the concrete traffic situations.

The database generated by our simulator is shown in Table 5. In this table, "$W1N1, W1N2$" and "$W1N1, W1N3$" represent sections in our simulator. The average traffic volume of each section is discretized to Low/Middle/High groups, e.g., the average traffic volume of section "$W1N1, W1N2$" has the value of 8 at time unit 0001, which means it belongs to the group High at time unit 0001 considering the thresholds.

**Table 5** Database generated by simulator

| Time | $W1N1, W1N2$ | | | $W1N1, W1N3$ | | |
|------|-----|--------|------|-----|--------|------|
| | Low | Middle | High | Low | Middle | High |
| 0001 | 0 | 0 | 1 | 1 | 0 | 0 |
| 0002 | 0 | 1 | 0 | 0 | 1 | 0 |
| 0003 | 0 | 0 | 1 | 1 | 0 | 0 |
| 0004 | 1 | 0 | 0 | 0 | 0 | 1 |

Middle threshold=4; High threshold=7;

The generation of cars is based on O/D (Origin / Destination) shown in Table 6. For example, in Table 6, the " #N1" is the name of a starting/end point, and the numerical value of 12 in the table means that the car traveling from the point named "#N2" to the point named"#N1" has the traffic flow of 12 vehicles per time unit. The car traveling from the starting point to itself is forbidden here.

The parameter setting of the proposed data mining is presented in Table 7. Attributes here are the judgment node functions, for example, an attribute named "$W4N6, W4N7(Low)$" can be interpreted as the section "$W4N6, W4N7$" has low traffic. We have $7 \times 8 \times 2 = 112$ sections here in our

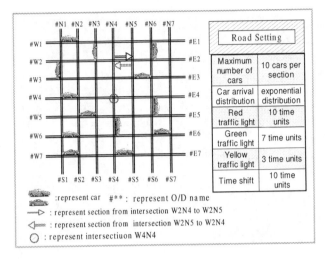

**Fig. 9** Road model used in simulations

**Table 6** Example of OD (Origin/Destination)

| O \ D | #N1 | #N2 | #N3 | #N4 |
|---|---|---|---|---|
| #N1 | ... | 7 | 1 | 8 |
| #N2 | 12 | ... | 7 | 5 |
| #N3 | 8 | 0 | ... | 6 |
| #N4 | 2 | 1 | 9 | ... |

**Table 7** Parameter setting for evolution

| Items | Values |
|---|---|
| Number of judgment nodes | 100 |
| Number of processing nodes | 10 |
| Number of attributes | 672 |
| Number of time units | 800 |
| Number of generations per Round | 50 |
| Sub attribute set size | 100 |

simulator and each section has two directions, so, there exist $112 \times 2 = 224$ sections in total. What's more, each section has 3 categories (Low/Mid/High), thus we have $224 \times 3 = 672$ attributes. Time units in Table 6 represents the number of total time units in our database. Simulator runs for 5000 time units, but, removes the first and last 500 time units for stabilization from the proposed data mining calculation. The samples are taken every 5 time units, i.e., (5000-500-500)/5= 800, thus it leads to 800 time units in the following simulations.

## 5.2  Simulation Results

The aim of our data mining is to extract rules on the time-related association relations among all of the sections on the map, and our method can be applied to the database with a large number of attributes, e.g., 672 sections in our simulations.

### Test Case 1

In test case 1, the number of rules stored in the Big Rule Pool(BRP) is compared between the proposed method and the conventional method without the attribute accumulation mechanism. Each round has the same number of generations of 50 and the selected set size is 100.

**Fig. 10** Comparison of the number of rules between the proposed method and conventional method

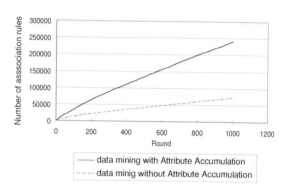

Fig.10 shows the number of rules obtained in the BRP versus round number. In the conventional method in Fig.10, GNP individuals are randomly initialized from the whole attribute set at the beginning of each "round". We can see from Fig.10 that the proposed method can extract important association rules efficiently, when compared with the conventional one.

The rule extracted is like the following:

$W4N5, W4N6, Low(t = 0) \Rightarrow$
$W4N5, W3N5, High(t = 9) \wedge W3N5, W3N6, High(t = 11)$

The above rule means that the section on the road named "$W4N5, W4N6$" has low traffic at time 0, then the section named "$W4N5, W3N5$" will probably have high traffic volume at time 9, and the section named "$W3N5, W3N6$" will also possibly have high traffic volume at time 11.

As Fig.11 describes, the above rule represents the trend that more cars will choose the section "$W4N5, W3N5$" and "$W3N5, W3N6$", which is triggered by the past low traffic volume of the section "$W4N5, W4N6$". According to this information, we can update the routing algorithm more accurately.

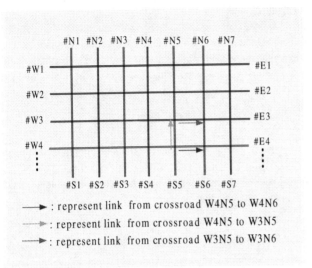

Fig. 11 Map and rule

## Test Case 2

The aim of the second simulation is to study the relations between the selected attribute set size and the performances of the algorithm. With other parameter setting being unchanged, the attribute set size of 25, 50, 100, 125, 150, 200, 400 and 672 has been studied. The result is shown in Fig.12.

Fig. 12 Number of rules in case of changing the attribute size

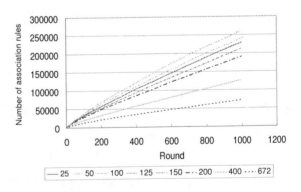

From Fig.12 we can see that the performance of the attribute set size 672 is almost the same as the result of the conventional method in Fig.10, since the number 672 is the total attributes size.

We can also see from Fig.12 that both too large or too small attribute set size will harm the performance of the algorithm and the best solution should be selected according to the concrete problem to solve. In our cases, the best solution is obtained when we use the attribute set size of 50.

## Test Case 3

The third simulation is to explore the relations between the generation size per round and the performances of the algorithm. With other parameter setting being unchanged, the performances of the algorithm with generation size of 25, 50, 100 and 150 are compared. The result is shown in Fig.13.

The average number of finding new rules per generation is shown in Fig.14 when changing the generation size per round. Since the time consumed in one generation is almost the same in four cases, the average number of extracted rules per generation in Fig.14. actually represents the time complexity of four cases.

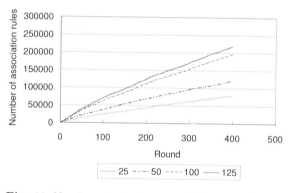

**Fig. 13** Number of rules in case of changing the generation size

The larger generation size certainly contributes to finding more rules, however, the time complexity increases. Considering this trade-off and that the different round has different convergence rate, the fixed size of generations per round is not a good method, which should be studied more in the future.

## Test Case 4

The main purpose of the last simulation in this paper is to study the effects of changing the accumulation percentage. With other parameter setting being unchanged, the top 10%, 20%, 40%, 50% and 70% attributes of the new rules in BRP are used as the attribute sub set of the next round. Here, the top attributes mean the attributes with the most frequent occurrence in BRP. The results are shown in Fig.15.

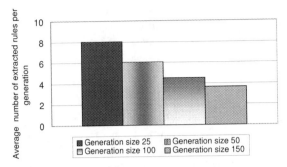

**Fig. 14** Average number of rules extracted per generation

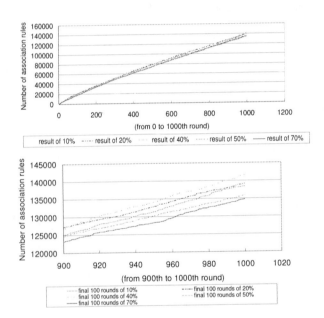

**Fig. 15** Number of rules in case of changing the accumulation percentage

The figure in the upper side of Fig.15 represents the number of rules obtained from 0 to $1000^{th}$ round. It is found from the figure that there is no huge difference among each other, since all of them are capable of obtaining a large number of association rules. However, we can examine the difference more clearly from the figure shown in the lower side of Fig.15, where the number of rules generated in the final 100 rounds is shown. We can see from the figure that both too large or too small accumulation percentage will harm the performance of the algorithm. And the best solution is obtained around 40% in our simulation.

# 6 Conclusion

In this chapter, a method of association rule mining using Genetic Network Programming with time series processing mechanism and attribute accumulation mechanism has been proposed. The proposed method can extract important time-related association rules efficiently. Extracted association rules are stored temporarily in Small Rule Pool(SRP) and finally all together in Big Rule Pool(BRP) through rounds of generations. These rules are representing useful and important time related association rules to be used in the real world. We have built a simple road simulator and examined the effectiveness and usefulness of our algorithm. The results showed that our proposed method extracts the important time-related association rules in the database efficiently and the attribute accumulation mechanism improves the performance considerably. These rules are useful in time-related problems, for example, traffic prediction. We are now studying the improvements of our method in terms of the combination of the proposed method with the optimal route algorithm.

# References

1. Zhang, C., Zhang, S.: Association Rule Mining: models and algorithms. Springer, Heidelberg (2002)
2. Agrawal, R., Srikant, R.: Fast Algorithms for Mining Association Rules. In: Proc. of the 20th VLDB Conf., pp. 487–499 (1994)
3. Brin, S., Motwani, R., Silverstein, C.: Beyond market baskets: generalizing association rules to correlations. In: Proc. of the 1997 ACM SIGMOD Conf., pp. 265–276 (1997)
4. Klemettinen, M., Mannila, H., Ronkainen, P., Toivonen, H., Verkamo, A.I.: Finding Interesting Rules from Large Sets of Discovered Association Rules. In: Proc. of the third Int'l Conf. Information and Knowledge Management, pp. 401–408 (November 1994)
5. Savasere, A., Omiecinski, E., Navathe, S.: An Efficient Algorithm for Mining Association Rules in Large Databases. In: Proc. of the 1995 Int'l Conf. Very Large Data Bases, pp. 432–443 (September 1995)
6. Park, J.S., Chen, M.S., Yu, P.S.: An Effective Hash-Based Algorithm for Mining Association Rules. In: Proc. of the 1995 ACM SIGMOD Conf., pp. 175–186 (1995)
7. Wu, X., Zhang, C., Zhang, S.: Efficient Mining of Both Positive and Negative Association Rule. ACM Transactions on Information Systems 22(3), 381–405 (2004)
8. Andrews, R., Diederich, J., Tickle, A.B.: Survey and critique of techniques for extracting rules from trained artificial neural networks. Knowlege-Based Systems 8(6), 373–389 (1995)
9. Rules and Networks. In: Andrews, R., Diederich, J. (eds.) Proc. of the Rule Extraction from Trained Artificial Neural Networks Workshop, AISB 1996, Queensland University of Technology (1996)

10. Holland, J.H.: Adaptation in Natural and Artificial Systems. University of Michigan Press, Ann Arbor (1975)
11. Goldberg, D.E.: Genetic Algorithm in search, optimization and machine learning. Addison-Wesley, Reading (1989)
12. Koza, J.R.: Genetic Programming, on the programming of computers by means of natural selection. MIT Press, Cambridge (1992)
13. Koza, J.R.: Genetic Programming II, Automatic Discovery of Reusable Programs. MIT Press, Cambridge (1994)
14. Tung, A.K.H., Lu, H., Han, J., Feng, L.: Efficient Mining of Intertransaction Association Rule. IEEE Transactions on Knowledge and Data Engineering 15(1), 43–56 (2003)
15. Wu, S., Chen, Y.: Mining Nonambiguous Temporal Patterns for Interval-Based Events. IEEE Transactions on Knowledge and Data Engineering 19(6), 742–758 (2007)
16. Kam, P.S., Fu, A.W.C.: Discovering Temporal Pattern for Interval-Based Events. In: Kambayashi, Y., Mohania, M., Tjoa, A.M. (eds.) DaWaK 2000. LNCS, vol. 1874, p. 317. Springer, Heidelberg (2000)
17. Omlin, C.W., Giles, C.L.: Extraction of Rules from Discrete-time Recurrent Neural Networks. Neural Networks 9(1), 41–52 (1996)
18. Mabu, S., Hirasawa, K., Hu, J.: A Graph-Based Evolutionary Algorithm: Genetic Network Programming (GNP) and Its Extension Using Reinforcement Learning. Evolutionary Computation 15(3), 369–398 (2007)
19. Eguchi, T., Hirasawa, K., Hu, J., Ota, N.: A study of Evolutionary Multiagent Models Based on Symbiosis. IEEE Trans. on Systems, Man and Cybernetics - Part B 36(1), 179–193 (2006)
20. Hirasawa, K., Eguchi, T., Zhou, J., Yu, L., Hu, J., Markon, S.: A Double-Deck Elevator Group Supervisory Control System Using Genetic Network Programming. IEEE Tran. on System, Man and Cybernetics -Part C 38(4), 535–550 (2008)
21. Shimada, K., Hirasawa, K., Hu, J.: Genetic Network Programming with Acquisition Mechanisms of Association Rules. Journal of Advanced Computational Intelligence and Intelligent Informatics 10(1), 102–111 (2006)

# Growing Self-Organizing Map for Online Continuous Clustering

Toby Smith and Damminda Alahakoon

**Abstract.** The internet age has fuelled an enormous explosion in the amount of information generated by humanity. Much of this information is transient in nature, created to be immediately consumed and built upon (or discarded). The field of data mining is surprisingly scant with algorithms that are geared towards the unsupervised knowledge extraction of such dynamic data streams. This chapter describes a new neural network algorithm inspired by self-organising maps. The new algorithm is a hybrid algorithm from the growing self-organising map (GSOM) and the cellular probabilistic self-organising map (CPSOM). The result is an algorithm which generates a dynamically growing feature map for the purpose of clustering dynamic data streams and tracking clusters as they evolve in the data stream.

## 1 Introduction

Artificial neural networks are a commonly used data mining technique which were inspired by the operation of biological neural networks in the mammalian brain [23, 12]. They are, in a very general sense, a network of interconnected, discrete adaptive processing elements (artificial neurons) with weighted inputs and outputs. Artificial neural networks acquire their intended functionality through repeated exposure to input and supervisory stimulus however the precise functionality of the networks is an emergent property arising from the architecture of the network and the behavioural characteristics of the particular neurons. There have been neural network architectures

Toby Smith

Cognitive and Connectionist Systems Lab, Monash University, Australia

e-mail: `tobin.smith@infotech.monash.edu.au`

Damminda Alahakoon

Cognitive and Connectionist Systems Lab, Monash University, Australia

e-mail: `damminda.alahakoon@infotech.monash.edu.au`

A. Abraham et al. (Eds.): Foundations of Comput. Intel. Vol. 4, SCI 204, pp. 49–83.
springerlink.com © Springer-Verlag Berlin Heidelberg 2009

designed with applications in all areas of data mining including trend prediction, unsupervised clustering, supervised classification, and image recognition. This chapter is concerned with a type of neural network architecture
and algorithm called a Self-Organising Map (SOM) which has broad application in the data mining area of unsupervised clustering. SOMs as they are
known today owe most of their development to the work of Teuvo Kohonen
in 1982 [16] whose research is in turn based upon a model of the visual cortex
of higher vertebrates developed by Von der Malsburg in 1973 [30].

The SOM algorithm maps input vectors to a lower dimensional lattice of
neurons which is typically 2 or 3 dimensional (as this is easily visualised).

After repeatedly exposing a SOM network to input stimuli, the SOM adjusts its internal state variables (synaptic weightings) so that similar input
vectors map close together on the output lattice, and gradually this builds a
lower dimensional, granular representation of the input data. The purpose of
SOMs is essentially to perform a type of topological vector quantisation of
the input data with the aim of reducing the dimensionality of the input data
but still preserve as much as possible of the spatial relationships within the
data.

Figure 1 is an example of the type of map generated by the growing
self-organising map (GSOM) algorithm. In this case, data about numerous animals has been acquired and used as input into the algorithm. This
map clearly delineates prominent clusters such as mammals, birds, and
fish as well as giving an indication of the spatial relationship between the
clusters.

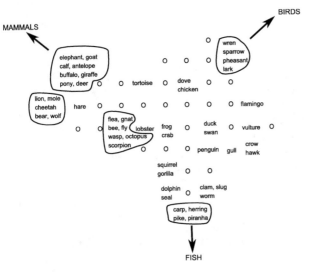

**Fig. 1** Feature map generated using the GSOM algorithm [3]

While SOMs have yielded many practical successes, SOMs have some limitations which have spawned many derivative algorithms. The limitations of the SOM algorithm that are of most interest to this research are:

- Size of the network has to be pre-specified at the start of the algorithm. Different applications require different map sizes. This can consume unnecessary computational resources during the training stage of the algorithm as different map sizes might have to be trialled.
- The SOM is an heuristically derived algorithm which can hinder rigorous mathematical analysis of a SOMs behaviour.
- SOMs produce clusters that essentially have the same level of detail (neuron resolution) however some clusters have a more complex structure than others and hence merit a higher level of map detail. This is especially important in hierarchical clustering (finding clusters within clusters).
- The original SOM algorithm can only learn static data sets since once a map learns and stabilises it loses its capacity to re-shape itself when new structures become manifest in the input data, although continuous learning alternatives have been developed [24, 9].
- The SOM algorithm can produce topologically twisted maps however this can usually be mitigated through appropriate tuning of the algorithms parameters or re-dimensioning the output lattice.

These concerns provide the primary motivation for the research reported in this chapter. Two algorithms which achieve some gains with regard to these issues are the growing self-organising map (GSOM) [3] and the cellular probabilistic self-organising map (CPSOM) [7].

The GSOM is a SOM algorithm which employs a dynamically growing neural network and as a result mitigates the issue regarding pre-specifying the map size as the map is dynamically grown according the needs of the data analyst and the complexity of the data. The GSOM algorithm also incorporates a parameter called the *Spread Factor* (SF) which controls the map resolution. By initially setting the SF to a lower value a data analyst may initially generate a fairly coarse map containing only the main clusters and filter out the data belonging to a cluster and run this data through the algorithm using a larger SF value to zoom in on the cluster and examine any subclusters. In this way a form of hierarchical clustering is achieved. Another advantage of the GSOM algorithm is that clusters tend to be more visually obvious to the data analyst since the GSOM generates maps with irregular geometries which adhere to the shape and location of prominent clusters.

The GSOM algorithm trains its network in a similar fashion to the SOM which has a largely heuristic derivation whilst the CPSOM is a more rigorously derived learning algorithm with a foundation in statistical mechanics.

The CPSOM is an online algorithm (making it suitable for large data sets) which incorporates a *forgetting* parameter which allows the network to forget stale patterns and adapt to new ones as they are presented. This perpetual maleability, made possible by the *forgetting factor*, allows the algorithm to better track the clusters which might be present in dynamic data streams whereas the maps produced by traditional SOMs will eventually stabilise and become inflexible.

This research seeks to unify the benefits of each algorithm by amalgamating the growing feature of the GSOM into the more rigorous learning framework of the CPSOM algorithm.

This chapter is divided into several sections. Following this introduction, section 2 will introduce the SOM in more detail and review research that is related to the CPSOM and GSOM algorithms as well as describing the algorithms in detail. In order to justify my choice of CPSOM and GSOM algorithms in this research, an overview of similar approaches which have had successes in overcoming some of the limitations of the SOM mentioned previously will also be briefly discussed. Section 3 will describe in more detail the advantages of combining the two algorithms and how this can be achieved, and introduce the new algorithm. In Section 4, the results of some experiments will be presented using the new algorithm and analysised such that meaningful comparisons to the CPSOM and GSOM can be made. In sections 5, the strengths and weaknesses of the proposed algorithm will be discussed in light of the experimental results. Section 6 will summarise how the algorithm performs with respect to the aims of this research and discuss applications and future research directions for the new algorithm.

## 2 Polymorphic Self-Organizing Maps and Clustering of Dynamic Data Streams

As alluded to in the introduction, the self-organising map is an artificial neural network consisting of an $l$ dimensional array of input neurons which are *fully connected* to a $p$ dimensional lattice of neurons ($p$ is typically 1, 2, or 3 as this facilitates easy visualisation). The architecture of the SOM (with $p = 2$) is illustrated in Fig. 2.

For each lattice neuron there is a corresponding $l$ dimensional weight vector. The SOM output layer is essentially a 'winner takes all' layer whereby the output neuron whose weight vector is closest to the input vector is

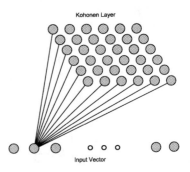

**Fig. 2** Kohonen SOM neural network topology

activated and the activation of all other lattice neurons are suppressed. The original SOM architecture is trained by gradually aligning the weight vector of the winning neuron (and to a lesser degree, neighbouring neurons) to lie closer to the input vectors as they are presented to the network.

A batch version of the algorithm (as opposed to the *online* version) has been developed which updates all the weight vectors after the presentation of a complete data set. Batch algorithms are not ideally suited to dynamic data clustering as the algorithm must be capable of continuous online learning (this would require the input data to be sampled for batch algorithms).

The training algorithm for the online version is enumerated as follows:

1. Initialise all the weight vectors. Typically, initialised to random vectors.
2. For each input vector, $x(t)$, calculate the winning neuron, $c$, in the Kohonen layer by

$$c = \arg\min_i \|x(t) - w_i\|, \qquad i = 1, 2, \ldots, N \tag{1}$$

   where N is the size of the feature map and $w_i$ is the weight of the $i$th neuron.
3. Calculate the new neural weightings by

$$w_i(t+1) = \begin{cases} w_i(t) + \varepsilon(t)h(c,i)(x(t) - w_i(t)), \forall i \in N_c \\ w_i(t), \qquad\qquad\qquad\qquad\qquad \text{otherwise} \end{cases} \tag{2}$$

   where $N_c$ is the set of neurons which lie in the neighbourhood of $i$, $\varepsilon(t)$ is the learning rate at time $t$ (decreases with time), and $h(c,i)$ is the neighbourhood function centred at the winning neuron $c$. $h(c,i)$ decreases as the distance of neuron $i$ from neuron $c$ increases. The Gaussian function is most often used for the neighbourhood function.
4. Contract the neighbourhood function slightly in order to diminish the lateral influence that the winning neuron has on its neighbouring neurons. This will ensure that the map stabilises and converges over time, however, contracting the neighbourhood function too fast will tend to generate maps with topological twists in them.
5. Repeat from step 2 until a pre-specified number of epochs has elapsed or the mean squared error of neurons is sufficiently low.

$$\text{MSE} = \frac{1}{N}\sum_{i=1}^{N} \|x_i(t) - w_i(t)\|^2 \tag{3}$$

In this way the SOM partitions the input space into a set of convex Voronoi regions:

$$V_i = \{x| \ \| m_i - x \| < \| m_j - x \|, i \neq j\} \tag{4}$$

where $m_i$ is the midpoint of Voronoi region $V_i$

The SOM algorithm encompasses the twin concepts of competition and cooperation. Each neuron globally competes with each other for activation depending on how close they are to the input vector. Once a neuron has 'won', the weight vectors are moved towards the input vector with a strength dependant upon how close each neuron is to the winning neuron in the Kohonen layer (by way of the neighbourhood function). In this way, the winning neuron cooperates with the neurons in its local neighbourhood. In practice, not all the weight vectors need to be adjusted so long as neurons within a sufficiently large enough neighbourhood of the winning neuron have their weights adjusted (i.e. the neighbourhood function is effectively truncated).

Although this algorithm has proven capability in data clustering, Teuvo Kohonen has identified five properties [16] of SOMs which can affect the accuracy or clarity of the feature map:

Magnification Factor The magnification factor concerns the frequency that input events are mapped to a location on the feature map. This is typically not constant across mapping and as a result different map resolutions in the feature map may be required to generate accurate maps of some clusters but not others. It was found that the magnification (or scale) factor seemed to be approximately proportional to the occupied frequency. To quote Kohonen, "the network resources are utilised optimally in accordance with need".

Boundary Effects Boundary effect can sometimes be observed at the edge of maps as a consequence of the outer neurons having a truncated neural neighbourhood on one side. Typically this is usually a *contraction* of the map around the edges resulting from the distribution of weight vectors at the edges being rotated more inwards rather than outwards.

Pinch Phenomenon This phenomenon occurs when the distribution of weight vectors form a *ring* rather than spreading out into a planar arrangement. This typically results when the lateral range of interactions between the neurons in the map is too short (i.e., the neighbourhood function is too contracted).

Collapse Phenomenon This effect is an extension of the pinch effect and occurs when the range of lateral interactions are too long. This tends to *collapse* the weight distributions of the map as large groups of input vectors yield the same response.

Focusing Phenomenon This effect is the opposite of the collapse phenomenon and happens when a neuron (or group of neurons) become sensitised to a significant proportion of the training vectors. It can occur when the amount of lateral interactions between neighbouring neurons is too weak or if the normalisation of

> weight vectors is such that it becomes impossible to discriminate between the responses of neighbouring neurons. Neurons can become sensitised to training vectors if a significant number of similar vectors are presented in a sequence but this can be overcome by presenting the training vectors in random order.

Kohonen suggests that the last three effects are usually a result of poor design and can often be overcome through appropriate tuning of the learning rate and neighbourhood function.

In addition to these limitations and those listed in the introduction, SOMs do not provide complete topology preservation as quantified by Villmann et al [29].

## 2.1 SOM Inspired Algorithms

There have been an enormous amount of papers published in the area of SOMs since they were developed, in a wide range of disciplines, with many variants and tweaks to the algorithms having being developed. This section briefly discusses some of the algorithms which share common features with the new algorithm being developed as part of this research. Specifically, these features are: Capable of dynamic data clustering, polymorphic neural structure, and proven robustness in generating quality feature maps.

This section does not attempt to review other techniques that may be used for dynamic data clustering such as Bayesian networks, statistical techniques, etc.

### Growing Hierarchical Self-Organising Map

The growing hierarchical self-organising map (GHSOM) [25] is a technique which starts by using a small SOM to determine the main clusters than iteratively generate new SOMs to *break apart* clusters into their subclusters thereby exposing their hierarchy. Each map has the capability to grow new rows or columns in order to generate mappings at a particular detail level. Once a specified level of granularity is reached in a SOM, any nodes which have too diverse inputs mapped to them are expanded into small SOMs at a lower layer. The GHSOM algorithm contains two parameters for controlling the growth process and the granularity of the generated maps.

### Growing Neural Gas Model

The growing neural gas (GNG) network developed by Bernd Fritzke [9] is a technique which uses competitive Hebbian learning in a similar way to the neural gas (NG) model [22] with the advantage that it is an incremental model capable of continuous learning and only constant parameters. In this model, the network is initialised with two randomly placed neurons in the output space. The network is iteratively evaluated using local statistical metrics obtained in prior adaptation steps and new nodes are added (or deleted) were appropriate as training progresses. This process is illustrated in the growing cell structures model by the same author [8], although that model only has a fixed dimensionality. GNG is more targeted towards generating an optimal topology than the NG model.

As this algorithm is capable of continuous online learning and nodes tend to be pruned in a Hebbian fashion, the model is capable of dynamic data clustering although the output maps tend to be hard to visualise due to a topology which can have differing local dimensionalities.

## Dynamic Adaptive Self-Organising Hybrid Model

The dynamic adaptive self-organising hybrid algorithm (DASH) algorithm [13] was primarily designed for dynamic data clustering (although hierarchical clustering was another prime consideration). DASH is essentially a hybrid of GNG and the GHSOM. DASH has the following features: dynamic structure (including node pruning), hierarchical clustering, parameter self adaptation, and dynamic data learning. The authors of the original paper have shown that DASH outperforms SOM and GNG on static data and is much superior at tracking dynamic data in comparison to the SOM, However, its efficacy against other online algorithms has yet to be established.

## Multiple Self-Organising Maps

Multiple Self-Organising Maps (M-SOMs) is a technique developed by Nils Goerke et al [10] for the purpose of extracting information out of nonlinear dynamic systems. The M-SOM architecture consists of a set of independent partner SOMs (with their own topology, size, and dimensionality) which collectively cover the feature space. This approach has the advantage that each partner SOM will be less susceptible to twisting (due to their size [17]) than would a normal SOM covering the entire feature space. It was noted by the authors that as the number of maps approaches 1, the M-SOM becomes equivalent to the classical SOM whilst if the size of each partner SOM shrinks to 1, the M-SOM becomes a Neural Gas. With this algorithm, each output class can be represented by a whole SOM rather than a discrete neuron as is the case with the classical SOM.

Because each map is topologically distinct, there is greater scope for parallel computation.

The Authors demonstrated the M-SOMs ability to efficiently Map / Classify the behaviour of chaotic attractors (in this case, the Lorenz and Roessler attractors).

## Plastic Self Organising Map

There has been relatively little research into using SOMs for the purpose of dynamic data clustering with the notable exception of CPSOM. Another notable exception is the Plastic Self Organising Map (PSOM) developed by Lang & Warwick [18]. The PSOM is a continuously learning dynamic neural network which adds new neurons in order to capture new information and incorporates an ageing process to remove neurons that encode stale information. According to Lang et al, the PSOM organises "itself into a visual graph structure displaying high dimensional data in a temporal submanifold domain" at the expense of producing a more topographic representation of the data. It was shown that the PSOM can track and classify non-stationary data sets although its effectiveness as compared with other algorithms is not entirely known. One drawback of the PSOM algorithm is that it has parameters controlling the growth and pruning nature of network which need to be set without knowing anything of the patterns present in the input data (this is a common issue with many dynamic SOMs). Tuning of these parameters was achieved via trial and error over multiple simulations. A self-tuning process for network parameters has been suggested as a topic for further investigation.

## 2.2 *Deterministic Annealing*

Deterministic annealing [26] is not a technique that is directly related to SOMs however it is the basis of the CPSOM algorithm (described in a subsequent section) which is pertinent to the algorithm presented in this chapter. The technique is described here in a fairly cursory way and is only included so that the reader has more of an appreciation for the derivation of the CPSOM algorithm.

Deterministic Annealing is a technique derived from statistical physics whereby data is probabilistically assigned to classes such that their distributions minimise a cost constrained *randomness* (Shannon's entropy [28]).

In physical chemistry, materials often have many crystalline defects caused by an inability of atoms to self-organise into a regular lattice that has minimal internal energy. Annealing refers to the process of heating a material giving its atoms sufficient energy to transition between states of higher energy and then gradually cooling the material giving the atoms a chance of finding a configuration with a lower internal energy than the initial one (i.e., the atoms are less likely to be trapped in local energy minima). Through this process the size of the crystals in the material increases and the number of defects is reduced.

Deterministic annealing essentially replaces the need for stochastic simulation (as in simulated annealing) by use of an expectation measure. An energy function (incorporating a pseudo temperature parameter) is derived from this expectation and is deterministically optimised as its temperature parameter is slowly reduced.

Let $x$ denote a source vector, $y(x)$ denote denote the Voronoi region that $x$ belongs in, $p(x)$ denote the probability that $x$ should be mapped to Voronoi region $y(x)$, and $d(x, y(x))$ be the distance between input vector $x$ and Voronoi region $y(x)$. The cost criterion we want to minimise is the total quantisation error for the feature map (expected distortion)

$$D = \sum_x p(x) \mathrm{d}(x, y(x)) \approx \frac{1}{N} \sum_x \mathrm{d}(x, y(x)) \tag{5}$$

assuming that the source distribution can be approximated by a finite training set of $N$ independent vectors.

Deterministic annealing is a probabilistic framework defined by randomisation of the input space into partitions with input vectors being assigned to partitions based on an *assignment probability*. The randomisation of the partition leads us to rewrite the expected distortion as

$$D = \sum_x \sum_y p(x, y) \mathrm{d}(x, y)$$
$$= \sum_x p(x) \sum_y p(y|x) \mathrm{d}(x, y) \tag{6}$$

where $p(x, y)$ is the joint probability distribution and $p(x|y)$ is the conditional probability relating input vector $x$ to Voronoi region $y$. We seek to minimise $D$ subject to a specified level of randomness (as measured by the Shannon entropy [28])

$$H(X, Y) = -\sum_x \sum_y p(x, y) \log p(x, y) \tag{7}$$

This can be reformulated as a minimisation of the Lagrangian

$$F = D - TH \tag{8}$$

where $T$ is the Lagrange multiplier. Alternatively this is equivalent to maximising the lagrangian corresponding to $H - \beta D$ with the lagrange multiplier $\beta = 1/T$.

The joint entropy can be rewritten as $H(X, Y) = H(X) + H(Y|X)$ with $H(X) = -\sum p(x) \log p(x)$ is the source entropy. Since $H(X)$ is independent of clustering (and essentially constant for our purposes), we may eliminate it from the lagrangian and use only the conditional entropy

$$H(Y|X) = -\sum_x p(x) \sum_y p(y|x) \log p(y|x). \tag{9}$$

Minimising $F$ with respect to $p(y|x)$ yields

$$p(y|x) = \frac{\exp\left(-\frac{d(x,y)}{T}\right)}{\sum_y \exp\left(-\frac{d(x,y)}{T}\right)} \tag{10}$$

where the denominator normalises the distribution.

With further derivation (see [26]), we obtain the following condition which minimises $F$

$$\frac{1}{N} \sum_x p(y|x) \frac{d}{dy} d(x,y) = 0 \quad \forall y \in Y. \tag{11}$$

The process of deterministic annealing can be summarised as

1. Initialise the codevectors
2. calculate the assignment probabilities using (10)
3. Optimise the codevectors via (11)
4. Decrease the temperature parameter gradually
5. repeat from 2

## 2.3 The Cellular Probabilistic Self-Organising Map

A useful way of describing vector quantisation is as a technique for transforming information from a higher dimensional space to a lower dimensional space in such a way that a given cost function is minimised based on an appropriate distortion measure. This definition of vector quantisation is at the core of the development of the cellular probabilistic self-organising map (CPSOM), which has a more rigorous mathematical basis than most other SOM related algorithms which are often developed heuristically.

Luttrell [19, 20, 21] developed the topological vector quantisation (TVQ) algorithm by modelling topological mapping using a vector quantisation model with a noisy transmission medium. The author showed that the SOM is equivalent to an efficient approximation of gradient descent optimisation performed on the TVQ cost function and hence SOMs are related to noisy vector quantisation. Luttrell also developed a batch version of the TVQ algorithm which is a predecessor to the CPSOM.

In order to prevent the TVQ algorithm becoming stuck in local minima, Graepel et al [11] applied deterministic annealing to the optimisation of the TVQ cost function and derived the soft topological vector quantisation (STVQ) algorithm. By considering what happens when the STVQ neighbourhood function tends towards the SOM neighbourhood function, a *fuzzy* version of the SOM was obtained called the soft self-organising map (SSOM). Graepel et al showed that as the temperature parameter is increased towards infinity in the SSOM algorithm, a batch version of the SOM is obtained.

It was subsequently shown that the STVQ algorithm (due to it being equivalent to the fuzzy VQ algorithm [27]) is a variant of expectation-maximisation (EM) optimisation as illustrated by Kloppenburg et al [15], this enabled Chow to derive the CPSOM [7] algorithm from the STVQ algorithm by replacing the batch EM algorithm in the STVQ with an online EM algorithm [1] thus allowing the algorithm to learn incrementally from data (a requirement for clustering dynamic data sets). The algorithm was further adapted for dynamic data clustering through the inclusion of a *forgetting* parameter which acts to reduce the weight of old input data stored in the neuron state variable, $B_i(t)$ (see algorithm later in this section). Another feature of the CPSOM algorithm is that it is more memory efficient than the STVQ algorithm and hence can utilise larger data sets. Fig. 3 illustrates the relationships between the various algorithms mentioned.

**Fig. 3** CPSOM related
algorithms

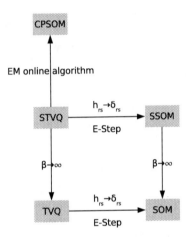

By way of motivation for developing the CPSOM algorithm, Chow asserts that the STVQ has a "slow convergence rate due to the batch EM algorithm" and "is easily trapped in local minima such that topological maps among neurons may be disordered".

The "cellular" in the CPSOM name derives from the fact that each neuron is locally connected to other neurons (like in biological cells) and similarly the "probabilistic" in CPSOM derives because there is a soft probability assignment associated with each neuron calculated from the input data.

The CPSOM was tested using two synthetic static data sets, one real static data set, and two synthetic dynamic data sets. The following features were demonstrated by the author:

- The CPSOM shows faster convergence than the STVQ with small map sizes.
- The CPSOM is less likely to be trapped in local minima with large map sizes and maps tend to be more ordered with lower quantisation errors.
- The effect of the forgetting factor was demonstrated successfully on the synthetic dynamic data sets.

The complete algorithm is illustrated as follows:

1. Initialise variables $\beta = \beta^{\text{start}}$, $\gamma = \gamma^{\text{start}}$, and static neighbourhood function $h_{ij}$.
2. Select an input vector x from the training data.
3. Update weightings and state variables for all neurons according to the following two steps

   Expectation Step:

$$P_i(x(t)) = \frac{\exp\left(-\beta \sum_{j=1}^{N} h_{ij} D(x(t), w_j)\right)}{\sum_{n=1}^{N} \exp\left(-\beta \sum_{j=1}^{N} h_{nj} D(x(t), w_j)\right)}, \quad i = 1, 2, \ldots, N. \quad (12)$$

   Maximisation step:

$$w_i(t) = w_i(t-1) + \frac{1}{B_i(t)} \sum_{j=1}^{N} h_{ij} P_j(x(t))(x(t) - w_i(t-1)), \quad i = 1, 2, \ldots, N \quad (13)$$

   where

$$B_i(t) = B_i(t-1) + \sum_{j=1}^{N} h_{ij} P_j(x(t)), \quad \text{for } i = 1, 2, \ldots, N. \quad (14)$$

   Here, $N$ is the size of the output map and $D(x(t), w_j)$ is the quantisation error between $x(t)$ and the weight $w_j$ of neuron $j$.

4. If current iteration $t$ is an integer multiple of parameter $\lambda_1$ and if $\beta < \beta^{\text{final}}$ then increment $\beta$ by positive incremental factor $\Delta_1$.
   If the current iteration $t$ is an integer multiple of parameter $\lambda_2$, set $B_i(t) = B_i(t)/\gamma$ where $\gamma \leftarrow \gamma - \Delta_2$, $\gamma \geq 1$ and $\Delta_2 > 0$. $\gamma$ is also called the *forgetting* factor as it reduces the weightings of old input data stored in the neuron state variables $B_i(t)$.
5. This step is used only for dynamic data sources. If the current iteration is a multiple of parameter $\lambda_3$ where $\lambda_3$ is an integer multiple of $\lambda_2$ and $\lambda_3 \gg \lambda_2$, set $\gamma = \gamma^{\text{start}}$.
6. Repeat procedure from step 2 or until a pre-specified epoch has been reached (for static data sets).

The variable $\beta$ is the inverse temperature parameter which is gradually increased from $\beta^{\text{start}}$ to $\beta^{\text{final}}$. The neuron state variable $B_i(t)$ can be considered an inverse learning rate for the weight updation. $B_i(t)$ is gradually increased per iteration according to the assignment probability of the input vector and gradually slows down the trainability of the network. However, every $\lambda_2$ iterations, $B_i(t)$ is decreased by a factor of $\gamma$ which increases the malleability of neural weightings allowing the network to adapt to new patterns as they are presented. In this way, $\gamma$ acts as a *forgetting factor* for the algorithm. After $\lambda_3$ iterations, $\gamma$ is reset to $\gamma^{\text{start}}$, allowing the network to maintain it's retrainability which is needed for clustering dynamic data streams.

The parameters $\beta^{\text{start}}$, $\Delta_1$, and $\lambda_1$ should be selected such that $\beta$ reaches $\beta^{\text{final}}$ before $\frac{2}{3}T^{\text{final}}$ epochs have been presented for static data sets (or $\frac{2}{3}\lambda_3$ iterations since $\gamma$ was reset to $\gamma^{\text{start}}$ for dynamic data sets). Similarly, the parameters $\gamma^{\text{start}}$, $\Delta_2$, and $\lambda_2$ should be selected such that $\gamma$ reaches $\gamma^{\text{final}}$ before $\frac{2}{3}T^{\text{final}}$ epochs have been presented for static data sets (or $\frac{2}{3}\lambda_3$ iterations since $\gamma$ was reset to $\gamma^{\text{start}}$ for dynamic data sets). The forgeting factor, $\lambda_3$, should be determined by the pattern-varying rate of the dynamic data set.

## 2.4   The Growing Self-Organising Map

The growing self-organising map (GSOM) developed by Alahakoon et al [3, 2] is a heuristically derived variant of the SOM. The GSOM overcomes the requirement of having to pre-specify the size of the output map by growing nodes at the boundary of the map when the total accumulated error distance for a neuron exceeds a pre-calculated growth threshold. The advantage that this scheme has over the standard SOM is that the output maps tend to be only as large as is needed and the shape of the output map is determined as well which can aid visualisation. Growth is controlled by a *spread factor* parameter which can be utilised to zoom in on parts of the map to ascertain sub clusters. The following advantages are identified by the authors:

- The GSOM is able to generate maps that are comparable to SOM with a smaller number of nodes and same level of spread resulting in more efficient use of computer resources.
- Weightings of generated nodes are initialised so that they "fit in with existing map".
- The GSOM is more suitable for higher dimensional data than the SOM due to greater algorithmic efficiency resulting from less nodes at the start of training and localised weight adaptation.

The GSOM starts with a small map to begin with (typically, a 2x2 lattice for a 2D map) which are randomly weighted, initially. The GSOM keeps track of the accumulated error of each *winning* node via the following equation.

$$E_i(t+1) = E_i(t) + D(x(t), w_i) \tag{15}$$

where $D(x(t), w_i)$ is the quantisation error between $x(t)$ and the weight $w_i$ of neuron $i$.

If the neuron with the greatest accumulated error is greater than a pre-specified growth threshold, then the feature map is seen to be under represented at the neurons loci and new nodes are grown to fill up any vacancies in the lattice around the neuron.

New nodes can only be grown at the boundary of the map with all available positions being filled as it is easier than calculating exactly which positions should be filled. Any redundant nodes can be identified and removed periodically as they should accumulate zero "hits". New nodes are weighted so that they fit into the local neighbourhood. This negates the need for an ordering phase as is the case in the SOM.

Since node growth can not proceed from a non-boundary node, a mechanism is provided so that when a winning non-boundary neuron has the highest accumulated error greater than the growth threshold, its accumulated error is distributed to its neighbouring neurons. In this way, non-boundary nodes will indirectly contribute to node growth around their nearest boundary node(s) due to their errors being propagated outwards. Specifically, the new errors are calculated as:

$$E^c_{t+1} = GT/2 \tag{16}$$
$$E^{n_i}_{t+1} = E^{n_i}_t + FD \times E^{n_i}_t \tag{17}$$

where $E^{n_i}_{t+1}$ is the error of the $i$th neighbour of the winning neuron $c$ and FD is a constant termed the *factor of distribution*.

At some point, node growth will saturate (identified by low frequency of node growth). At this stage, a smoothing phase is entered into (characterised by reduced learning rate, neighbourhood restricted to immediate neighbours, and no new new node growth).

During training, weights are updated in a similar way to the original SOM:

$$w_j(t+1) = \begin{cases} w_j(t), & j \notin N_c \\ w_j(t) + LR(t) \times (x_t - w_j(t)), & j \in N_c \end{cases} \tag{18}$$

where the learning rate reduction parameter, $\alpha$, has a constant value $0 < \alpha < 1$ and the learning rate, $LR(t)$, has the form

$$LR(t+1) = LR(t) \times \alpha \times \psi(n) \tag{19}$$

where n is the number of neurons in the network. $\psi(n)$ should be selected such that $\psi(n) \rightarrow 1$ as $n \rightarrow \infty$. This has the function of reducing learning fluctuations as new nodes are grown. $\psi(n) = 1 - \frac{R}{n(t)}$ has been suggested as a possible candidate. The training neighbourhood is large to begin with but shrinks to 1 during the smoothing phase.

The complete algorithm can be summarised as follows:

1. Seed neural network with four nodes and calculate the growth threshold (GT) parameter which controls the growth of the self-organising map:.

$$GT = -D \times \ln(SF) \tag{20}$$

where $D$ is the dimensionality of the input space and SF is a parameter called the *spread factor*.

2. After presenting input to the network a winning neuron, $c$, is calculated in a similar fashion to the SOM by:

$$c = \arg\min_i \|x_t - w_i\|, \qquad i = 1, 2, \ldots, N \tag{21}$$

3. All the neural weightings in the neighbourhood of $c$ are calculated by:

$$w_j(t+1) = \begin{cases} w_j(t), & j \notin N_c \\ w_j(t) + \text{LR}(t) \times (x_t - w_j(t)), & j \in N_c \end{cases} \tag{22}$$

where $N_c$ represents the set of all neurons in the neighbourhood of $c$ and $LR$ is the learning rate at the $t^{\text{th}}$ iteration and is given by $\text{LR}(t) = \text{LR}(t-1) \times \alpha$ where $\alpha$ is a constant parameter called the learning rate reduction $(0 < \alpha < 1)$.

4. Update the cumulative error value for the winning neuron

$$E_i(t+1) = E_i(t) + \| x_t - w_i \| \tag{23}$$

The Euclidean distance is a common metric to use in this equation.

The highest accumulated error is recorded. i.e, if $E_i^{\text{new}} > H_{\text{err}}$ then $H_{\text{err}} = E_i^{\text{new}}$.

5. If $H_{\text{err}} > \text{GT}$, grow new nodes if the winning neuron is a boundary node and distribute neighbouring neural weightings to the new nodes. If $H_{\text{err}} > \text{GT}$ but the winning neuron is not a boundary node, distribute the neurons accumulated error its neighbouring neurons as per the following equations

$$E_{t+1}^c = \text{GT}/2 \tag{24}$$
$$E_{t+1}^{n_i} = E_t^{n_i} + \text{FD} \times E_t^{n_i} \tag{25}$$

6. Reinitialise LR and repeat from step 2 until node growth has attained a minimum level.

7. Subsequent to the *growing* phase of the map is a *smoothing* which is characterised by a lowered learning rate (and reduced $\alpha$), a more contracted neighbourhood function and no new node growth. In all other respects it is a repetition of the algorithm from step 2.

There have been a few variants of the GSOM developed including:

BGSOM.  The BGSOM [31] variant is a batch version of the GSOM algorithm and yields improved performance over the online version.

HDGSOM.  The higher dimensional GSOM [5] was developed to address a limitation of the GSOM which is that the growth can be adversely affected when trained with very large data sets. The HDGSOM accomplishes this with a calibration phase that precedes the growing phase and a variable growth threshold which is reduced with each epoch.

HDGSOMr. The HDGSOMr [4] algorithm introduces an element of randomness into the self-organising process which improves the quality of feature maps generated by the HDGSOM in fewer epochs and hence improves its computational efficiency.

# 3 Introducing a New Algorithm: The GCPSOM

## 3.1 Motivation

The original motivation for this research is to develop a polymorphic self-organising map algorithm for tracking clusters in dynamically changing data. The expected benefits of the hybrid algorithm of the CPSOM and the GSOM algorithms include:

- Map which grows in accordance with need rather than having to be pre-specified before training
- A perpetually learning algorithm which can track new pattern as they are present in a data stream
- Greater efficiency than the CPSOM due to having a minimally sized neural lattice
- A more rigorously derived weight updating rule than the SOM or GSOM

The CPSOM algorithm has been to shown to converge fast at small map sizes, but there is a tendency for the map to start twisting with large map size (say, greater than 10x10). An anticipated advantage that the new algorithm would have is less of a tendency to twist as the map would start very small (say 2x2) which will allow the algorithm to untwist and stabilise before new neurons are grown. Provided there is sufficient time to order the map between new neurons being added, the map should remain relatively untwisted.

Aside from the previous advantages, this research will investigate the use of deterministic annealing as a self-organising process in a dynamically growing map.

## 3.2 Growing the CPSOM

The basis of this research is to combine the growing logic of the GSOM with the weight updating mechanism of the CPSOM algorithm. The following sections will examine various facets of the new algorithm and provide solutions to any incompatibilities between the operation of the GSOM and CPSOM.

### Growing Logic

At the core of the growing mechanism behind the GSOM, is the heuristically derived growth threshold. New neurons are grown around neuron $i$ when its total accumulated error exceeds the growth threshold. i.e.,

$$TE_i \geq GT \tag{26}$$

When the CPSOM algorithm is being trained from dynamic data, the progression of training is characterised by regular periods of *forgetting* whereby the map forgets some of the information it has learned and becomes more flexible and able to adapt to new patterns (every $\lambda_3$ iterations). This has implications for the adaptaion of the GSOM growth logic.

In the GSOM, the rate that errors are accumulated will be dependant on how many neurons are used to quantise the input data (assuming a static data set is used as input). Eventually, the accumulated error will increase at such a low level that the $TE_i$ will take longer and longer to exceed the growth threshold and neuron growth will effectively cease. This method cannot be blindly applied to the CPSOM algorithm as the accumulated error will not be bounded (i.e., there will be continual neuron growth) due to sudden increases in the accumulated error at the start of each forgetting cycle as the map reorganises itself. To prevent neurons from continually accumulating higher and higher errors in the CPSOM, there has to be a mechanism to reset the accumulated error of each neuron based on the pace of change in the clusters present in the data.

Since the *forgetting factor*, $\gamma$, of the CPSOM is a parameter which is set in accordance with how rapidly the input patterns are changing, it is convenient to reduce accumulated errors at this point. Because the map becomes a lot more dynamic when the $\gamma$ parameter is reset, reducing the errors of all neurons at this time delays new neuron growth and allows the map to re-stabilise a bit before new neurons are added otherwise the map might begin to develop a twist or fold. The accumulated errors should ideally preserve some of their original value but also be scaled back in a way that ensures that the growth will be bounded. One such proposal is to scale the total accumulated errors using:

$$TE_i = TE_i \left(1 - \frac{N}{\chi}\right) \tag{27}$$

where $N$ is the number of neurons and $\chi$ is a growth inhibitor parameter which forces $TE_i$ to zero as $N$ approaches $\chi$.

Similarly to the GSOM, new nodes are only grown from *winning* neurons which are also boundary neurons. In the GSOM, the winning vector is simply the neuron with closest weight vector to that of the input vector. In the CPSOM, a winning vector can be obtained by determining which neuron has the greatest probability assignment for the given input vector using 12. This is the strategy chosen for this algorithm.

Should a non-boundary neuron have the greatest probability assignment, then some of the cumulative error of the neuron is redistributed to its immediately neighbouring neurons. In this way it will indirectly contribute to the growth of the boundary neurons closest to it like the GSOM.

$$E_{t+1}^c \quad = GT/2 \tag{28}$$

$$E_{t+1}^{n_i} = E_t^{n_i} + (\text{FD})E_t^{n_i} \tag{29}$$

where $E^{n_i}$ is the error of the $i$th neighbour of the winning neuron, $c$, and FD is the *factor of distribution*.

### New Node Initialisation

As the map should be partially organised by the time new nodes are grown, it doesn't make sense to initialise the weight vectors to a random value (as is the case with the initial map nodes). The GSOM algorithm seeks to initialise the weight vectors of new nodes in such a way that the new nodes are grown away from their immediate neighbours. In doing this, the self-organisation process is helped slightly and the map is less likely to fold back on itself when the new nodes migrate during the self-organisation process.

The weight updating rule of the CPSOM produces a map which is more dynamic (i.e., neurons move around more) during much of the self-organisation process than that of the original SOM or the GSOM. It is deemed that simply initialising the weight vector of new nodes with that of the parent node would be just as effective as the GSOM style initialisation and any more attempt at trying to assist the self-organisation process would be swamped during subsequent weight updating.

The $B_i$ variable for all new nodes is initialised to that of their parent node. This ensures that the new nodes acquire the learning rate of their parent nodes and hence learn at rate appropriate to their local neighbourhood.

### Weight Updating

Weight updating can proceed in a similar way to the CPSOM algorithm with the assignment probability (using (12)) being calculated for each neuron on presentation of an input vector. After which, the new weights can be calculated using (13) and (14) respectively.

The weight updating rule requires that the neighbourhood function $h_{ij}$ be normalised and hence every addition of new nodes must be followed by a renormalisation of the neighbourhood function.

## 3.3  The GCPSOM Algorithm

1. **Initialise Step**

   Initialise annealing parameter and forgetting factor:

   $$\beta = \beta^{\text{start}} \tag{30}$$
   $$\gamma = \gamma^{\text{start}} \tag{31}$$

   Initialise a small starting map with randomly weighted neurons. For a 2D map, a 2x2 lattice is sufficient. Initialise neuron inverse learning rates, cumulative errors, and highest error parameters:

$$B_i = 0 \; \forall i \in N \tag{32}$$

$$E_i = 0 \; \forall i \in N \tag{33}$$

$$H = 0 \tag{34}$$

Calculate the neighbourhood function $h_{ij}$ such that $\sum_{j=1}^{N} h_{ij} = 1$.
Calculate the Growth Threshold using

$$GT = -D \times \ln(SF) \tag{35}$$

where $SF$ is the *spread factor*.

2. Select a vector $x$ randomly from the input data set (or sequentially if the input is a dynamic data set).

3. **E Step** Calculate the probability assignment for all neurons using (12)

$$P_i(x(t)) = \frac{\exp\left(-\beta \sum_{j=1}^{N} h_{ij} D(x(t), w_j)\right)}{\sum_{n=1}^{N} \exp\left(-\beta \sum_{j=1}^{N} h_{ij} D(x(t), w_j)\right)} \tag{36}$$

where $D$ is the quantisation error between input vector $x(t)$ and neuron weight vector $w_j$, defined as

$$D(x(t), w_j) = \frac{\|x(t) - w_j\|^2}{2} \tag{37}$$

4. **M Step** Adjust neural weightings via (13) and (14):

$$w_i(t) = w_i(t-1) + \frac{1}{B_i(t)} \sum_{j=1}^{N} h_{ij} P_j(x(t))(x(t) - w_i(t-1)) \tag{38}$$

$$B_i(t) \qquad = B_i(t-1) + \sum_{j=1}^{N} h_{ij} P_j(x(t)) \tag{39}$$

5. **Growth Step**
   The neuron with the greatest probability assignment is determined (call it $c$). Calculate the accumulative error for neuron $c$ using

$$E_c = E_c + D(x(t), w_c) \tag{40}$$

If $E_c > H$, Assign $H = E_c$
   If $H > GT$ and $c$ is a boundary node, grow new neurons around $c$ such that it now has four immediate neighbouring neurons.

After growing new neurons, the neighbourhood function must be renormalised such that $\sum_{j=1}^{N} h_{ij} = 1$ and any new neurons should have their weight vectors and $B_i$ variable initialised to that of their parent neurons. If $D(c) > \text{GT}$ but $c$ is not a boundary neuron, the cumulative quantisation error of $c$ is halved and redistributed to its neighbours as in the GSOM algorithm.

6. **Parameter Adjustment Step**

Adjust parameters according to the following conditions:

| Condition | Adjustment |
|---|---|
| $\beta < \beta^{\text{final}}$ $t \bmod \lambda_1 = 0$ | $\beta \leftarrow \beta + \Delta_1$ |
| $\gamma > 1$ $t \bmod \lambda_2 = 0$ | $\gamma \leftarrow \gamma - \Delta_2$ $B_i(t) = B_i(t)/\gamma$ |
| $t \bmod \lambda_3 = 0$ | $\gamma = \gamma^{\text{start}}$ $E_i = E_i \left(1 - \frac{N}{\chi}\right) \qquad \forall i \in N$ $H = H \left(1 - \frac{N}{\chi}\right)$ |

## 4 Experimental Results

The performance of topology preserving vector quantisation algorithms can essentially be evaluated according to how well they quantise input data or how well they preserve the topology of the input data manifold. This section will illustrate the capabilities of the GCPSOM in relation to these two quantities using both synthetic and *real world* data.

The aim of this section is to demonstrate that the GCPSOM has a performance comparable with other related algorithms and can dynamically grow the neural network to accommodate new patterns that may occur within dynamic data sets. To achieve this aim, the following performance measures are used:

Quantisation Error    Calculated as the average Euclidean distance between the input vectors and their corresponding winning neurons' weight vector, over the entire training set:

$$E_q = \frac{1}{N} = \sum_{i=1}^{N} \|x_i + w_i\| \tag{41}$$

Topographic Error    The topographic error [14] essentially evaluates the proportion of sample vectors which indicate a local discontinuity in the map:

$$\varepsilon_t = \frac{1}{N} \sum_{i=1}^{N} u(x_i) \tag{42}$$

where $u(x_i) = 1$ if the best- and second-best-matching neurons for input vector $x_i$ are non-adjacent, otherwise $u(x_i) = 0$.

## 4.1  Static Data Set

The synthetic data used in these experiments is 2 dimensional as this facilitates easy visualisation of the neural weight vectors.

### Synthetic Data

The data set used in this section consists of 8648 2d vectors which are randomly distributed in such a way to form a *smiley face*. The parameters for the CPSOM were selected such that $\gamma$ decays to $\gamma^{\text{final}}$ and $\beta$ decays to $\beta^{\text{final}}$ within 1 epoch. Fig. 4 illustrate the clustering capabilities of the GSOM, CPSOM and GCPSOM on the face data set. The algorithm parameters used in the experiments are given as follows:

| | |
|---|---|
| GSOM | Spread factor = 0.9, pruning interval = 2000 iterations |
| CPSOM | $\beta^{\text{start}} = 20$, $\beta^{\text{final}}$, $\gamma^{\text{start}} = 10$, $\gamma^{\text{final}} = 1$, $\lambda_1 = 100$, $\lambda_2 = 100$, $\Delta_1 = 10$, $\Delta_2 = 0.1$, neighbourhood radius, $\sigma = 0.8$. |
| GCPSOM | Same parameters as the CPSOM. Spread factor = 0.9. |

One observation that can be made immediately (see Fig. 4) is that the CPSOM algorithm tends to produce maps with more strongly delineated clusters irrespective of map size. This comes at the expense of some topographic preservation. It can be seen that the final quantisation errors of each algorithm are roughly comparable, however, the topographic error of the GSOM algorithm jumps up at each interval of 2000 iterations corresponding to the node pruning interval. This is one of the reasons that the GSOM node pruning strategy was not introduced into the GCPSOM algorithm. Also, irregular concave map geometries tend to decrease map stability making the map susceptible to twisting. This is not so much of a problem in the GSOM algorithm as the weightings of new neurons are initialised to fit in with the map and the learning rate is not as great as in the GCPSOM algorithm. Since the GCPSOM algorithm has a greater learning rate than the GSOM, the GCPSOM algorithm tends to provide better coverage of the input manifold during its growing phase. The effect of this is that neuron growth is not as highly directed as in the GSOM algorithm and as a result of this, more rounded (convex) map geometries are produced which adds stability to the map which is needed when clustering dynamic data sets. This is illustrated in Fig. 6 which is the map geometries resulting from clustering of the data set in Fig. 5.

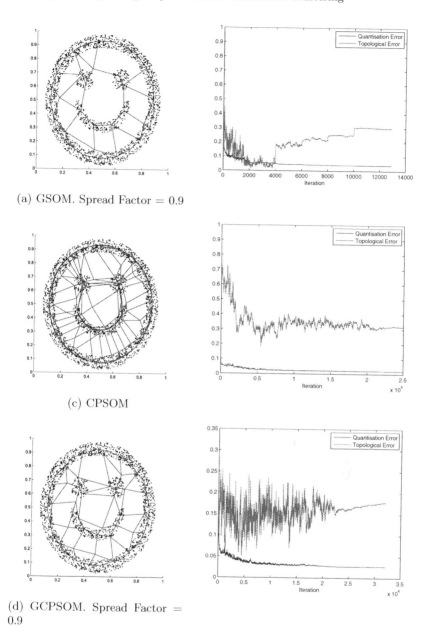

(a) GSOM. Spread Factor = 0.9

(c) CPSOM

(d) GCPSOM. Spread Factor = 0.9

**Fig. 4** Comparison of topological mapping of an 8648 point smiley face data set. (a) Final map size of 36 Neurons; (c) Map size 16x16; (d) Final map size of 52 neurons

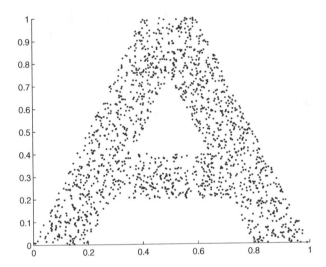

**Fig. 5** 'A' Data set

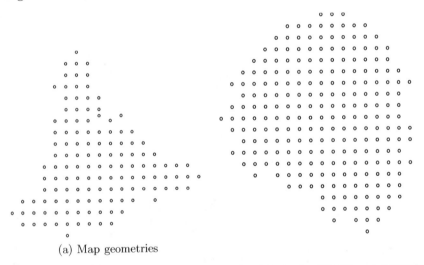

(a) Map geometries

**Fig. 6** Comparison of map geometries generated from the GSOM and GCPSOM algorithms. (a) Final map size = 70; (b) Final map size = 235

## Real-World Data

The real world data used to test the algorithm, although contrived, is of a form that is typically encountered in real world applications. That is to say, the data set is fairly high dimensional (16 dimensional in the experiments conducted herein) and the attributes are predominantly categorical in nature.

The data set is taken from the UCI machine learning repository at the University of California [6] and contains zoological information on approximately 100 animals.

Fig. 7 and 8 are the feature maps generated using the GSOM and GCP-SOM algorithms respectively. The neurons have been annotated with their respective taxonomic categories including any exceptions. Both algorithm would consistently group the animals correctly.

The GCPSOM is clearly capable of clustering static, real world data with comparable performance to the GSOM, although, the ability of the GCPSOM to accomplish hierarchical clustering using the spread factor parameter wasn't evaluated.

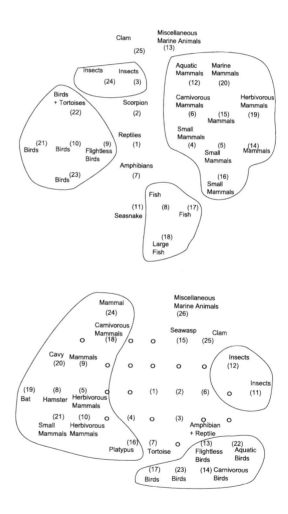

**Fig. 7** Static zoological data set clustered with the GSOM. $SF = 0.9$

**Fig. 8** Static zoological data set clustered with the GCPSOM. $SF = 0.9$

## 4.2   Dynamic Data Set

### Synthetic Data

To test the clustering capabilities of the GCPSOM algorithm against a dynamic data set, a dynamic data set was synthesised that gradually builds up from a circle to a smiley face and back again over 4000 iteration intervals (the data set is illustrated in Fig. 9).

**Fig. 9** Face data set. Each plot represents 4000 consecutive iterations in time

The Spread factor chosen for this data set is 0.6. However, in the GCPSOM, the spread factor doesn't affect the overall spread of the map when clustering dynamic data sets. It determines the growth rate of the neural network and as such, should be set depending on how dynamic the data set is. That is to say, if the network grows to quickly, neurons will be added before the network is sufficiently organised and twists / folds might develop. The final map size is principally determined by the growth inhibitor parameter, $\chi$, which in this experiment was set to 144. The factor of distribution was set to 0.4. The other parameters of the GCPSOM were set as follows: $\beta^{start} = 100$, $\beta^{final} = 1000$, $\gamma^{start} = 10$, $\gamma^{final} = 1$, $\lambda_1 = 100$, $\lambda_2 = 100$, $\lambda_3 = 4000$, $\Delta_1 = 35$,

$\Delta_2 = .35$, $\sigma = 0.5$. The forgetting factor, $\lambda_3$, is selected such that the algorithm can relearn new patterns every 4000 iterations and the parameters $\Delta_1$, $\Delta_2$ were selected such that $\beta$ and $\gamma$ would decay to their final values within $\frac{2}{3}$ of each forgetting cycle (2667 iterations).

Fig. 10 demonstrates the progression of the clustering generated by the GCPSOM algorithm. The CPSOM can achieve similar results contingent on being able to form a relatively ordered map (i.e., without any significant twists or folds) initially which is hard to achieve with large map sizes. A significant advantage that the GCPSOM has over the CPSOM is that a topologically accurate map is typically generated initially due to a small initial map size and the map tends to remain fairly ordered throughout the growing phase (provided the growth rate isn't too rapid).

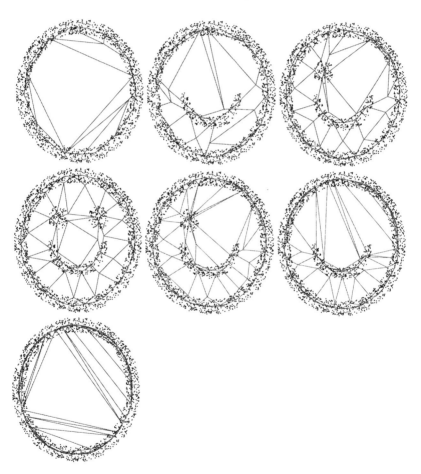

**Fig. 10** Clustering of dynamic face data set. Each plot is taken at 4000 iteration intervals

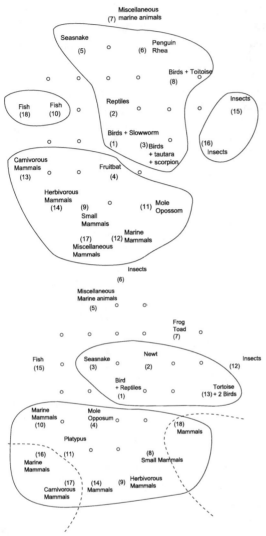

**Fig. 11** Dynamic zoological data set. New nodes are indicated by the dashed line

## Real-World Data

A real-world dynamic data set was manufactured from the zoological data set by removing the group of birds for the first 800 input vectors. Fig. 11 shows the feature map after the first 800 input vectors have been presented to the algorithm. Although the birds were omitted during the training, the group of birds were included in the simulation for these two maps. The birds are principally clustered together with the reptiles and a few miscellaneous animals.

**Fig. 12** Dynamic zoo-
logical data set. After
1300 iterations

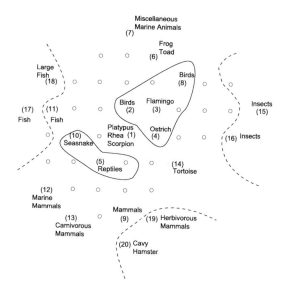

The birds are still fairly strongly grouped. On inspection of the data set, the birds only have significant variation in three of the sixteen attributes which explains why they are clustered reasonably well despite being omitted from the training data. After 1300 iterations (Fig. 12), the birds have predominantly separated from the reptiles into their own cluster

## 5   Analysis

The advantages that are often cited for dynamically growing self-organising maps are:

1. They produce maps which are sized according to need and data complexity. This reduces computational requirements needed to generate maps.
2. They may produce maps which are shaped according to the contours of the input data manifold.
3. In the case of the GSOM, the shape of the map has the ability to attract the attention of the data analyst to prominent clusters.
4. The dimensionality of the map can be made to better reflect the topology of the input data manifold so that there is reduced folding of high dimensional input spaces.

Point 4 is not relevant to this algorithm as growth is constrained to grow within the dimensionality specified in the output map. The GCPSOM doesn't succeed as well as the GSOM in reference to point 3 although often the data analyst does not rely heavily on the raw output map and some auxiliary processing is often done to make visualising feature maps easier and to attract the attention of data analysts to interesting features. This leaves points 1

**Fig. 13** Dynamic zoo-
logical data set clustered
by the CPSOM algorithm

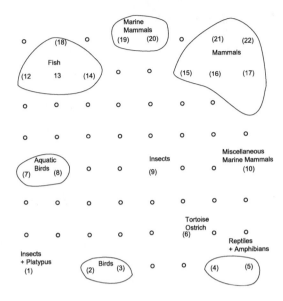

and 2 as the primary motivations for developing growing self-organising map
algorithms.

In order to evaluate how well the GCPSOM meets these criteria, it is
necessary to investigate the growth heuristics which form the basis of any
growing SOM algorithm. I have identified four such properties present in such
algorithms which are: growth rate, growth conditions, growth inhibition, and
neuron pruning.

Before elaborating on the four properties just mentioned, it is worth making a few comments regarding what is an optimal map size. This depends
largely on how the feature map is going to be used. Some applications are
only concerned with locating clusters within a map while other applications
may require greater topology preservation so that cluster are more easily
navigated. The latter will obviously require more neurons to encode the topographic information of the input data manifold but may also require some
form of post processing to make clusters more obvious. Considering that the
purpose of these algorithms is essentially topological vector quantisation, it
could be argued that the raw feature maps are better utilised/interpreted by
some auxiliary computational technique rather than directly by a data analyst. In this case, the optimal map size should be as large as possible subject
to computational/time constraints. This would yield the most accurate and
detailed maps

## 5.1   Growth Rate

The self-organising process can be divided into two phases. The first phase is
a very dynamic organisation phase which tries to reduce the number of major

twists and folds. The second phase is primarily concerned with stabilising the map and fine tuning the neuron weightings to reduce the quantisation error of the map.

The growth of neurons should be driven by the need to cluster data to a sufficient level of detail. If the network grows too fast, neurons will be added before major topological defects have been ironed out. This will exacerbate the topological deformity of the map as it evolves. Ideally the growth rate should be as high as possible subject to growing neurons only during times of map stability.

An advantage of the CPSOM style learning is the presence of a per neuron learning rate which can be used to adjust the local maleability of a map depending on local map dynamics.

## 5.2 Growth Conditions

It has been my observation that the more neighbours a neurons has, the greater local stability it will have during the self-organisation process. Conversely, regions which have relatively few number of neurons in their neighbourhood (such as near the edge of maps) are places where twisting and folding are most likely to occur. After neuron growth, new neurons and their neighbouring neurons should be allowed to reorganise themselves before growing new neurons, otherwise additional neurons will continue to grow in the direction of new data leading to concave shaped maps making them susceptible to twisting or folding. The GSOM does this to a certain degree (see Fig 6), but new neurons have their weights initialised in such a way that the map requires little reorganisation to accommodate the new neurons. Whilst the GSOM yields good quantisation of the input data, the topology preservation doesn't seem as accurate as other algorithms (This is based purely on visual inspection of maps generated during the experimentation. The measurement of topological error in Sect. 4 only really reveals the degree of twisting or folding of the maps and not how accurate the topological preservation is).

A sound criterion for growing new neurons is if by the addition of new neurons, the total quantisation error (TQE) of the map decreases. This is achieved in the GSOM and GCPSOM by maintaining a record of each neurons relative contribution to total error and growing new neurons when the neuron with the greatest error exceeds a specified growth threshold. In this way neurons are grown at locations which are likely to yield the maximum decrease in TQE for the map.

## 5.3 Growth Inhibition

Obviously the growth must be bounded in some way. In the GSOM (and GCPSOM), the growth is limited by gradually reducing the rate of accumulation of quantisation error for boundary neurons until they are unable to exceed the growth threshold. The final number of neurons present in the map is an indirect byproduct of the selection of the *spread factor* (and consequently, growth threshold). A significant advantage to growing self-organising maps is the ability to efficiently generate feature maps subject to computational constraints. Hence the number of neurons in the final map should be determined by the available resources (i.e., time, computer memory, computer architecture) and so it may be more useful to specify a hard limit on the total number of neurons in the final map directly rather than through an indirect parameter to control growth.

## 5.4 Neuron Pruning

Once a network has reached its final size, the network may need to reshape itself to better define the contours of new patterns in the data (This is more important when it comes to mapping dynamic data sets). It is problematic to prune nodes from the interior of the map since this tends to destroy topographic information (although the quantisation error may not be affected appreciably) as shown in Fig. 4a. Several methods of node pruning were trialled for the GCPSOM, however, matching the pruning rate to the growth rate such that the map structure would stabilise proved to be difficult. On reflection, I believe node pruning should only start once the network has reached its final size and every node pruned should be accompanied by a reciprocal growth of new nodes elsewhere in the map. This style of node pruning will only affect the shape of the map and consequently will have negligible impact on the total quantisation error of the map. Hence the node pruning condition should be made such that the topology preserving nature of the map is increased.

## 6 Conclusion

The viability of hybridising the GSOM algorithm with the CPSOM has been demonstrated. The resulting algorithm, christened the Growing Cellular Probabilistic Self-Organising Map (GCPSOM), is more capable at mapping data sets at a higher map size than the CPSOM since the CPSOM has difficulty untwisting large maps whilst the GPCSOM resolves any twists or folds when the map is small and gradually grows the map thereafter. However, the main purpose of this research was to develop a growing SOM type algorithm which can track dynamic data sets. This capability has been demonstrated for a 2D synthetic data set and a high dimensional data set of the kind likely to be encountered in the real world.

### 6.1 Extensions

The section describes a few extensions which would be valuable additions to the algorithm.

| | |
|---|---|
| Node Pruning | It would be very advantage to have some form of node pruning in the algorithm as after a period the node growth will plateau and any further self-organisation of the map will be constrained by the final map geometry. |
| HDGSOMr Randomisation | Adding an element of randomisation to the updating of the learning rate might help the algorithm escape local minima as in the HDGSOMr [5] algorithm. |
| Parameter Auto-tuning | The patterns in dynamic data sets often change at different rates. It would be advantageous if the GCPSOM algorithm incorporated some mechanism to *forget* patterns on a localised scale dependent on how dynamically the data patterns are changing on a local level. The algorithm has the capacity to incorporate this ability through the per neuron learning rate $\frac{1}{B_i(t)}$. |

## 6.2　Further Research Possibilities

The new algorithm can be considered a qualified success, although a few questions remain unanswered. The following is list of possible research extensions which would be valuable additions to the research contained within this thesis:

- The algorithm requires more thorough testing on real-world dynamic data streams instead of the pseudo realistic data sets used in the experiments in this thesis.
- Investigate the possible extensions to the algorithm as mentioned in Sect. 3.
- Re-design the algorithm in light of the analysis in Sect 5.

# References

1. Sato, M.A., Ishii, S.: On-line EM algorithm for the normalized gaussian network. Neural Computation 12(2), 407–432 (2000)
2. Alahakoon, D.: Controlling the spread of dynamic self-organising maps. Neural Comput. Appl. 13(2), 168–174 (2004)
3. Alahakoon, D., Halgamuge, S.K., Srinivasan, B.: Dynamic self-organizing maps with controlled growth for knowledge discovery. IEEE Transactions on Neural Networks 11(3), 601–614 (2000)
4. Amarasiri, R., Alahakoon, D., Smith, K., Premaratne, M.: Hdgsomr: A high dimensional growing self-organizing map using randomness for efficient web and text mining. wi, 215–221 (2005)

5. Amarasiri, R., Alahakoon, D., Smith, K.A.: Hdgsom: A modified growing self-organizing map for high dimensional data clustering. In: HIS 2004: Proceedings of the Fourth International Conference on Hybrid Intelligent Systems (HIS 2004), pp. 216–221. IEEE Computer Society, Washington (2004)
6. Asuncion, A., Newman, D.J.: UCI machine learning repository (2007)
7. Chow, T.W.S., Wu, S.: An online cellular probabilistic self-organizing map for static and dynamic data sets. IEEE Transactions on Circuits and Systems 51(4), 732–747 (2004)
8. Fritzke, B.: Growing cell structures-a self-organizing network for unsupervised and supervised learning. Neural Netw. 7(9), 1441–1460 (1994)
9. Fritzke, B.: A growing neural gas network learns topologies. In: Tesauro, G., Touretzky, D.S., Leen, T.K. (eds.) Advances in Neural Information Processing Systems, vol. 7, pp. 625–632. MIT Press, Cambridge (1995)
10. Goerke, N., Kintzler, F., Eckmiller, R.: Multi-soms: A new approach to self organised classification. In: ICAPR (1), pp. 469–477 (2005)
11. Graepel, T., Obermayer, K.: A stochastic self-organizing map for proximity data. Neural Computation 11(1), 139–155 (1999)
12. Hebb, D.O.: The Organization of Behaviour. Wiley, New York (1949)
13. Hung, C., Wermter, S.: A dynamic adaptive self-organising hybrid model for text clustering. In: ICDM 2003: Proceedings of the Third IEEE International Conference on Data Mining, p. 75. IEEE Computer Society, Washington (2003)
14. Kiviluoto, K.: Topology preservation in self-organizing maps. In: IEEE International Conference on Neural Networks, 1996, Washington, DC, USA, vol. 1, pp. 294–299 (June 1996)
15. Kloppenburg, M., Tavan, P.: Deterministic annealing for density estimation by multivariate normal mixtures. Phys. Rev. E 55(3), R2089–R2092 (1997)
16. Kohonen, T.: Self-organized formation of topologically correct feature maps. Biological Cybernetics 43(1), 59–69 (1982)
17. Kohonen, T.: Self organizing maps, 2nd edn. Springer, Heidelberg (1995)
18. Lang, R., Warwick, K.: The plastic self organising map. In: Proceedings of the 2002 International Joint Conference on Neural Networks, 2002. IJCNN 2002, Honolulu, HI, vol. 1, pp. 727–732 (May 2002)
19. Luttrell, S.P.: Self-organisation: a derivation from first principles of a class of learning algorithms. In: IJCNN: International Joint Conference on Neural Networks, vol. 2, pp. 495–498 (1989)
20. Luttrell, S.P.: Code vector density in topographic mappings: Scalar case. IEEE Transactions on Neural Networks 2(4), 427–436 (1991)
21. Luttrell, S.P.: A bayesian analysis of self-organizing maps. Neural Computation 6, 676–794 (1994)
22. Martinetz, T., Schulten, K.: A "neural-gas" network learns topologies. Artificial Neural Networks, 397–402 (1991)
23. Mcculloch, W., Pitts, W.: A logical calculus of the ideas immanent in nervous activity. Bulletin of Mathematical Biophysics 5, 115–133 (1943)
24. Menhaj, M.B., Jahanian, H.R.: An analytical alternative for som. In: International Joint Conference on Neural Networks, 1999. IJCNN 1999, vol. 3, pp. 1939–1942 (1999)
25. Merkl, D., Dittenbach, M., Rauber, A.: Uncovering hierarchical structure in data using the growing hierarchical self-organizing map. Neurocomputing 48(1-4), 199–216 (2002)

26. Rose, K.: Deterministic annealing for clustering, compression,classification, regression, and related optimization problems. Proceedings of the IEEE 86(11), 2210–2239 (1998)
27. Rose, K., Gurewitz, E., Fox, G.C.: Statistical mechanics and phase transitions in clustering. Phys. Rev. Lett. 65(8), 945–948 (1990)
28. Shannon, C.E.: A mathematical theory of communication. SIGMOBILE Mob. Comput. Commun. Rev. 5(1), 3–55 (2001)
29. Villmann, T., Der, R., Herrmann, M., Martinetz, T.M.: Topology preservation in self-organizing feature maps: Exact definition and measurement. IEEE Transactions on Neural Networks 8(2), 256–266 (1997)
30. von der Malsburg, C.: Self-organization of orientation sensitive cells in the striate cortex. Biological Cybernetics 14(2), 85–100 (1973)
31. Yu, Y., Alahakoon, D.: Batch implementation of growing self-organizing map. In: International Conference on Computational Intelligence for Modelling, Control and Automation, 2006 and International Conference on Intelligent Agents, Web Technologies and Internet Commerce, p. 162 (November 2006)

# Synthesis of Spatio-temporal Models by the Evolution of Non-uniform Cellular Automata

Ana L.T. Romano, Wilfredo J.P. Villanueva, Marcelo S. Zanetti, and Fernando J. Von Zuben

## 1 Introduction

Non-uniform or inhomogeneous cellular automata (NunCA) [28] are spatio-temporal models for dynamical systems in which space and time are discrete, and there is a distinct transition rule for each cell, with a finite number of states. The cells are in a regular lattice and the transition from one state to another is performed synchronously. The next state of a given cell will then be provided by a local and fixed transition rule that associates its current state and the current state of the neighbouring cells with the next state. The neighbourhood could also be specific for each cell, but will be considered the same, except for the cells at the frontiers of the regular lattice. So, the only distinct feature between NunCA and the traditional uniform cellular automata (CA) [29,34] is the adoption of a specific transition rule for each cell instead of a single transition rule for all the cells in the lattice.

Both CA and NunCA have been applied to a wide variety of scenarios, including (but not restricted to):

- CA: physical systems modeling [6,21], ecological studies [7,14], computational applications [29,35];

Ana L.T. Romano
Rio de Janeiro State University (UERJ)
Faculty of Oceanography - Physical Oceanograpy Department
Rio de Janeiro, RJ, Brazil
e-mail: romano@uerj.br

Wilfredo J.P. Villanueva and Fernando J. Von Zuben
Laboratory of Bioinformatics and Bioinspired Computing
School of Electrical and Computer Engineering
University of Campinas (Unicamp)
Campinas, SP, Brazil
e-mail: {wilfredo,vonzuben}@dca.fee.unicamp.br

Marcelo S. Zanetti
Department of Computer Science
ETHZ - Swiss Federal Institute of Technology
Bergstrasse 126, 8032 Zurich
e-mail: marcelos@student.ethz.ch

A. Abraham et al. (Eds.): Foundations of Comput. Intel. Vol. 4, SCI 204, pp. 85–104.
springerlink.com                                   © Springer-Verlag Berlin Heidelberg 2009

- NunCA: VLSI circuit design [15,32], computational applications [28,30,31,33].

NunCA has one predominant advantage over CA, i.e. the greater flexibility to define the transition rules, which can be explored to produce dynamic behaviors not (easily) obtainable by means of a single rule. So, the possibility of updating the state of each cell following local and distinct rules can be explored to conceive synthetic universes from simple rules, with the emergence of complex spatio-temporal structures. Section 2 will provide additional arguments to support NunCA, and section 3 will present the essential mathematical formalism.

Instead of investigating and/or exploring the computational power of NunCA, the purpose here is to provide a systematic procedure to achieve a mathematical model for the NunCA framework capable of reproducing a sequence of spatio-temporal behaviors. Two scenarios will be considered:

1. For one-dimensional lattices: given a desired sequence of state transitions, the aim is to determine one of the possibly multiple set of fixed, though distinct, rules that is capable of driving the sequential transition of states according to the desired profile. This has already been performed in the case of uniform CA ([19]), and the main purpose here is to indicate that some profiles can not be achieved when a single transition rule is defined for all cells in the lattice. In such a case, only the NunCA framework can fulfill the task.
2. For two-dimensional lattices: given the initial and final states of each cell, the aim is to determine one of the possibly multiple set of fixed, though distinct, rules that is capable of driving the sequential transition of states from the initial to the final one, with the final state as a stationary configuration. The trajectory between the initial and final states, denoted transitory phase, can be of interest or not. A case study will be considered in which the transitory phase is left unrestricted, and another case study will impose some restrictive conditions to the intermediary states.

The great challenge of such a formulation is the necessity of defining the whole set of transition rules, one rule for each cell in the regular lattice. The necessity of as many rules as cells has precluded a wider dissemination of similar approaches. Section 4 will present the necessary steps toward the synthesis of an evolutionary design of these transition rules, and Section 5 will comment on related approaches in the literature.

After a successful determination of an appropriate set of transition rules, the interpretation of the resulting NunCA may take place, even though the obtained set of rules is just one of the possible solutions. The interpretation is easier in the case of one-dimensional lattices, because the states are binary, but relevant information can be extracted from the resulting two-dimensional lattices too, where multivalued states are considered. The use of multivalued states can be interpreted as a quantization of the continuous state case, where the NunCA would then be equivalent to a non-uniform coupled map lattice [16] and a discrete-time cellular neural network [5].

Section 6 is devoted to the description and analysis of the obtained results. Concluding remarks are then outlined in Section 7.

## 2   The Motivation for Non-uniform Lattices

Every dynamical event whose description involves the evolution of variables in time and space is called a spatio-temporal phenomenon. Examples of these dynamics include dispersion, expansion, contraction, and local interrelation of groups of elements, and can be associated with living and other physical phenomena in nature [4,6,22].

Most of these spatio-temporal systems are continuous in space and time. However, a computational model will necessarily require the quantization of space, in the form of regular lattices, and the use of a discrete-time dynamics to update the state of the cells in the lattice, denoted a transition rule.

The use of cellular automata models is generally associated with one of two purposes:

1. Classification of the spatio-temporal behavior;
2. Reproduction of a predefined behavior from the simplest transition rule that can be defined.

The second approach is the one of interest here and has been explored in the literature in distinct ways, as:

- An architecture for fast and universal computation [35];
- An alternative paradigm for the investigation of computational complexity [35];
- Pattern recognition tools [18];
- Modeling devices [10].

Our primary concern here is the last item: cellular automata as powerful models of actual physical phenomena. The main motivation is the possibility of reproducing specific behaviors in space and time, always based on simple transition rules.

Gutowitz and Langton [11] have pointed out that lattices with interesting behavior are the ones that achieve a tradeoff between high-level and low-level of dependence among neighboring cells. With distinct transition rules for each cell, the level of inter-cell dependence may be established with much more flexibility, when compared with the existence of a single transition rule to be followed by every cell in the lattice. In fact, a uniform CA can be interpreted as a particular case of a non-uniform CA, here denoted NunCA.

To achieve a proper tradeoff capable of reproducing the desired spatio-temporal behavior, powerful search devices should be conceived to determine a proper set of transition rules. Evolutionary computation has already been demonstrated to provide effective procedures to optimize parameters of a single transition rule in uniform and binary-state cellular automata [19]. That is why we are going to extend the already proposed evolutionary approaches to search for an optimal set of parameters for each transition rule in the non-uniform case.

As a transition rule of a given cell will represent a local binding to the neighboring cells, multiple equivalent transition rules can be capable of reproducing the same global behavior, so that we will be interested in finding just one of them.

## 3 Mathematical Formalism

As already mentioned, cellular automata are idealizations of physical systems where time and space are discrete and the physical quantities, or cellular states, can assume a finite set of values. The transition rules can be seen as an expression of a microscopic dynamics that leads to a desired macroscopic behavior [8].

The existence of a regular lattice, a well-defined neighborhood, a transition rule and an initial state for the cells in the lattice, taken from a finite set of possible states, gives rise to the cellular automata mathematical model. The transition rule is a function that depends on the states of the neighboring cells and which leads each cell to the next state at each time step, synchronously or asynchronously [26] with the other ones in the lattice. So, each cell state is a fixed function of its previous state and of the previous state of the neighboring cells.

The dimension of the lattice will certainly depend on the spatial dimension of the physical phenomenon under investigation. In the literature, we can find one-dimensional [30], two-dimensional [24], and three-dimensional [2] lattices. The neighborhood of each cell is previously defined and is generally assumed to have the same shape for each cell in the lattice, even in the case of non-uniform cellular automata. Three usually adopted two-dimensional configurations for the neighborhood are illustrated in Figure 1.

**Fig. 1** Examples of neighborhoods for a two-dimensional lattice: (a) Moore; (b) von Neumann; and (c) Hexagonal

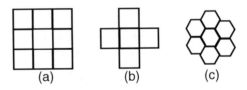

(a)          (b)          (c)

The simplest version of cellular automata is the binary-state one-dimensional CA, in which each cell can only assume the values 0 or 1. Examples of the neighborhood of this CA are presented in Figure 2.

**Fig. 2** One-dimensional lattice: (a) Neighborhood with radius 1; and (b) Neighborhood with radius 2

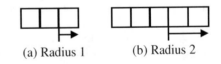

(a) Radius 1          (b) Radius 2

In the one-dimensional lattice, the number of cells involved in the updating of the state of a given cell is 2*Radius+1. For the binary-state one-dimensional cellular automata, the neighborhood can assume $2^{2*Radius+1}$ possible configurations, and the number of transition rules for a given cell is $2^{2^{2*Radius+1}}$. This is the size of the search space for the simplest cellular automata that can be conceived, supposing that each cell will obey the same transition rule. For the non-uniform case, the cardinality of the set of all possible transition rules is given by $\left(2^{2^{2*Radius+1}}\right)^n$, where $n$ is the number of cells in the one-dimensional lattice.

The experiments to be performed involving one-dimensional lattices will consider uniform and non-uniform binary-state cellular automata.

In the case of a two-dimensional $n \times m$ lattice, only von Neumann neighborhoods will be implemented, with multivalued states. Only non-uniform cellular automata will be considered, and the definition of the transition rule is based on the notation of Figure 3.

**Fig. 3** Parameters of the transition rules

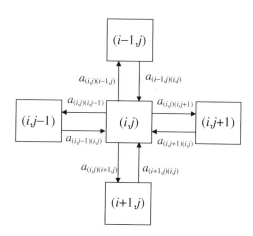

The spatio-temporal behavior will be associated with the flux of some material from cell to cell, according to the von Neumann neighborhood. The state of each cell is the concentration of that material and the set of four parameter values $\left\{a_{(i,j)(i,j+1)}, a_{(i,j)(i+1,j)}, a_{(i,j)(i,j-1)}, a_{(i,j)(i-1,j)}\right\}$ defines the amount that will be transferred from cell $(i,j)$ to cell $(k,p)$, where the indices $k$ and $p$ are defined according to the corresponding neighbor cell. Being interpreted as rate of flux, the following restrictions are imposed:

$$a_{(i,j)(i,j+1)} \geq 0;\ a_{(i,j)(i+1,j)} \geq 0;\ a_{(i,j)(i,j-1)} \geq 0;\ a_{(i,j)(i-1,j)} \geq 0;$$

$$a_{(i,j)(i,j+1)} + a_{(i,j)(i+1,j)} + a_{(i,j)(i,j-1)} + a_{(i,j)(i-1,j)} \leq 1;$$

$a_{(i,j)(k,p)} = 0$ when $(k,p)$ is an absent neighbor, motivated by the fact that $(i,j)$ is a cell at the frontier of the lattice.

When dealing with uniform cellular automata, the following additional restrictions are necessary:

$$a_{(i,j)(i,j+1)} = a_{(i,j-1)(i,j)};$$

$$a_{(i,j)(i+1,j)} = a_{(i-1,j)(i,j)};$$

$$a_{(i,j)(i,j-1)} = a_{(i,j+1)(i,j)};$$

$$a_{(i,j)(i-1,j)} = a_{(i+1,j)(i,j)}.$$

So, given that $c_{(i,j)}(t)$ is the concentration of material at cell $(i,j)$ in the instant $t$, the transition rule for cell $(i,j)$ is given by Eq. (1):

$$c_{(i,j)}(t+1) = \left(1 - a_{(i,j)(i,j+1)} - a_{(i,j)(i+1,j)} - a_{(i,j)(i,j-1)} - a_{(i,j)(i-1,j)}\right)c_{(i,j)}(t) +$$
$$a_{(i,j+1)(i,j)}c_{(i,j+1)}(t) + a_{(i-1,j)(i,j)}c_{(i-1,j)}(t) + a_{(i,j-1)(i,j)}c_{(i,j-1)}(t) + a_{(i+1,j)(i,j)}c_{(i+1,j)}(t)$$

$$(1)$$

where $i \in \{1,...,n\}$ and $j \in \{1,...,m\}$. Notice that $n$ can be taken equal to $m$ in a square lattice. When a non-toroidal neighborhood is considered, every time that $i=1$ and/or $j=1$, the terms involving indices $i-1$ and $j-1$ are null, and the same happens with the terms involving $i+1$ and $j+1$ when $i=n$ and/or $j=m$. For toroidal neighborhood, see Section 6.2.

## 4 Methodology for Evolutionary Design

### 4.1 Genetic Algorithms for One-Dimensional and Binary-State Cellular Automata

Genetic algorithms (GAs) [9] have been successfully applied to the synthesis of uniform cellular automata [20,23]. Inspired by the process of natural selection, a GA maintains a population of candidate solutions in a genotypic representation, and mutation and recombination operators [9] are then conceived to promote a proper exploration of the search space in a population-based mechanism. Selection is performed to implement the principle of the survival of the fittest, and individuals with higher fitness values have a high probability of being selected to spread their genetic material to the next generation of individuals. The recursive application, generation after generation, of selection and genetic operations, together with local search procedures when available, tends to promote an increase in the average fitness of the population, at least in the fitness of the best individual at each generation. Better fitness means a candidate solution with better quality. Every problem will have its own fitness function.

In one-dimensional lattices composed of $n$ cells, each individual will be a binary vector describing the single transition rule, in uniform cellular automata, and the whole set of transition rules, in non-uniform cellular automata. In fact, the codification will interpret each possible configuration of the neighborhood (given by a sequence of 2*Radius+1 bits, where Radius is the order of the neighborhood) as an integer index, and this index will indicate the position of its corresponding next state in the transition rule.

**Table 1** An example of transition rule (3rd column) for a neighborhood of Radius = 1

| Configuration | Index | Next State |
|---|---|---|
| 000 | 0 | 0 |
| 001 | 1 | 1 |
| 010 | 2 | 1 |
| 011 | 3 | 0 |
| 100 | 4 | 1 |
| 101 | 5 | 0 |
| 110 | 6 | 0 |
| 111 | 7 | 1 |

As an example, taking Radius = 1, Table 1 presents in the third column a possible transition rule, so that every configuration of neighborhood has an indication of next state, e.g. when the neighborhood achieve 100 then the next state of the cell under analysis will suffer a transition from 0 to 1, and for a neighborhood 010, the state remains the same (equal to 1).

In the non-uniform case, the genetic codification of a transition rule will be given by a binary vector whose size is $n$ times the size of the binary vector in the uniform case, because each cell can have a distinct next state for each configuration of the neighborhood.

The fitness function will be simply given by the inverse of 1 plus the Hamming distance between ⟨the observed evolution of states in time⟩ and ⟨the desired one⟩. When a given transition rule is capable of exactly reproducing the spatio-temporal behavior, then the Hamming distance will be zero and the fitness will achieve the maximum value. The highest the Hamming distance, the smallest the fitness value, so that the fitness is restricted to fit in the range (0,1]. The initial condition of the automata is arbitrarily defined and is considered fixed.

**Fig. 4** Simple GA Flowchart

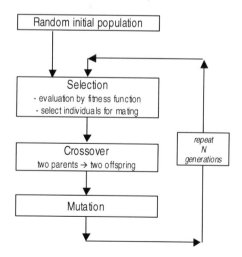

The flowchart in Figure 4 depicts the main steps of the adopted GA. A local search is also applied every time a new individual is obtained. This local search consists in definitely changing one of the next states suggested by the transition rule if this change turns to improve the overall performance of the cellular automata.

## 4.2 Evolution Strategy for Two-Dimensional and Multivalued-State Cellular Automata: The Non-uniform Case

Evolution Strategies (ESs) [1,27] have primarily been proposed to serve as a searching device for the optimization of continuous-valued parameters in a wide variety of applications. The mutation operator is the basic genetic operator and the next generation is obtained from the current population by means of one of two strategies: $(\mu,\lambda)$ or $(\mu+\lambda)$. In the $(\mu,\lambda)$ conception, the population is composed of $\mu$ individuals and $\lambda$ new individuals are generated from each one of the $\mu$ ancestors. Then $\mu$ individuals are selected solely from the offspring. On the other hand, in the $(\mu+\lambda)$ framework, the same happens except for the way the $\mu$ individuals are selected to compose the next generation: the ancestors and the $\lambda$ offspring are candidates to the next generation, and the $\mu$ individuals with the highest fitness are then selected. The (1+1) is the simplest version of evolution strategy, where one parent creates one single offspring via Gaussian mutation. Parameters of the Gaussian distribution may be evolved together with the individuals, incorporated into the genetic code. The recombination may be implemented as done in genetic

**Fig. 5** Flowchart for the Evolution Strategy

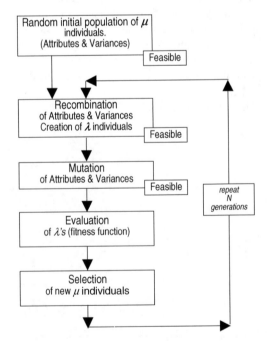

algorithms, and here we have adopted uniform crossover [9]. In the flowchart of Figure 5, describing the basic steps of the algorithm, the individuals are formed by the attributes of the candidate solution and the variance to be used by the Gaussian mutation operator [3]. Every time that the search space is composed of feasible and unfeasible candidate solutions, additional procedures should be incorporated to deal with feasibility issues.

In terms of codification of the attributes, Figure 3 indicates that, in a two-dimensional $n \times m$ lattice, each cell $(i,j)$, $i=1,...,n$ and $j=1,...,m$, will require four parameters in the genetic codification. So, in the NunCA framework, the size of the chromosome will be $4*n*m$.

The fitness will be given by the inverse of one plus the sum of the squared difference between the desired final state of each cell and the obtained final state. When intermediary states are of concern, additional terms will be included in the fitness function. As in the one-dimensional lattice, here the initial condition of the automata is arbitrarily defined and is considered fixed.

## 5   Related Approaches and Possible Extensions

Sipper [28] proposes an evolutionary-like and local procedure to update transition rules for binary states, including the possibility that one cell changes the state of a neighbor cell and copies itself onto that neighbor cell. Vacant cells, i.e. cells without a transition rule, are also accepted. However, the applicability was restricted to binary NunCA and requires specific operators to evaluate the fitness of individual rules, according to its local success, when applied to updating the state of its corresponding cell. Such methodology can hardly be directly extended to deal with global description of the intended spatio-temporal behavior.

Vassilev et al. [33] proposed a co-evolutionary procedure to deal with transition rules for binary states, and the spatio-temporal event under investigation was global synchronization.

Li [17] investigated partially and totally wiring (non-local CAs) and pointed out that the connection profile is decisive in the emergence of certain dynamical behaviors, like edge of chaos and attractors of convergent dynamics.

Structurally dynamic cellular automata (SDCA) are generalizations of uniform and non-uniform CA such that the lattice itself is part of the optimization process [12,13].

## 6   Experimental Results

Our experiments aim to show the flexibility of the NunCA approach when compared to the conventional uniform CA. We are going to consider two scenarios: a one-dimensional lattice with binary-state cells, and a two-dimensional lattice with multivalued-state cells.

In the former case, the purpose is to reproduce a sequence of state transitions in time, and in the latter case the intent is to obtain a non-uniform cellular automata capable of converging to a predefined final state, starting from an initial state and having the intermediary states submitted to some restrictive conditions or not. In both cases, the initial condition was set arbitrarily.

In the two-dimensional lattice, the transition rules admit an interpretation in terms of a local pattern of dispersion of a given material. This is one of the possible physical interpretations of the spatio-temporal dynamical model.

The set of binary rules for each cell in the one-dimensional CA was obtained via GA (see Figure 4) with binary representation, and the real values of the rules for each cell in the two-dimensional CA were provided by an evolution strategy (see Figure 5).

The individuals in the population are transition rules, and to evaluate each individual the corresponding cellular automata should be implemented and executed along time. Every discrete instant of time is relevant in the one-dimensional lattice, but in the two-dimensional lattice the final state may be the only relevant information or it may be considered together with the intermediary states. With the restrictions imposed to the parameter values of cell $(i,j)$, presented in section 3, the dynamic of the two-dimensional non-uniform cellular automata is guaranteed to be convergent.

The fitness function for the one-dimensional case will be given by Eq. (2):

$$F = \frac{1}{1 + d_{Hamm}(SS_{des}, SS_{obt})}, \tag{2}$$

where $d_{Hamm}(\cdot,\cdot)$ is the Hamming distance between matrices $SS_{des}$ and $SS_{obt}$, which contain respectively the desired and obtained sequence of states of the one-dimensional cellular automata. The number of columns equals the number of cells in the lattice, and the number of rows equals the number of state transitions along time. Though you will see two-dimensional pictures in Figures 6, 7, 8 and 9, they are just the representation of matrices $SS_{des}$ and $SS_{obt}$, with the time evolution being represented by the sequence of rows. The gray represents state 0 and the black corresponds to state 1.

On the other hand, in the two-dimensional lattice the fitness has two alternative expressions. When the final state is the only relevant information, the fitness function is expressed as follows:

$$F = \frac{1}{1 + \sum\limits_{i=1}^{n} \sum\limits_{j=1}^{m} \left( c_{(i,j)}^{end} - \hat{c}_{(i,j)}^{end} \right)^2}, \tag{3}$$

where $c_{(i,j)}^{end}$ and $\hat{c}_{(i,j)}^{end}$ are respectively the desired and the obtained final states of cell $(i,j)$ in the lattice, with the state being associated with the concentration of a given material.

When intermediary states are also relevant, one possibility is to express the fitness function in the form of Eq. (4):

$$F = \frac{1}{\left( 1 + \sum\limits_{k=1}^{N} \sum\limits_{i=1}^{n} \sum\limits_{j=1}^{m} \left( c_{(i,j)}^{k} - \hat{c}_{(j,i)}^{k} \right)^2 \right)}, \tag{4}$$

where $c_{(i,j)}^k$ and $\hat{c}_{(i,j)}^k$ are respectively the desired and the obtained states of cell $(i,j)$ in the lattice, at instant $k$, and $N$ is the number of intermediary states under consideration. In the experiments to be presented in what follows, we will adopted an alternative fitness function that emphasizes the necessity of a symmetrical dispersion, so that cells in opposite sides of a two-dimensional lattice have similar concentrations along time. The expression is given by Eq. (5):

$$F = \frac{1}{\left(1 + \sum_{i=1}^{n}\sum_{j=1}^{m}\left(c_{(i,j)}^{end} - \hat{c}_{(i,j)}^{end}\right)^2\right)\left(1 + \sum_{k=1}^{N}\sum_{i=1}^{n-1}\sum_{j=i+1}^{m}\left(\hat{c}_{(i,j)}^k - \hat{c}_{(j,i)}^k\right)^2\right)} \tag{5}$$

## 6.1  One-Dimensional Lattice

All the results in this subsection have been obtained with a genetic algorithm.

**Experiment 1: One-dimensional CA – "synchronization"**

In the synchronization task the objective is to alternate the one-dimensional lattice between states 1 and 0, so that every cell in the lattice share the same state at a given instant, and simultaneously change to the complementary state in the next instant. As already emphasized along the text, the initial configuration of states is arbitrary, though fixed. The *Radius* of the neighborhood is 1.

Figure 6 presents the results obtained with the NunCA approach, with Figure 6.b depicting the set of rules for each cell in the lattice. The set of transition rules for each cell is represented in each column of Figure 6.b. Given that the neighborhood is 1, we have eight possible configurations for the binary states of neighbor cells, and the transition rules should indicate the next state to every possible configuration. Figure 7 presents the best result obtained with a uniform CA. Figure 7.b shows the unique transition rule for the uniform CA, and Figure 7.a indicates that the uniform CA was incapable of solving the task.

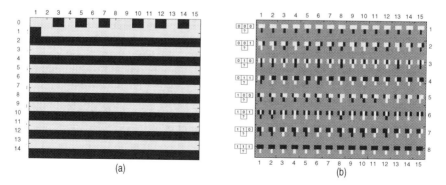

**Fig. 6** Results for Experiment 1 using NunCA

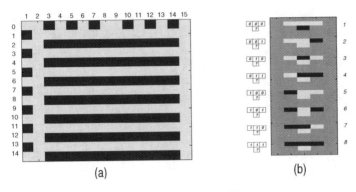

**Fig. 7** Results for Experiment 1 using uniform CA

Returning to Figure 6.b, which represents a successful implementation of the synchronization effect, we can see that no pair of cells shares the same transition rule, indicating that non-uniformity is a necessity here. Notice that this set of non-uniform transition rules may not be the only one capable of reproducing the desired behavior.

### Experiment 2: One-dimensional CA – "waves"

Experiment 2 consists in reproducing the temporal pattern that resembles the behavior of sinusoidal waves. Again, Figure 8 shows a successful performance of NunCA, and Figure 9 indicates that uniform CA fails to achieve the desired spatio-temporal behavior, because the best obtained behavior is far from the desired one.

Figure 8.b shows that each cell is associated with a distinct transition rule, again a strong indication of the complexity of the task and of the flexibility inherent to the NunCA framework.

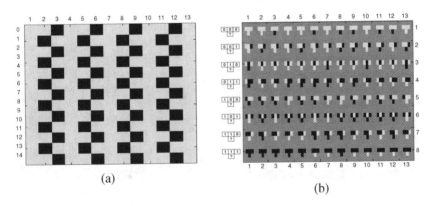

**Fig. 8** Results for Experiment 2 using NunCA

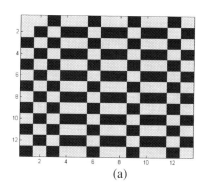
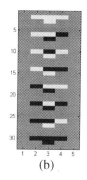

(a)                                          (b)

**Fig. 9** Results for Experiment 2 using uniform CA

In Experiment 2, using a neighborhood with *Radius*=2, a uniform CA gains enough representation power to accomplish the task that was successfully executed by a NunCA with *Radius*=1.

## 6.2 Two-Dimensional Lattice

Now we will analyze some experiments involving two-dimensional lattices and multivalued states. The synthesis of the desired spatio-temporal behavior will now be implemented by an evolution strategy.

The motivation for such experiments is the possibility of emulating dispersion phenomena in a great range of applications. In Experiments 3 and 4 we adopted a toroidal neighborhood, i.e. right-most cell is a neighbor of the left-most one, in the same row, and the up-most cell is a neighbor of the bottom-most one, in the same column. However, in Experiments 5 and 6 cells at the frontier of the lattice can not promote dispersion to the outside world, so that the dispersion is restricted to happen in a compact two-dimensional space.

### Experiment 3: Two-dimensional CA – "homogeneous distribution"

Experiment 3 is illustrated in Figure 10 and the purpose is to start with a maximum concentration of material at the cell in the centre of the lattice. The final convergent state will be a homogeneous distribution of concentration along the cells. So, if we start with 100 in the central cell of the lattice (see Figure 10.a), and the lattice has a 5×5 dimension, the desired final concentration per cell will be 4.

Here, each candidate NunCA should be put in operation and the convergence of the dynamics is measured by means of a threshold. When the sum of the square distance between two consecutive states (each term in the summation corresponds to a cell in the lattice) is below a predefined threshold, the convergence is detected and the fitness of that proposal is then evaluated. The best NunCA, obtained by the evolutionary search procedure based on an evolution strategy, produces the behavior illustrated in Figure 10 when put in operation.

Figure 14.a shows the gradient of the dispersion for this experiment, extracted from the interpretation of the resulting set of parameters for each transition rule.

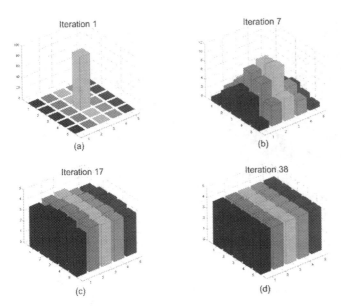

**Fig. 10** Convergence of the dynamic for Experiment 3, produced by the best evolved NunCA

As expected, there is no preferential direction of dispersion. Even with an unbalanced profile for the obtained gradient of dispersion, we have the emergence of a homogeneous equilibrium.

### Experiment 4: Two-dimensional CA – "contour"

Experiment 4 is illustrated in Figure 11. In this experiment, the objective was to equally distribute all the initial mass at the frontier of the lattice, so dividing the initial mass by 16. As a consequence, starting with 100 at the central cell, we want to obtain 6.25 in each of the sixteen cells at the frontier. Figure 11.d presents the convergent state, indicating the ability of the best evolved NunCA to reproduce the desired spatio-temporal behavior.

Figure 14.b shows the gradient of dispersion for this experiment. We can see that there is a preferential direction of dispersion pointing from the centre to the borders of the lattice.

### Experiment 5: Two-dimensional CA – "barrier"

Experiments 5 and 6 are the most complex to be considered here, and Experiment 5 is illustrated in Figure 12. In this experiment, the purpose was to move all the initial mass in cell (1,1), the one at the top-left corner, to cell (5,5), the one at the bottom-right corner. However, there is a barrier at cells (4,2), (3,3) and (2,4), so that the gradual transfer of mass must be accomplished avoiding the obstacle at the centre of the lattice.

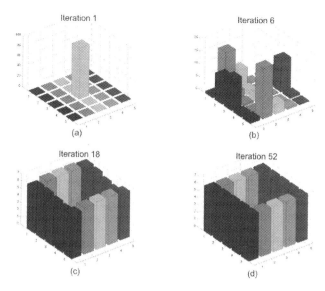

**Fig. 11** Convergence of the dynamic for Experiment 4, produced by the best evolved NunCA

Notice that cells (4,2), (3,3) and (2,4) has no transition rule and can not receive or deliver any amount of mass. Figure 12 shows the result and Figure 14.c shows the gradient of dispersion for this experiment. It can be inferred from Figure 14.c that

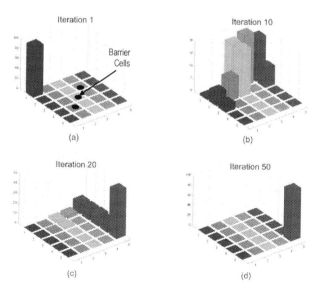

**Fig. 12** Convergence of the dynamic for Experiment 5, produced by the best evolved NunCA

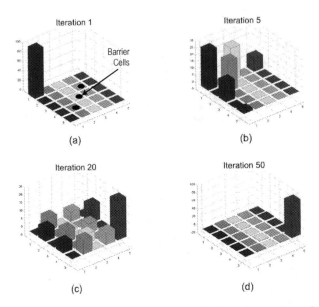

**Fig. 13** Convergence of the dynamic for Experiment 6, produced by the best evolved NunCA

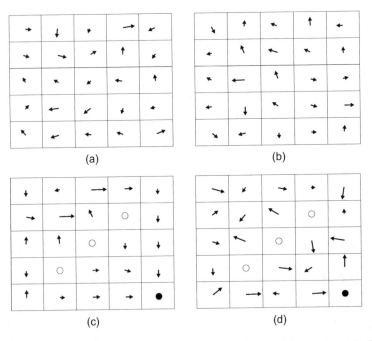

**Fig. 14** The gradient of dispersion throughout the lattice. (a) Experiment 3; (b) Experiment 4; (c) Experiment 5; (d) Experiment 6

the obtained solution forces the dispersion to follow the path through the top-right corner only. Of course, the bottom-left corner could have been considered as well, and Experiment 6 will impose an additional restriction requiring that the dispersion be symmetrical between both corners.

**Experiment 6: Two-dimensional CA – "barrier with symmetrical dispersion"**

Experiment 6 involves the same scenario already presented in Experiment 5, with the additional restriction of having a symmetrical dispersion along both sides of the barrier. Figure 13 indicates that the best evolved NunCA was capable of producing the intended spatio-temporal behavior, and Figure 14.d shows the gradient of dispersion for this experiment.

# 7  Concluding Remarks

Non-uniform cellular automata (NunCA) have been proposed here as mathematical models capable of reproducing desired spatio-temporal behaviors. The necessity of defining one transition rule per cell in the regular lattice motivated the application of evolutionary algorithms, due to the impossibility of performing an exhaustive search.

Evolutionary algorithms have already been proposed to design uniform and non-uniform cellular automata. However, none of these previous applications were devoted to spatio-temporal modelling using NunCA. When the purpose was the same, the cellular automata were taken to be uniform [19]. When the cellular automata were non-uniform, other purposes were involved in the evolutionary design, as the straight classification of the obtained transition rules according to the qualitative nature of the spatio-temporal behavior produced by the cells in the regular lattice [28].

When the transition rules involve a binary codification, a genetic algorithm has been designed to properly search for a feasible solution. In this scenario, the cellular automata are restricted to be one-dimensional lattices, and the purpose is to reproduce some specific and periodic profiles along time. The increment in flexibility provided by the NunCA framework was demonstrated to be essential to allow the reproduction of the intended spatio-temporal behavior. Very simple profiles have been defined, and even under these favorable circumstances (including a fixed initial condition for the cells in the lattice) there is no uniform CA capable of accomplishing the task, while multiple equivalent solutions have been obtained with NunCA.

A more challenging scenario is characterized by two-dimensional lattices with transition rules that implement dispersion of a given material, where the state of each cell is associated with the concentration of material at that position in space and at a given instant of time. Here, the cellular automata is characterized by transition rules each one obeying a difference equation with 4 parameters to be independently determined, once a set of physical restrictions is not violated. Due to the continuous nature of the parameters to be optimized, an evolution strategy has been conceived. Four distinct experiments have been implemented, and the last two ones incorporate spatial restrictions to the dispersion process. The spatial restrictions may be interpreted as a physical barrier to the flux of material. The single objective in the first three experiments was to design a two-dimensional NunCA capable of achieving a predefined final state from a predefined initial state, no matter the transitory behavior between the two configurations. The fourth experiment incorporates a temporal restriction associated with symmetrical flux, and here the intermediary states do matter, besides the initial and final states. The obtained results are promising in the sense of favoring the application of NunCA as mathematical model for arbitrary two-dimensional spatio-temporal behaviors.

The NunCA framework represents a significant increment in the computational demand of the design phase. However, the additional flexibility in implementing a distinct transition rule per cell in the regular lattice opens the possibility of multiple solutions and gives rise to an additional step in the investigation of means to reproduce spatio-temporal phenomena: the obtained transition rules for the NunCA can be interpreted and can be used to raise hypothetical explanations for complex spatio-temporal events in nature. This is exactly what has been done in Romano et al. [25], where real datasets obtained from experiments of larval dispersal have been considered to synthesize generalized versions of the NunCA model presented here.

**Acknowledgments.** This work has been supported by grants from CNPq, Capes, and Fapesp.

# References

1. Bäck, T., Hoffmeister, F., Schwefel, H.-P.: Survey of Evolution Strategies. In: Proceedings of the 4th International Conference on Genetic Algorithms, pp. 2–9 (1991)
2. Basanta, D., Miodownik, M.A., Bentley, P.J., Holm, E.A.: Investigating the Evolvability of Biologically Inspired CA. In: Proceedings of the Ninth International Conference on the Simulation and Synthesis of Living Systems, ALIFE9 (2004)
3. Beyer, H., Schwefel, H.: Evolution strategies: A comprehensive introduction. Natural Computing (2002)
4. Camazine, S., Deneubourg, J.-L., Franks, N.R., Sneyd, J., Theraulaz, G., Bonabeau, E.: Self-Organization in Biological Systems. Princeton University Press, Princeton (2001)
5. Chua, L.O., Yang, L.: Cellular Neural Networks: Theory. IEEE Transactions on Circuits and Systems 35(10), 1257–1272 (1988)
6. Chopard, B.: Cellular automata modeling of physical systems. Cambridge University Press, Cambridge (1998)
7. Collasanti, R.L., Grime, J.P.: Resource dynamics and vegetation process: a deterministic model using two-dimensional cellular automata. Functional Ecology 7, 169–176 (1993)
8. Frisch, U., Hasslacher, B., Pomeau, Y.: Lattice-gas automata for the Navier-Stokes equation. Physical Review Letters 56, 1505–1508 (1986)
9. Golberg, D.E.: Genetic Algorithms in Search, Optimization & Machine Learning. Addison-Wesley, Reading (1989)

10. Gregorio, S.D., Serra, R.: An empirical method for modelling and simulating some complex macroscopic phenomena by cellular automata. Future Generation Computer Systems 16, 259–271 (1999)

11. Gutowitz, H., Langton, C.: Methods for Designing 'Interesting' Cellular Automata. CNLS News Letter (1988)

12. Hillman, D.: Combinatorial Spacetimes, Ph.D. Thesis, University of Pittsburgh (1995)

13. Ilachinski, Halpern, P.: Structurally dynamic cellular automata. Complex Systems 1, 503–527 (1987)

14. Jai, A.E.: Nouvelle approche pour la modélisation des systèmes en expansion spatiale: Dynamique de végétation, Tendences nouvelles en modélisation pour l'environnement, pp. 439–445. Elsevier, Amsterdam (1999)

15. Kagaris, D., Tragoudas, S.: Von Neumann Hybrid Cellular Automata for Generating Deterministic Test Sequences. ACM Trans. On Design Automation of Electronic Systems 6(3), 308–321 (2001)

16. Kaneko, K. (ed.): Theory and Applications of Coupled Map Lattices. Wiley, Chichester (1993)

17. Li, W.: Phenomenology of Non-Local Cellular Automata, Santa Fe Institute Working Paper 91-01-001 (1991)

18. Maji, P., Ganguly, N., Saha, S., Roy, A.K., Chaudhuri, P.P.: Cellular Automata Machine for Pattern Recognition. In: Bandini, S., Chopard, B., Tomassini, M. (eds.) ACRI 2002. LNCS, vol. 2493, pp. 270–281. Springer, Heidelberg (2002)

19. Mitchell, M., Hraber, P., Crutchfiled, J.: Revisiting the edge of chaos: Evolving cellular automata to perform computations. Complex Systems 7, 89–130 (1993)

20. Mitchell, M., Crutchfield, J.P., Das, R.: Evolving Cellular Automata with Genetic Algorithms: A Review of Recent Work. In: Proceedings of the First International Conference on Evolutionary Computation and Its Applications (1996)

21. Nagel, K., Herrmann, H.J.: Deterministic models for traffic jams. Physica A 199, 254–269 (1993)

22. Nicolis, G., Prigogine, I.: Self-organization in non-equilibrium systems. Wiley, Chichester (1977)

23. Oliveira, G.M.B., de Oliveira, P.P.B., Omar, N.: Definition and Application of a Five-Parameter Characterization of Unidimensional Cellular Automata Rule Space. Artificial Life 7, 277–301 (2001)

24. Rabino, G.A., Laghi, A.: Urban Cellular Automata: The Inverse Problem. In: Bandini, S., Chopard, B., Tomassini, M. (eds.) ACRI 2002. LNCS, vol. 2493, pp. 349–356. Springer, Heidelberg (2002)

25. Romano, A.L.T., Gomes, L., Gomes, G., Puma-Villanueva, W., Zanetti, M., Von Zuben, C.J., Von Zuben, F.J.: Evolutionary Modeling of Larval Dispersal in Blowflies Using Non-Uniform Cellular Automata. In: IEEE Congress on Evolutionary Computation, pp. 3872–3879 (2006)

26. Schönfisch, B., de Roos, A.: Synchronous and asynchronous updating in cellular automata. Biosystems 51(3), 123–143 (1999)

27. Schwefel, H.-P.: Numerical Optimization of Computer Models. Wiley, Chichester (1981)

28. Sipper, M.: Non-Uniform Cellular Automata: Evolution in Rule Space and Formation of Complex Structures. In: Artificial Life IV, pp. 394–399. MIT Press, Cambridge (1994)

29. Toffoli, T., Margolus, N.: Cellular automata machines – A new environment for modeling. MIT Press, Cambridge (1987)

30. Tomassini, M., Perrenoud, M.: Cryptography with cellular automata. Applied Soft Computing 1(2), 151–160 (2001)
31. Tomassini, M., Sipper, M., Zolla, M., Perrenould, M.: Generating high-quality random numbers in parallel by cellular automata. Future Generation Computer Systems 16, 291–305 (1999)
32. Tsalides, P.: Cellular automata based build-in self test structures for VLSI systems. IEE-Electronics Letters 26(17), 1350–1352 (1990)
33. Vassilev, V.K., Miller, J.F., Fogarty, T.C.: The Evolution of computation in co-evolving demes of non-uniform cellular automata for global synchronization. In: Proceedings of the 5th European Conference on Artificial Life, Berlin (1999)
34. Von Neumann, J.: The General and Logical Theory of Automata, John von Neumann: Collected Works. In: Taub, A.H. (ed.) Design of Computer, Theory of Automata and Numerical Analysis, vol. 5. Pergamon Press, Oxford (1961)
35. Wolfram, S.: Cellular Automata and Complexity – Collected Papers. Addison-Wesley Publishing Company, Reading (1994)

# Part II
# Bio-Inspired Approaches in Classification Problem

# Genetic Selection Algorithm and Cloning for Data Mining with GMDH Method

Marcel Jirina and Marcel Jirina Jr.

**Summary.** The Group Method Data Handling Multilayer Iterative Algorithm (GMDH MIA) is modified by use of the selection procedure from genetic algorithms while including cloning of the best neurons generated to obtain even less error. The selection procedure finds parents for a new neuron among already existing neurons according to the fitness and also with some probability from the network inputs. The essence of cloning is slight modifying the parameters of the copies of the best neuron, i.e. the neuron with the largest fitness. The genetically modified GMDH network with cloning (GMC GMDH) can outperform other powerful methods. It is demonstrated on some tasks from the Machine Learning Repository.

## 1 Introduction

Classification of multivariate data into two or more classes is an important problem of data processing in many fields of data mining. For classification of multivariate data into classes (groups, categories etc.) the well-known GMDH MIA (group method data handling multilayer iterative algorithm) is often used. This approach – in difference to others – can even provide a closed form polynomial solution when needed.

Although GMDH MIA is one of relatively standard methods, there are new findings in the application of genetic optimization for enhancing GMDH

Marcel Jirina

Institute of Computer Science, Pod vodarenskou vezi 2, 182 07 Prague 8 – Liben, Czech Republic

e-mail: marcel@cs.cas.cz

Marcel Jirina Jr.

Faculty of Biomedical Engineering, Czech Technical University in Prague, Nam. Sitna 3105, 272 01, Kladno, Czech Republic

e-mail: jirina@fbmi.cvut.cz

A. Abraham et al. (Eds.): Foundations of Comput. Intel. Vol. 4, SCI 204, pp. 107–125.

springerlink.com          © Springer-Verlag Berlin Heidelberg 2009

behavior and there is the idea of using the cloning principle borrowed from immune networks to further optimize GMDH based algorithms.

The standard GMDH MIA method has been described in many papers since 1971 e.g. in [1], [4], [5], [6], [7], [10]. The basis of the GMDH MIA is that each neuron in the network receives input from exactly two other neurons from previous layer; the first layer is formed using neurons representing the input layer. The two inputs, $x$ and $y$ are then combined to produce a partial descriptor based on the simple quadratic transfer function (the output signal is $z$):

$$z = a_1 + a_2x + a_3y + a_4x^2 + a_5y^2 + a_6xy \tag{1}$$

where coefficients $a,..f$ are determined by linear regression and are unique for each neuron. The coefficients can be thought of as analogous to weights found in other types of neural networks.

The network of transfer functions is constructed one layer at a time. The first network layer consists of functions of each possible pair of n input variables (zero-th layer) resulting in $n(n+1)/2$ neurons. The second layer is created using inputs from the first layer and so on. Due to exponential growth of the number of neurons in a layer, a limited number of best neurons are selected and the other neurons are removed from the network after finishing the layer. For illustration see Fig. 1. In the standard GMDH MIA algorithm all possible pairs of neurons from the preceding layer (or inputs when the first layer is formed) are taken as pairs of parents. The selection consists of selecting a limited number of the best descendants, "children", while the others are removed after they arose and were evaluated. In this way, all variants of GMDH MIA are rather ineffective as there are a lot of neurons generated, evaluated and then immediately removed with no other use.

In this Chapter we solve the classification task by use of GMDH MIA, modified by a selection algorithm common in genetic algorithms and by cloning the best neurons generated up to a given moment of the learning process. In our approach, when generating a network of neurons, for each new neuron, it's parents are selected from all presently existing neurons and network inputs also. Then six parameters of the new neuron are computed using the least mean squared error method. No explicit layered structure arises because of the random selection of the parents.

The number of neurons grows during learning of one neuron at a time. No neuron is deleted during the learning process and in the selection procedure its chance to become a parent for a neuron is proportional to its fitness. If a new neuron appears to be the best, its clones are generated. Clones are inexact copies of the parent neuron, which was found to be the best neuron generated up to now. The true minimum is searched in the neighborhood of the minimum given by the parameters found by standard linear regression. Therefore, the clones have mutated, i.e. have slightly changed parameters of the parent individual to cover that neighborhood. In this respect we use the idea that a clone is an inexact copy of the parent GMDH neuron.

Note that GMDH is principally a function approximator. When used as a two class classifier, the network approximates the class mark, often 1 and 0 or 1 and –1. To decide to which class an applied sample belongs to, the network output is compared with the threshold $\theta$. If the output is equal to or larger than this threshold, the sample belongs to one class or else to the other class. The value of the threshold can be optimized to get a minimal classification error. We use this approach here too.

**Fig. 1** An example of
GMDH MIA structure

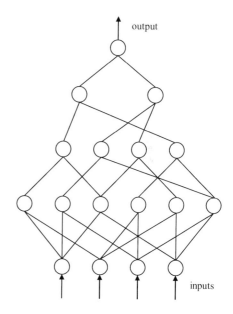

It is shown here that a new algorithm, including cloning, allows tuning the GMDH neural network more effectively than is possible in genetically optimized GMDH networks.

We tested the genetically modified GMDH network with cloning (GMC GMDH) on some tasks from the Machine Learning Repository and compared it with other widely accepted methods. It is demonstrated that GMC GMDH can outperform other powerful methods.

## 2  Related Works

There are several works dealing with genetic optimization of the GMDH MIA method [3], [7], [8], and [9]. These approaches use the GMDH networks of limited size as the number of neurons in a layer and the number of layers. Some kind of randomization must be used to get a population of GMDH networks because the original GMDH MIA neural network is a purely deterministic. Individuals in a population are subjects of genetic operations of selection, crossover and mutation. As each GMDH network represents a rather general graph, there must be procedures for crossover of graphs similarly as in other kinds of genetically optimized neural networks, e.g., NNSU [10]. For example, in [18] the NN architecture is built by adding hidden layers into the network, while configuration of each neuron's connections is defined by means of GA. An elitist GA with binary encoding, fitness proportional selection, standard operators of crossover and mutation are used in the algorithm. Different genetically optimized GMDH networks in literature differ in the way how a population, especially the first population of networks, is formed and by variants of the genetic procedures used. It seems, however, that up to now no approach in genetically optimized GMDH networks essentially brings better results than the standard GMDH MIA algorithm. On the other hand, genetically optimized GMDH networks eliminate the necessity to set up, in advance, the number of the best neurons left in each layer at least. In this way, such GMDH networks become even less "parameter-less" than before.

The difficult problem with genetically optimized GMDH method lies in the fact, that a population of GMDH networks must be generated. There must be crossover of individuals. Individuals are networks, which generally have different graphs. The crossover of two different graphs is a rather complex task. Its solution is known and also used in other genetically optimized neural networks [10], [15], [16], and [17].

A very interesting and principally simple application of the selection process to GMDH MIA was published by Hiasaat and Mort in 2004 [8]. Their method does not remove any neuron during learning. Thus it allows unfit individuals from early layers to be incorporated at an advanced layer where they generate fitter solutions. Secondly, it also allows those unfit individuals to survive the selection process if their combinations with one or more of the other individuals produce new fit individuals, and thirdly, it allows more implicit non-linearity by allowing multi-layer variable interaction. The GMDH algorithm is constructed in the same manner as the standard GMDH algorithm except for the selection process. In order to select the individuals that are allowed to pass into the next layer, all the outputs of the GMDH algorithm at the current layer are entered as inputs in the genetic algorithm. It was shown in [8] that this approach can outperform the standard GMDH MIA when used in the prediction of two daily currency exchange rates. No other test of this approach classification ability was performed in the literature cited. The GMDH network [8] has a layered structure where the input to the neuron can be the output of any already existing layer or even network input. To keep the layered structure in this context seems rather complicated. One can use generalization where a new neuron input can be the output of any already existing neuron or even network input. Thus, the strict layered structure disappears.

Immunity-based models including cloning are new optimization techniques. Note first that the authors dealing with artificial immune systems, e.g. [11], [12] use different terminology than used in the neural network community and the genetic algorithm community. So, some translation or mapping is needed. Here especially, antibody – neuron, affinity – fitness, antigen or antigenic stimulus – signal. Leukocytes or white blood cells are divided into three classes, one of which is lymphocyte. Lymphocytes include B cells, which mature in bone marrow, and T cells, which mature in the thymus. White blood cells produce antibodies – proteins that each respond to a specific antigen. Antigens are enemy agents, i.e. viruses or bacteria dangerous for an organism. There, various mechanisms or processes in the immune system appear, which are investigated in the development of artificial immune systems (AIS) [11].

1. Negative selection is the process that happens during the development of T cells in the thymus. Immature T cells go through a censoring process so those that recognize self-cells are eliminated. Only the rest are deployed into the immune system so T cells will not attack the self-cells. Artificial negative selection algorithms were proposed to mimic that procedure: generating detector candidates randomly, and then eliminating those that match the self-samples. It suits the need of the so-called "one-class classification" problem, e.g. anomaly detection, in which training data from only one of the two classes are available.

2. The immune network model is another widely used model in the AIS field. It was based on the idiotypic[1] network concepts proposed by Jerne in 1974, in which the dynamics of the immune system can be described as a network of antibodies where activation by antigens, and activation and suppression between antibodies co-exist.

3. The relatively new models in AIS include the danger theory, which emphasizes discriminating the danger posed by the invaders instead of whether it is self or nonself.

4. Besides the immune network, the other theoretical framework of immunology, clonal selection theory, which is in fact more widely accepted, is also often used in AIS. The clonal selection principle describes the basic features of an immune response to antigenic stimulus. It establishes the idea that only those cells that recognize the antigen proliferate, thus being selected against those that do not. The main features of the clonal selection theory are that:

   - The new cells are copies of their parents[2] (clone) subjected to a mutation mechanism with high rates (somatic hypermutation);
   - Elimination of newly differentiated lymphocytes carrying self-reactive receptors;
   - Proliferation and differentiation on contact of mature cells with antigens.

---

[1] Idiotype - individual genotype.

[2] In fact, of the original - parent cell, as there are no parents in the standard sense of word.

From these ideas we use results of clonal selection theory, especially a cloning procedure derived from the Simple Clonalg algorithm [12]:

```
BEGIN
   Construct the initial population of antibodies
   REPEAT
      Evaluate antibodies to calculate their affinities
      Select the n best affinity antibodies and clone them
      Maturate the clones and evaluate them
      Allow the best antibody from each subpopulation to survive
      Replace the d lowest affinity antibodies with new
      antibodies randomly produced
   UNTIL a terminal criterion is satisfied or the maximum
   generation number of clones is reached
END
```

In biological systems clones are not exact copies of the parent cell because some mutations are in effect. In artificial systems, clones are also not exact copies of the parent cell or neuron, and therefore some mutation must be in effect. The clone, to be a true clone, must have the same parents, i.e. input signals. So, the basic parameters – the two parents are not changed. The problem is, how to change the parameters $a, .. , f$ of the parent neuron. These changes should be small enough to keep a sufficient similarity of the clone to the original individual, and, at the same time, sufficiently large enough to reach the necessary changes for searching the data space in the neighborhood of the parent neuron.

The application of the cloning process to GMDH MIA was not proposed up to now. Usually – and we do it as well – the parameters of the new neuron are set up by linear regression, i.e. with a least mean squared error method. This method uses the following Gauss-Markov assumptions [19]:

The random errors have an expected value of 0.

The random errors are uncorrelated.

The random errors are homoscedastic, i.e., they all have the same variance.

The errors are not assumed to be normally distributed, nor are they assumed to be independent, nor are they assumed to be identically distributed.

We expect that due to the nonlinearity of the problem as well as the GMDH network, the assumption of homoscedasticity is not met especially for the classification problem. In classification problems each class has a different origin from which a different distribution of regression errors may arise. Even such a weak, but essential, assumption as homoscedasticity is not met. We show it in a detailed analysis in the following paragraph Linear regression. Thus, the true minimum may lie somewhere in the neighborhood of the minimum given by the parameters found by standard linear regression.

# 3   Linear Regression

In this chapter we show, on an example, that the Gauss-Markov assumptions of an expected value 0 and of homoscedasticity need not be fulfilled in GMDH network. The solution obtained by the least mean squared error method then need not be the optimal solution. Thus, the truly optimal solution may lie somewhere in the neighborhood of the solution found by the LMS method.

To prove the assertion that homoscedasticity need not be fulfilled in GMDH network, a simple example will suffice. We used the standard GMDH MIA method for the Brest CW data from the UCI Machine learning repository [13]. The Brest CW problem is a two-class classification task. We analyzed the input and output values of the neurons and found that zero mean and homoscedasticity of residuals is often not fulfilled. It is shown on the example of one neuron.

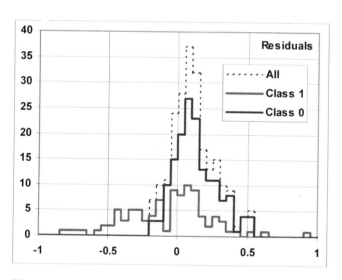

**Fig. 2** Histograms of residuals for both classes and both classes together

In Fig. 2, histograms of residuals, i.e. histograms of errors for one neuron and for both classes separately and for all data are depicted. There, it is seen that first, the expected value apparently is not zero, and second, residuals are heteroscedastic. The heteroscedasticity originates in the fact that there is data of two classes, each with different statistics. It is shown in Figs. 3 and 4, where histograms of input signals to the neuron analyzed are depicted.

It is seen that classes have very different statistics. After polynomial transformation according to (1) the difference increases because the target of the transformation is to get the output as close as possible to the value 0 for class zero and to the value 1 for class one. From this difference, the difference in

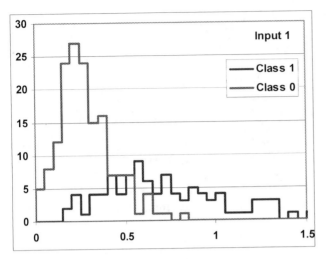

**Fig. 3** Histograms of input signals for both classes for input 1 of the neuron

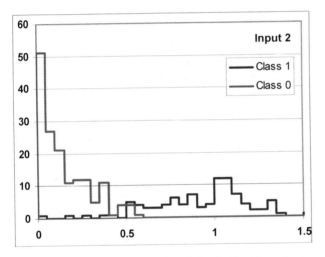

**Fig. 4** Histograms of input signals for both classes for input 2 of the neuron

statistics of residuals also follows - see Fig. 2 - after transformation (1) where coefficients were set up by linear regression.

   Different approaches can be used to find a better solution than the solution obtained by linear regression. The solution obtained by linear regression can be used as the first approximation. We use cloning, i.e. we generate neurons with the same inputs and with parameters $a, .. f$ slightly modified with respect to their original values.

## 4 Genetically Modified GMDH

Here we describe the approaches which result in our construction of a genetically modified GMDH network with cloning.

### 4.1 The Learning Set

The learning set consists of $n + 1$ dimensional vectors $(x_i, y_i) = (x_{1i}, x_{2i}, ... x_{ni}, y_i)$, $i = 1, 2, ... N$ where $N$ is the number of learning samples (patterns or examples). The learning set can be written in the matrix form

$$[X, Y].$$

The matrix $X$ has $n$ columns and $N$ rows; $Y$ is a column vector of $N$ elements. In the GMDH, the learning set is usually broken into two disjoint subsets, the training set (or construction or setup set) and the so-called validation set. In the learning process the former one is used for setting up parameters of neurons of the newly created neuron, the latter for evaluation of an error of the newly created neuron. Thus $N = N_s + N_v$, where $N_s$ is the number of rows used for setting up the parameters of neurons (the training set), and $N_v$ is the number of rows used for error evaluation during learning (the validation set).

### 4.2 New Genetically Modified GMDH Network Algorithm

The standard quadratic neuron is an individual of the genetically modified GMDH network. Its parents are two neurons (or possibly one or two network inputs) from which two input signals are taken. A selection of one neuron or input as one parent and of another neuron or input as the other parent can be made by the use of different criteria. In genetic algorithms, in the selection step, there is a common approach that the probability to be a parent is proportional to the value of the fitness function. Just this approach is used here. The fitness is simply a reciprocal of the mean absolute error on the validation set.

An operation of a crossover in the genetically modified GMDH is, in fact, no crossover in the sense combining two parts of parents' genomes. In our approach, Equation (1) gives us a symmetrical procedure of mixing the parents' influence but not their features, parameters. The parameters $a, .. f$, see (1), are stated separately.

## 4.3   Selection Procedure

The initial state form $n$ inputs only, there are no neurons. If there are alreadyk neurons, the probability of a selection from inputs and from neurons is given by

$$p_i = n/(n + k),$$
$$p_n = k/(n + k)$$

for $n/(n + k) > p_0$, where $p_0$ is the minimal probability that one of the network inputs will be selected; we found $p_0 = 0.1$ as optimal. Otherwise,

$$p_i = p_0,$$
$$p_n = (1 - p_0).$$

The fitness function is equal to the reciprocal error on the verification set. Let $\varepsilon(j)$ be the mean error of the j-th neuron on the validating set. The probability that neuron $j$ will be selected is:

$$p_n(j) = (1 - p_n)\frac{1/\varepsilon(j)}{\sum\limits_{s=1}^{N_T} 1/\varepsilon(s)}.$$

Moreover, it must be assured that the same neuron or the same input is not selected as the second parent of the new neuron.

After the new neuron is formed and evaluated it can immediately become a parent for another neuron. Thus, the network has no explicit layers. Each new neuron can be connected to any input or up to any existing neurons at that point.

The computation of six parameters $a, .. f$, see (1), of the new neuron is the same as in the GMDH MIA algorithm.

## 4.4   Best Neuron

The new neuron added need not be better than all others. Therefore, the index and error value of the best neuron is stored as long as a better neuron arises. Thus every time there is information about the best neuron, i.e. the best network's output without need of sorting. After learning, this output is used as a network output in the recall phase.

## 4.5   Cloning Mechanism

There are lots of ideas how to do the cloning. From these ideas, we use cloning in the form close to the SIMPLE CLONALG algorithm [12] in this way:

```
BEGIN
    Given the Best GMDH Neuron with parents (i.e. input signals
    from) In₁, In₂ and with six parameters a, b, .. f.
    REPEAT
        Produce a copy of the Best Neuron. A copy has the same
        inputs In₁ and In₂ but mutated parameters a, .. f, i.e.
        parameters slightly changed (details see below)
        Evaluate fitness of this clone neuron.
        If this neuron happens to be better than the Best
        Neuron, break this clone generating cycle (and start
        this cloning algorithm from the beginning with
        new Best Neuron again).
    UNTIL a terminal criterion is satisfied or the maximum
    number of clones is reached
END
```

## 4.6  Mutation

It has no sense for the clones to be the exact copies of the Best Neuron. Therefore, some mutation must be in effect. The clone, to be a true clone, must have the same parents. So, the basic parameters – the two parents are not changed. A problem is how to change the parameters $a, .. f$. These changes should be small enough to keep a sufficient similarity of the clone to the original individual (the Best Neuron) and, at the same time, a sufficiently large amount to reach the necessary changes for searching the data space in the neighborhood of the Best Neuron.

The simplest approach spreads value of each parameter randomly around its value in the Best Neuron. For each parameter $a, .. f$ one can use normal distribution with the mean equal to the value valid for the best neuron and standard deviation equal to $1/6$ of this value. There is also a multiplicative constant for setting up the standard deviation, which a user can set up appropriately; the default value is 1.

## 4.7  Error Development and Stopping Rule

A stopping rule different from searching for the minimal error on the validating set like in the original GMDH MIA method also follows from the new strategy of network building in the GMC-GMDH method. In our case, the error on the validating set for the best neuron monotonously decreases having no minimum. On the other hand, the indexes of the best neurons became rather distant. For illustration see Figs. 5 and 6. The process can be stopped either when very small change in error is reached, or too many new neurons are built without appearance of a new best neuron or when a predefined number of neurons is depleted.

## 4.8  Pruning

After learning, the best neuron and all its ancestors have their role in the network function. All others can be removed.

Pruning reduces the size of network graph to the necessary neurons (nodes) and edges. It may appear that some input variables are not used and have no influence to the network's output. This phenomenon appears in the standard GMDH-MIA algorithm as well as in our algorithm. Input variables not used can be omitted and thus, we have an effective means for dimensionality reduction of the data space. If the final GMDH network is not too complex, one can even obtain a closed form of the high order polynomial of the network's function. It was shown already in [2] for the standard GMDH-MIA.

## 4.9  Network Structure

It can be seen here that resulting network consists of individual neurons connected more or less randomly forming an oriented graph with leaves in network inputs and the root formed by the network output, the final best neuron. When generating the network, the notion of layer was not defined or used. In the end, the network has no clear layered structure like the network generated by the original GMDH MIA algorithm and most of other GMDH algorithms including those genetically optimized.

## 4.10  Recall

After learning, the resulting network is a feed-forward network. When a sample is applied to inputs, the outputs of the individual neurons are computed successively according to their order numbers. Thus at any time all information needed for the neuron's output computation is known. The last neuron is the best neuron and its output is the output of the whole network. If the network is used for approximation or prediction, the output gives just the approximation of the value desired. If the network is used as a two class classifier, one must set up a proper threshold $\theta$ and an output value larger than or equal to this threshold means that the sample applied belongs to one class or else it belongs to the other class. The value of the threshold can be tuned with respect to the classification error or to the other features of the classifier.

## 5  Performance Analysis

The experiments described below show that in most cases our genetically modified GMDH algorithm with cloning (GMC GMDH) outperforms 1-NN method in most cases, and in many cases it outperforms naïve Bayes method and also the $k$-NN method where $k$ equals to the square root of the number of training set points.

The classification ability of the genetically modified GMDH algorithm with cloning (GMC GMDH) was tested using real-life tasks from the UCI Machine Learning Repository [13]. We do not describe these tasks in detail here as all can be found in [13]. For each task, the same approach to testing and evaluation was used as described in [13] for other methods. We also show convergence of the learning process.

For running GMC GMDH program, default parameters were used as follows for all tasks: The no. of neurons generated for stopping computation was 10000. The probability that the new neuron's input was one of the input signals was 10 %; the probability that new neuron's input was one of the already existing neurons was 90 %. The maximal number of clones generated from one parent neuron was limited to int(sqrt(No. of neurons generated up to now)). For each method, an optimal threshold $\theta$ for the minimal error was used. The fitness function was the reciprocal of the mean absolute error. An experiment was also made with the fitness function equal to the reciprocal of the square of the mean absolute error to make a lower relative probability that the bad neuron is selected as a parent for a new neuron.

**Table 1** Classification errors for four methods on some data sets from UCI MLR. GMC GMDH is with the fitness function equal to the reciprocal of the mean absolute error, and GMC GMDH 2 is with the fitness function equal to the reciprocal of the squared mean absolute error

| Data set | Algorithm | | | | | |
|---|---|---|---|---|---|---|
| | 1-NN | sqrt-NN | bay1 | LWM1 | GMC GMDH | GMC GMDH 2 |
| German | 0.4077 | 0.2028 | 0.2977 | 0.7284 | 0.2947 | 0.2617 |
| Adult | 0.2083 | 0.2124 | 0.1637 | 0.1717 | 0.1592 | 0.1562 |
| Brest CW | 0.0479 | 0.0326 | 0.0524 | 0.0454 | 0.0419 | 0.0436 |
| Shuttle-small | 0.0259 | 0.0828 | 0.1294 | 0.0310 | 0.0259 | 0.0465 |
| Spam | 0.0997 | 0.1127 | 0.1427 | 0.1074 | 0.1008 | 0.0917 |
| Splice | 0.4035 | 0.3721 | 0.2866 | 0.2587 | 0.1309 | 0.1339 |
| Vote | 0.1053 | 0.0602 | 0.0977 | 0.0741 | 0.0667 | 0.0741 |

In Table 1, the results are shown together with the results for four other well-known and very often used classification methods. In the second column, the cross validation factor is given. The methods for comparison are:

1-NN – standard nearest neighbor method

Sqrt-NN – the $k$-NN method with $k$ equal to the square root of the number of samples of the learning set

Bayes – the naïve Bayes method using ten bins histograms

LWM1 – the learning weighted metrics method [14] modified with nonsmooth learning process.

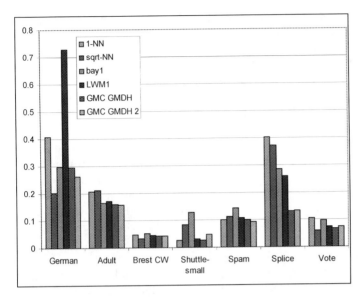

**Fig. 5** Classification errors for four methods on some data sets from the UCI MLR. Note that for Shuttle, small data the errors are enlarged ten times in this graph. In the legend, GMC GMDH 2 means the GMC GMDH method with fitness equal to the reciprocal of the square of the mean absolute error

These results are also depicted in a more distinct form in Fig. 5.

The learning process of the GMC GMDH network is stable and convergent. In Figs. 6 and 7, the successive lowering of the error is depicted. The error is stated on the validating set for the data VOTE from UCI MLR. In difference to the GMDH MIA algorithm, there is no minimum and the stopping rule here is based on the exhausting of the total number of neurons given in advance.

In Fig. 6, it is shown how an error of the best neuron on the validating set decreases with the number of best neurons successively found during learning. It is seen that in this figure, the order number of the best neuron, i.e. true number of neurons generated during learning process, is not seen.

The dependence on the true number of neurons generated is shown in Fig. 7. There, on the horizontal axis are a number of neurons generated and the points on the line show the individual best neurons successively generated. For each point, the corresponding value on horizontal axis is the order number of the best neuron. Note a logarithmic scale on the horizontal axis. It is seen that a new Best Neuron appears in successively longer and longer intervals of neuron generation bringing smaller and smaller improvement in error on the validating set.

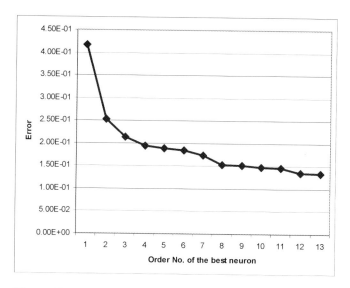

**Fig. 6** Error on the validating set count of best neurons successively found during learning

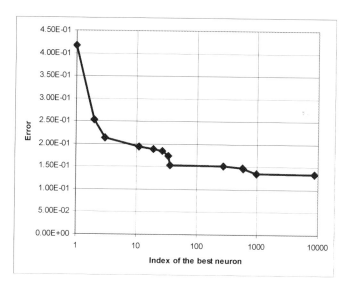

**Fig. 7** Error on the validating set vs. the index of the best neuron, i.e. the number of neurons generated

# 6 Conclusion and Discussion

The target of this Chapter was to generalize the idea of the genetically modified GMDH neural network for processing multivariate data appearing in data mining problems and to extend this type of network by cloning. Clones are close, but not identical copies of original individuals. An individual in our case is a new neuron, which appears during the learning process.

The new genetically modified GMDH method with cloning (GMC GMDH) has no tough layered structure. During learning, when a new neuron is added it is randomly connected to two already existing neurons or to network inputs with some probability derived from the fitness and keeping some minor probability that an input is selected. When the inputs are assigned to a new neuron, the six parameters $a, ..f$ of the quadratic polynomial (1) are computed. For it, the least mean squared method is used as in the basic GMDH MIA approach using the training part of the learning set. The fitness, i.e. the reciprocal of the mean absolute error, is evaluated using the validating set. If a new neuron generated is found to be the best neuron, the clones are derived to reach even better fitness. The clones have the same two "parent" signals as the best neuron. The mutation operation slightly changes values of six parameters of the best neuron and thus the clones are similar to, but not exact copies of the best neuron. When a clone is found to be the best neuron, the clone generating process is broken and immediately starts again with this new best neuron.

It was found that our expectation that the true optimum may lie somewhere in the neighborhood of parameters of the best neuron computed by linear regression holds. We have shown that some assumptions for least squares method for linear regression are not met and thus the solution does not represent an exact optimum. Cloning is a useful technique to get closer to the true optimum. Cloning with large changes of parameters has a small effect, but with small changes a new best neuron often arises. From it, one can deduce that in practice differences between pairs of parameters corresponding to a minimum found by linear regression and a minimum found by cloning is not too large but not negligible either.

Classification errors for four methods on some data sets from the UCI MLR are depicted in Fig. 5. Note that for Shuttle, small data the errors are enlarged ten times in this graph. In Table 1 and in Fig. 5 it is seen that the GMC GMDH method outperforms other methods in the tasks Adult, Shuttle small, and Splice or nearly outperforms Brest CW, Spam, and Vote. The GMC GHMDH is the second best with a very small difference with respect to the best method considered. It is the second best in the task German. The experiments described above show that the GMC GMDH approach outperforms 1-NN method in most cases, and in many cases outperforms the naïve Bayes method and also the k-NN method where k equals the square root of the number of training set samples.

It is also seen here that for a fitness function equal to the second power of the mean absolute error, the error may be slightly different from the case of the fitness function equal to the mean absolute error. Then, the sensitivity to fitness function definition is rather small in these cases.

In Figs. 6 and 7 it is seen that the learning process converges rather fast, i.e. for a relatively small number of neurons generated, the error on the validating set decreases fast and then for a large number of neurons generated, the error decreases only slightly. Practical tests show that a further enlargement of the number of neurons generated up to the order of a hundred thousand has no practical effect. As there is no searching or sorting like in the nearest neighbor-based methods or in the classical GMDH MIA algorithm, the GMC GMDH is much faster than the methods mentioned, especially for large learning sets.

The genetically modified GMDH method with cloning (GMC GMDH) presented here appears to be an efficient approach giving reliably good results better or comparable to the best results obtained by other methods. The Genetically modified GMDH method is an elegant idea how to improve the efficiency of the popular GMDH MIA method. It is based on the usage of a selection principle of genetic algorithms instead of systematic assignment of all pairs formed by neurons of the last layer. Thus, all neurons once generated remain potential parents for new neurons during the whole learning process. Also, each input signal may be used with some probability as a parent signal for a new neuron. Thus, the strictly layered structure of the GMDH algorithm disappears, as any new neuron can be connected to the output of any already existing neuron or even to any input of the network. The essential advantage of the genetically modified GMDH with cloning is that one need not set up the number of the best neurons selected in the newly generated layer and thus indirectly control the learning time and size of the network as in the standard GMDH MIA algorithm. The only limitations of the GMC GMDH method are the learning time or the memory size.

Here, we presented a new method of building the GMDH network with genetic selection of parents for each new neuron and with cloning of the best neuron. We have shown efficiency and good behavior for two class classification tasks. As GMDH networks also serve as an approximator and predictor, there is the open possibility to use GMC GMDH for approximating and predicting tasks.

**Acknowledgements.** This work was supported by the Ministry of Education of the Czech Republic under the project Center of Applied Cybernetics No. 1M0567, and No. MSM6840770012 Transdisciplinary Research in the Field of Biomedical Engineering II.

# References

1. Ivakhnenko, A.G.: Polynomial Theory of Complex System. IEEE Trans. on Systems, Man and Cybernetics SMC-1(4), 364–378 (1971)
2. Farlow, S.J.: Self-Organizing Methods in Modeling. GMDH Type Algorithms. Marcel Dekker, Inc., New York (1984)
3. Tamura, H., Kondo, T.: Heuristics-free group method of data handling algorithm of generating optimal partial polynomials with application to air pollution prediction. Int. J. Systems Sci. 11(9), 1095–1111 (1980)
4. Ivakhnenko, A.G., Müller, J.A.: Present State and New Problems of Further GMDH Development. SAMS 20, 3–16 (1995)
5. Ivakhnenko, A.G., Ivakhnenko, G.A., Müller, J.A.: Self-Organization of Neural Networks with Active Neurons. Pattern Recognition and Image Analysis 4(2), 177–188 (1994)
6. Ivakhnenko, A.G., Wunsch, D., Ivakhnenko, G.A.: Inductive Sorting/out GMDH Algorithms with Polynomial Complexity for Active neurons of Neural network. IEEE 6/99, 1169–1173 (1999)
7. Nariman-Zadeh, N., et al.: Modeling of Explosive Cutting process of Plates using GMDH-type neural network and Singular value Decomposition. Journ. of material processes technology 128(1-3), 80–87 (2002)
8. Hiassat, M., Mort, N.: An evolutionary method for term selection in the Group Method of Data Handling. Automatic Control & Systems Engineering, University of Sheffield, www.maths.leeds.ac.uk/statistics/workshop/lasr2004/Proceedings/hiassat.pdf
9. Oh, S.K., Pedrycz, W.: The Design of Self-organizing Polynomial Neural Networks. Information Sciences 141(3-4), 237–258 (2002)
10. Hakl, F., Jiřina, M., Richter-Was, E.: Hadronic tau's identification using artificial neural network. ATLAS Physics Communication, ATL-COM-PHYS-2005-044 (August 26, 2005), http://documents.cern.ch/cgi-bin/setlink?base=atlnot&categ=Communication&id=com-phys-2005-044
11. Ji, Z.: Negative Selection Algorithms: From the Thymus to V-Detector. Dissertation Presented for the Doctor of Philosophy Degree. The University of Memphis (August 2006)
12. Guney, K., Akdagli, A., Babayigit, B.: Shaped-beam pattern synthesis of linear antenna arrays with the use of a clonal selection algorithm. Neural Network world 16, 489–501 (2006)
13. Merz, C.J., Murphy, P.M., Aha, D.W.: UCI Repository of Machine Learning Databases. Dept. of Information and Computer Science, Univ. of California, Irvine (1997), http://www.ics.uci.edu/~mlearn/MLrepository.html
14. Paredes, R., Vidal, E.: Learning Weighted Metrics to Minimize Nearest-Neighbor Classification Error. IEEE Transactions on Pattern Analysis and Machine Intelligence 20(7), 1100–1110 (2006)
15. Zhang, M., Cieselski, V.: Using Bask propagation Algorithm and Genetic Algorithm to Train and refine neural networks for Object Detection. In: Bench-Capon, T.J.M., Soda, G., Tjoa, A.M. (eds.) DEXA 1999. LNCS, vol. 1677, pp. 626–635. Springer, Heidelberg (1999)
16. Lu, C., Shi, B.: Hybrid back-propagation/genetic algorithm for multilayer feed forward neural networks. In: 5th International Conference on Signal Processing Proceedings, 2000. WCCC-ICSP 2000, vol. 1, pp. 571–574. IEEE, Los Alamitos (2000)

17. Kalous, R.: Evolutionary operators on ICodes. In: Hakl, F. (ed.) Proceedings of the IX PhD. Conference. Institute of Computer Science, Academy of Sciences of the Czech Republic, Prague 2004, September 29–October 1, pp. 35–41. Matfyzpress, Paseky nad Jizerou (2004)
18. Vasechkina, E.F., Yarin, V.D.: Evolving polynomial neural network by means of genetic algorithm: some application examples. Complexity International 09, 1–13 (2001), http://www.complexity.org.au/vol09/vasech01/
19. Wikipedia - Gauss–Markov theorem (2007), http://en.wikipedia.org/wiki/Gauss-Markov_theorem

# Inducing Relational Fuzzy Classification Rules by Means of Cooperative Coevolution

Vahab Akbarzadeh, Alireza Sadeghian, and Marcus V. dos Santos

**Summary.** An evolutionary system for derivation of fuzzy classification rules is presented. This system uses two populations: one of fuzzy classification rules, and one of membership function definitions. A constrained-syntax genetic programming evolves the first population and a mutation-based evolutionary algorithm evolves the second population. These two populations co-evolve to better classify the underlying dataset. Unlike other approaches that use fuzzification of continuous attributes of the dataset for discovering fuzzy classification rules, the system presented here fuzzifies the relational operators "greater than" and "less than" using evolutionary methods. For testing our system, the system is applied to the Iris dataset. Our experimental results show that our system outperforms previous evolutionary and non-evolutionary systems on accuracy of classification and derivation of inter-relation between the attributes of the Iris dataset. The resulting fuzzy rules of the system can be directly used in knowledge-based systems.

## 1 Introduction

In the context of machine learning, classification is the task of predicting the value of one attribute (goal attribute) based on the value of other attributes (predicting attribute). Most common classification algorithms work

Vahab Akbarzadeh

Department of Computer Science, Ryerson University, Toronto, Ontario, Canada
e-mail: vahab.akbarzadeh@ryerson.ca

Alireza Sadeghian

Department of Computer Science, Ryerson University, Toronto, Ontario, Canada
e-mail: asadeghi@ryerson.ca

Marcus V. dos Santos

Department of Computer Science, Ryerson University, Toronto, Ontario, Canada
e-mail: m3santos@ryerson.ca

A. Abraham et al. (Eds.): Foundations of Comput. Intel. Vol. 4, SCI 204, pp. 127–147.
springerlink.com                                    © Springer-Verlag Berlin Heidelberg 2009

as follows; the dataset to be mined is divided into two mutually exclusive sets, namely the training and the test set. Then the classification algorithm is applied to the training set, which includes the value of the predicting and the goal attributes. When the classification process is finished, prediction accuracy of the classification algorithm is tested with the test set [1]. Different algorithms have been proposed for classification tasks, some of which includes: Multi Layer Perceptron Artificial Neural Network, Bayesian Networks, and Decision Trees.

IF-THEN rules are one of the most popular ways of expressing explicit classification knowledge in data mining applications. Rules can express implicit knowledge of datasets in a form that is comprehensible for humans. High level comprehensible rules can help an expert through the decision making process. This process can be disease diagnosis, credit evaluation or any other classification process. The expert can combine the system rules with his own knowledge to make well-informed decisions. Classification methods based on IF-THEN rules are more comprehensible for the end-user than the black box methods, *e.g.*, Artificial Neural Networks.

Real-world datasets are filled with inconsistent and imprecise information. Fuzzy logic, which is based on the concept of fuzzy sets, is a form of logic that uses approximate reasoning method. By integrating fuzzy logic with a rule derivation system, the system will manage the vagueness and uncertainty of the underlying information with more flexibility. The resulting fuzzy rule based system is an extension of classical rule based system that incorporates fuzzy logic statements in the antecedent and consequent of the IF-THEN rules. Fuzzy rule based systems include linguistic identifiers, *i.e.*, "low", "high", "very", that humans use for reasoning. Derivation of fuzzy rules with linguistic identifiers as terminals, improves the comprehensibility of rules for humans. Mamdani [2] and TSK [3] are two of the most well-known models of fuzzy rule based systems. In the Mamdani model, the consequent of each rule is a single linguistic identifier, while the consequent of TSK rules is a function of the input variables. Comparing these methods, Mamdani type system is more interpretable and provides a better mechanism for integration of the expert's knowledge into the system, while TSK system is more accurate and derives a set of more compact rules [4].

Determination of the system parameters is a crucial component of any fuzzy system design process. The basic structural parameters of the fuzzy system which deal with the inference mechanism are usually determined by an expert, *e.g.*, reasoning method and defuzzification mechanism. An automatic mechanism is more suitable for derivation of operational parameters, *e.g.*, Membership function values, for the following reasons: (a) System experts generally have less accurate assumption for the proper values of the operational parameters, and even if they have, the complexity of the problem increases exponentially with the number of attributes and fuzzy sets of each attribute. (b) Derivation of the operational parameters can easily be encoded as a search problem, where different operational parameters of the fuzzy system constitutes the search space.

Evolutionary computation (EC), [5, 6] inspired from the principles of natural evolution date backs to the principle of Darwin's theory of evolution "The survival of the fittest". This field of study has become an active research area in Computational Intelligence with enormous applications. Although EC is the simplified evolutionary model of its biological counterpart, it is a powerful search and optimization technique. There are different variations of evolutionary techniques namely genetic algorithm, evolutionary strategies, evolutionary programming, gene expression programming ,and genetic programming.

Derivation of operational parameters of a fuzzy rule based system can be considered as exploration of a high dimensional search space with lots of local optimas. As a search strategy, evolutionary computation is a good candidate for this problem for the following reasons: First, it has less chance of getting stuck in a local optima [7]. Second, evolutionary methods need less prior knowledge of the search problem which, as we mentioned before, is one of the characteristics of the fuzzy classification rule design problem. Finally, different evolutionary strategies can be combined with each other under the concept of coevolutionary strategies. This ability gives the designer the freedom to model each part of the system with the type of evolutionary method that best meets it needs,and then combine them all together.

Genetic programming (GP) [8] is a powerful search method, which is applicable when a direct search is impossible. Classification of a dataset based on its attribute values is an example of a large search space. Holland [9] and Smith [10] initiated the application of evolutionary algorithms in classification problem. Their rule representation methods are referred to as Michigan and Pittsburgh approaches, respectively. In the Michigan approach, each individual includes only one rule, while Pittsburgh approach encodes a set of rules in one individual. It is now understood that the Pittsburgh approach has higher accuracy for complicated problems, while the Michigan approach converges faster. The following hybrid approaches are also worth mentioning: COGIN [11], REGAL [12], and G-Net [13]. More information regarding application of evolutionary algorithms for classification task can be found in [14].

Some authors have integrated fuzzy logic with evolutionary based classifiers [15, 16, 17, 18]. Chien et al. [16] proposed a classification algorithm based on discrimination functions and fuzzification of numerical attributes. The main problem with their discrimination function approach is that the end result is less understandable for the end-user compared to methods based on classification rules. Dounias et al. [19] designed a system to derive fuzzy classification rules from medical data. Kishore et al. [20] proposed a binary classification algorithm for multicategory classification problem. In that approach, each run of the system is dedicated to classify one of the goal attribute classes, so the rest of the goal attributes are considered as one undesired class. At the end, all the classifiers compete with each other to classify the test set.

Coevolutionary Computation is a form of evolutionary computation in which the evaluation of fitness is based on interactions between multiple individuals. As such, an individuals ranking in a population can change depending on other individuals.

Coevolutionary Algorithms (CEAs) are similar to traditional evolutionary algorithm (EA) methods in the sense that individuals encode some aspect of potential solutions, individuals are altered during search by genetic operators, and the search is directed by selection based on fitness. CEAs differ from traditional EAs in the following ways: evaluation requires interaction between multiple individuals, and such interaction requires more complex notions of cooperation and competition. As well, CEAs can use a single population or multiple populations.

Single population CEAs (*e.g.*, [21, 22]) employ a single EA, often using traditional operators. Individuals represent candidate solutions to a problem, as well as "tests" for other individuals. An individual is evaluated by having it interact with individuals in the same population. In this form of interaction, individuals compete in a game-theoretic sense, but also compete as resources for evolution.

Multi-population CEAs (*e.g.*, [23]) employ multiple EAs, one for each population. Like in single population CEAs, individuals represent candidate solutions to a problem, as well as "tests" for other individuals. Evaluation of an individual is accomplished by having it interact with individuals from other populations. Typically individuals in different populations do not compete directly for EA resources.

Multi-population CEAs that use cooperative-species models are of particular relevance to the work presented here. In this type of model two populations (species) have a symbiotic relationship in that the second species uses the representations coevolved by the first species.The multi-population CEA for fuzzy classification presented in [24] is an example of coevolutionary system that employs this type of model. Their system fuzzified each continues attribute of the system using a membership function and specification of all the membership functions co-evolve with GP-based classification rule induction system. Delgado et al. [25] encoded the parameters of a TSK fuzzy model into a four level hierarchical cooperative coevolutionary system. These four levels include membership functions, individual rules, rule-bases and the fuzzy systems. Comparing single population and coevolutionary methods, Potter [23] showed that the coevolutionary methods need less computation power and converge faster while maintaining a comparable accuracy level.

In this paper, we propose a classification rule induction evolutionary system that derives relational fuzzy classification rules from the defined dataset. Our system is based on two populations. One population encodes the antecedent of inductive classification fuzzy rules, and the other one keeps a population of membership functions. The membership function population is used for the fuzzification of continuous attributes of the dataset and fuzzification of relational operators between the attributes of the dataset that has the same type. The first population is evolved using a GP algorithm whereas the second population is evolved based on a mutation-based evolutionary algorithm. These two populations co-evolve to better adapt to each other.

The main characteristic of our system that makes it different from previous works is that it derives relational fuzzy rules from dataset. We will fuzzify the relational operators of "greater than" and "less than" for two variable valued operands, *i.e.*, dataset attributes in the dataset. As mentioned in [4] one of drawbacks of the traditional Mamdani type systems arise when there is a dependency between the attributes of the dataset. Derivation of relational fuzzy rules, help the user to recognize and handle these dependencies. As will be shown, this improves the classification performance of our algorithm and helps us to derive more information about the interrelation between the attributes of the mined dataset.

## 2   Relational Fuzzy Classification Rule Derivation (RFCRD) System

Our system derives fuzzy classification rules from any defined dataset. This system is based on the evolutionary computation methods. A genetic programming algorithm evolves a set of fuzzy classification rules, and a simple mutation-based evolutionary algorithm calculates two sets of membership functions to fuzzify continuous attributes and relational operators between the attributes of the dataset, as will be explained below. These two sets of rules and membership function definitions will co-evolve to derive a set of rules and membership function definitions that fit together on predicting the class of each example.

### 2.1   Fuzzy Classification Rule Population

Classification is the task of assigning a class label $C_i$ from the set of class labels $\{C_1 \ldots C_n\}$ to each example of the dataset. Each example of the dataset can be represented as a point in an $N$ dimensional space, where $N$ represents the number of predicting attributes of each example. The main task of the classifier system is to find a mapping function $S$ such that:

$$S : R^N \to C_i \tag{1}$$

Classification rules are one of the common methods to represent the mapping function of a classifier system. The classification rules have the form IF (antecedent)-THEN (consequent). The antecedent of the classification rule includes a logical expression among the predicting attributes of the dataset, and the consequent has the form $Class = C_i$. If the predicting attribute values of a record in the dataset satisfy the antecedent of a rule, then the class in the consequent is assigned to that example.

There has always been a trade off between accuracy and interoperability of a classification system. Fuzzy logic is a good candidate to be integrated into a classification rule based system, as the resulting system's rules are more comprehensible for human. Moreover, the system can better manage the uncertainty of the underlying dataset.

To better understand the concepts discussed in the rest of this paper, we will give a brief introduction to fuzzy set theory. Consider a dataset $S$ (shown in Fig.1 ) with $M$ records, where each record $Record_i$ has $m$ predicting attributes and one goal attribute $G$.

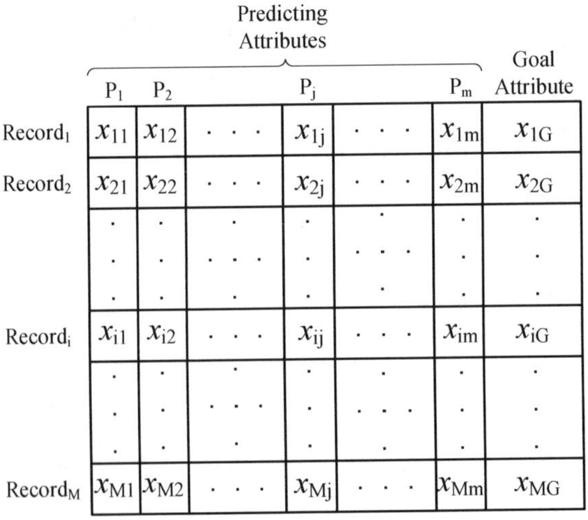

|  | Predicting Attributes | | | | | | Goal |
|  | $P_1$ | $P_2$ | | $P_j$ | | $P_m$ | Attribute |
|---|---|---|---|---|---|---|---|
| $Record_1$ | $x_{11}$ | $x_{12}$ | $\cdots$ | $x_{1j}$ | $\cdots$ | $x_{1m}$ | $x_{1G}$ |
| $Record_2$ | $x_{21}$ | $x_{22}$ | $\cdots$ | $x_{2j}$ | $\cdots$ | $x_{2m}$ | $x_{2G}$ |
| $\cdot$ | $\cdot$ | $\cdot$ | $\cdot$ | $\cdot$ | $\cdot$ | $\cdot$ | $\cdot$ |
| $\cdot$ | $\cdot$ | $\cdot$ | $\cdot$ | $\cdot$ | $\cdot$ | $\cdot$ | $\cdot$ |
| $\cdot$ | $\cdot$ | $\cdot$ | $\cdot$ | $\cdot$ | $\cdot$ | $\cdot$ | $\cdot$ |
| $Record_i$ | $x_{i1}$ | $x_{i2}$ | $\cdots$ | $x_{ij}$ | $\cdots$ | $x_{im}$ | $x_{iG}$ |
| $\cdot$ | $\cdot$ | $\cdot$ | $\cdot$ | $\cdot$ | $\cdot$ | $\cdot$ | $\cdot$ |
| $\cdot$ | $\cdot$ | $\cdot$ | $\cdot$ | $\cdot$ | $\cdot$ | $\cdot$ | $\cdot$ |
| $\cdot$ | $\cdot$ | $\cdot$ | $\cdot$ | $\cdot$ | $\cdot$ | $\cdot$ | $\cdot$ |
| $Record_M$ | $x_{M1}$ | $x_{M2}$ | $\cdots$ | $x_{Mj}$ | $\cdots$ | $x_{Mm}$ | $x_{MG}$ |

**Fig. 1** Dataset $S$

Let us assume that each predicting attribute $P_j$ has been divided into $K$ fuzzy sets $\{F_{j1}, F_{j2}, \ldots, F_{jK}\}$, and each fuzzy set is defined through the $\mu_{jk}$ membership function. More precisely, $\mu_{jk}(x_{ij})$ represents the degree which $Record_i$ belongs to fuzzy set $F_{jk}$. In this notation, a fuzzy classification rule has the form:

**IF**
> $((P_1$ is $F_{1p})$ and $(P_j$ is $F_{jp'})$
> and $\ldots$ and $(P_m$ is $F_{mp''}))$

**THEN**
> $(x_G$ belongs to $C_s)$
>
> $s = 1,2,\ldots,n$ ; $p = 1,2,\ldots,K$

The fuzzy classification rule based system is composed of a Knowledge Base (KB) and a fuzzy reasoning method [4]. The Rule Base (RB) and the

Data Base (DB) constitute the KB. As their name suggest, the RB contains all the classification rules and the DB keeps the definition of membership functions of the fuzzy system.

The fuzzy reasoning method is the mechanism which uses the information presented in the KB to assign the class label to an unclassified input record. In this paper, we used the fuzzy reasoning method proposed in [26] to calculate the output of the fuzzy rule based system. Assume that the input record $x_i = \{x_{i1}, x_{i2}, \ldots, x_{im}\}$ and a Rule Base $R = \{R_1, R_2, \ldots, R_o\}$ are given. The input vector's goal attribute $x_{iG}$ is classified as the output of the single winning fuzzy rule from the rule base which has the highest compatibility degree with the input record. This process is shown in Fig.2 .In this figure the recored belongs to the class which is indicated in the consequent of the winning rule.

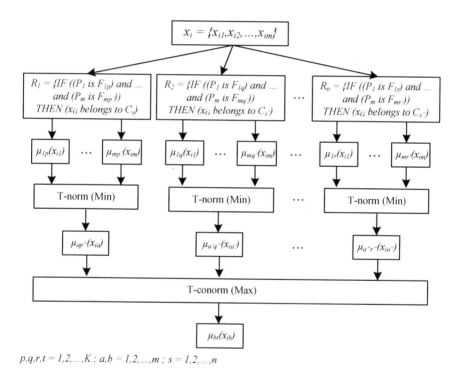

$p,q,r,t = 1,2,\ldots,K \; ; \; a,b = 1,2,\ldots,m \; ; \; s = 1,2,\ldots,n$

**Fig. 2** The fuzzy reasoning method

Implementation of a fuzzy classification rule based system can be broken down into two phases. First, the basic fuzzy system characteristics (such as reasoning method, fuzzy operators, and defuzzification method) needs to be selected. This phase is usually done by the system designer based on the experience on problem characteristics [7]. Secondly, the KB of the system should be generated. This problem consists of two distinct problems of deriving the DB and the RB. In DB, the membership function parameters associated to each linguistic variable need to be specified, while in RB the specific composition of the antecedent and the consequent of each classification rule should be defined. Automatic learning methods are a good candidate for this problem as the search space of the problem is huge and there exists little or no a priori knowledge about the problem. Among different learning methods that have been proposed for KB derivation, the simplest ones deal only with one of the KB components at a time. In other words, these methods try to generate the RB from a predefined DB or vice versa. Although these methods are computationally efficient, they ignore the interdependence that exists between the DB and RB of the KB [27]. On the other hand, methods which try to derive the RB and the DB at the same time are computationally expensive, as these methods need to deal with a much bigger search space.

In the fuzzy classification rule based system the antecedent of each rule is a fuzzy logic expression that contains a set of fuzzy expressions connected by fuzzy operators. So far, only logical operators of $\{AND, OR, NOT\}$ have been used as fuzzy operators. We fuzzified the relational operators $\{>, <\}$ for two variable valued operands. More precisely, the fuzzy expression $Att_i > Att_j$ return a value between $[0,1]$. This value shows the amount that $Att_i$ is bigger than $Att_j$.

## Rule Representation

We followed the Pittsburgh approach for rule representation [10]. In this population the function set contains the fuzzy logic operators $\{AND, OR, NOT, >, <, =\}$. We used the standard fuzzy operators for $\{AND, OR, NOT\}$ [28]. More precisely, if $\mu_A(x)$ represents the membership degree of element $x$ in fuzzy set $A$, then the fuzzy operators are defined as follows:

$$\mu_A(x) \ AND \ \mu_B(x) = \min(\mu_A(x), \mu_B(x)) \tag{2}$$

$$\mu_A(x) \ OR \ \mu_B(x) = \max(\mu_A(x), \mu_B(x)) \tag{3}$$

$$NOT \ \mu_A(x) = 1 - \mu_A(x) \tag{4}$$

Disjunctive Normal Form (DNF) is a logical form which is the disjunction of conjunction clauses. In our system the antecedent of the classification rules are represented in DNF. In order to show the antecedent of the formulas in DNF, we propose some restrictions on the structure of the individuals. These restrictions are: (a) The root of each tree is an $OR$ node. Each $OR$ node can have one or more than one $AND$ node (rule) as its child(ren). Although the generality of the rules could be increased by integration of other connectives in the rule antecedent, it was shown in [29] that the rules with only conjunctive connectives are generic enough to cover other connectives.(b) The second level of the tree includes $AND$ nodes (rules). Each $AND$ node can have one or more than one terminal node as its child(ren). (c) The third level is the *Terminal* level. Terminals of the system are defined using the following rules:

$< terminal >::= NOT < literal >$
$< terminal >::=< literal >$
$< literal >::=< op >< relational\ operator >< op >$
$< literal >::=< op >=< linguistic\ value >$
$< relational\ operator >::= \{>, <\}$
$< linguistic\ value >::= \{Low, Med, High\}$
$< op >::= \{Att_1 \ldots Att_N\}$

Note that in the case of relational operator, two operands of the operator should have the same type, so the values of them can be compared with each other. (d) No more than one child of an *AND* node can refer to a terminal with the same attribute as the first operand and the $=$ as the operator. For example, two terminals of $Temp = Low$ and $Temp = High$ cannot be the children of the same *AND* node. A sample rule set of the system is shown in Fig. 3. This rule set is equal to: *IF((Att$_1$=Low) AND (Att$_2$=High)) OR ((Att$_3$ < Att$_1$) AND (NOT Att$_1$ > Att$_2$) AND (Att$_3$=Med))*

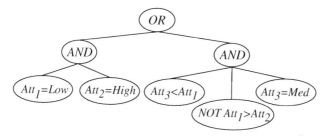

**Fig. 3** Sample rule set of the system

## Fitness Function

We used binary classification in our system, so at each run of the system, all the rules of the system are trying to predict only one class of the goal attribute. Fitness of each individual is calculated using the following formula [30]:

$$fitness = (\frac{TP}{TP + FP}) \times (\frac{TN}{TN + FN}) \qquad (5)$$

where,

- TP (true positive) is the number of examples that satisfied the antecedent of one of the rules of the individual and the examples have the same class as the consequent of the rule.
- FP (false positive) is the number of examples that satisfied the antecedent of one of the rules of the individual but the examples do not have the same class as the consequent of the rule.

- TN (true negative) is the number of examples that did not satisfy the antecedent of any of the rules of the individual and the examples do not have the same class as the consequent of the rule.
- FN (false negative) is the number of examples that did not satisfy the antecedent of any of the rules of the individual but the examples have the same class as the consequent of the rule.

It is generally assumed that the above values are crisp numbers, but since we used fuzzy classification rules in our system, antecedents of each classifying rule return a fuzzy value when trying to classify each example of the dataset. In other words, each example satisfies the antecedent of each rule to a certain degree between [0, 1]. We could either defuzzify the returning values of the rules or fuzzify value of TP, TN, FP and FN. We fuzzified value of TP, TN, FP and FN with the same approach proposed in [24].

**Selection**

We used 2-player tournament selection [31] in which the higher objective function value wins and is chosen for reproduction. The tournaments are played as follows: For each member of the population, one tournament is held. The second player for each tournament is randomly chosen from the population (sampled independently with replacement, *i.e.* being chosen as a second player in one tournament does not alter the probability of being chosen as a second player in another tournament. And an individual may be chosen to play itself). The winners of the tournaments are collected into a "parent pool" to which the genetic operators are applied. The major steps of the system is shown in Fig.4

**Crossover**

We used a modified version of cut and splice method [32] for crossover. The method works as follows:
A node is randomly selected in parent one.

- If the selected node is an *AND* node, one other *AND* node is randomly selected in parent two, and all the *AND* nodes beyond the two selected *AND* nodes are swapped between the two parents.
- If the selected node is a *Terminal* node, one other *Terminal* node is randomly selected in parent two, and all the *Terminal* nodes beyond the two selected *Terminal* nodes are swapped between the rules in two parents.

**Mutation**

Regarding mutation, a *Terminal* node is randomly selected in the individual and its value is swapped with a randomly selected terminal value.

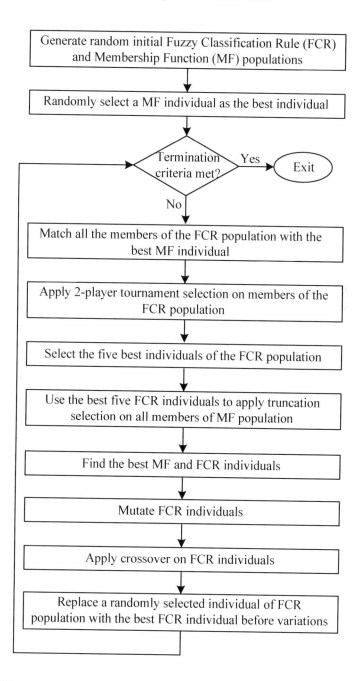

**Fig. 4** Major steps of the system

## 2.2   Membership Function Population

The second population keeps the encoding of the membership functions that
the first population uses for its rule interpretation. Each individual has a
predefined static structure that encodes the membership function for all continuous variables and relational operators between each two attributes of the
dataset that has the same type. Each continuous variable is divided into
three linguistic variables namely, $\{Low, Med, High\}$. This process is shown
in Fig.5. Each individual of the population keeps the values for the parameters of $mc1, mc2, mc3$ and $mc4$. These values evolve to better adapt to
their corresponding attributes. The system ensures that after each evolution
$mc1 < mc2 < mc3 < mc4$.

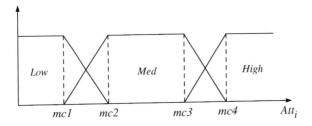

**Fig. 5** Membership function for continuous attributes

As mentioned before, the main characteristic of our system is fuzzification of relational operators. For this purpose, the relation between each two
variables of the system that have the same type is fuzzified through the
definition of membership function for "greater than" operator. This process
is shown in Fig.6. In this figure, the difference between the values of two
attributes ( *i.e.* $Att_i - Att_j$) is the main parameter that determines the degree that $Att_i$ is bigger than $Att_j$. Note that in our system the expression
$Att_i > Att_j$ represents the degree that $Att_i$ is bigger than $Att_j$, so the value
of $NOT\, Att_i > Att_j$ shows how much $Att_i$ is "not much bigger" or "less bigger" than $Att_j$. Also note that the "less than" operator is only a different
representation of "greater than" operator. More precisely, the value that $Att_i$

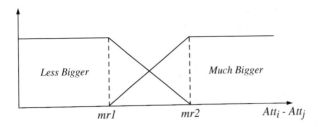

**Fig. 6** Membership function for "greater than" operator

is bigger than $Att_j$ is equal to the amount that $Att_j$ is less than $Att_i$, so there is no need to explicitly fuzzify the less than operator in the system. Like the membership function of continuous variables, the system evolves the value of $mr1$ and $mr2$ and ensures that $mr1 < mr2$.

Our approach for modeling the "greater than" operator is a simplified version of a more general binary fuzzy relation concept [33]. More precisely, to model a relationship between two attributes, one needs to define a two dimensional membership function over their universe of discourse. Let us assume that $Att_i$ and $Att_j$ are two attributes of the system, where $I$ and $J$ be their universes of discourse, respectively. A binary fuzzy relation $R$ on $I \times J$ can be defined as:

$$R = \{((Att_i, Att_j), \mu_R(Att_i, Att_j)) | (Att_i, Att_j) \in I \times J\} \tag{6}$$

Where $\mu_R(Att_i, Att_j)$ represents the two dimensional membership function. For example, the membership function of the "greater than" operator between $Att_i$ and $Att_j$ $(Att_i > Att_j)$ can be specified as:

$$\mu_R(Att_i, Att_j) = \begin{cases} \frac{Att_i - Att_j}{Att_i + Att_j} & Att_i \geq Att_j \\ 0 & Att_i < Att_j \end{cases} \tag{7}$$

We used a mutation based evolutionary algorithm for the evolution of the membership functions [24]. More specifically, we used the $(\mu + \lambda)$-evolutionary algorithm [34] where $\mu = 1$ and $\lambda = 5$. In this algorithm, each individual of the second population is cloned 5 times, and then a relatively high degree of mutation is applied to each clone. The mutation function increases or decreases the value of each parameter of the membership functions with a small value. Once all 5 clones have undergone mutation process, the fitness function of each clone is calculated in the process discussed below.

As we mentioned before the fitness value of each membership function individual depends on the predictive accuracy of the fuzzy rule individuals, so, in order to calculate the fitness value of each membership function individual, the individual is considered as the membership function definition for a group of the best fuzzy rule individuals (best 5 in this project). At the end, the fitness value of the membership individual is equal to the sum of the fitness values of the fuzzy rule individuals. This process is shown in Fig. 7.

## 3   Results and Discussion

Recall that our system should be run $c$ times for a dataset that has $c$ classes. As a result, we would have $c$ fuzzy classification rule sets and each rule set would have its own membership function definition. Each example of the test set will be classified with all resulting classification rule sets and the example will be assigned to the rule set that has the highest degree of membership value. The classification accuracy of the system is calculated as the

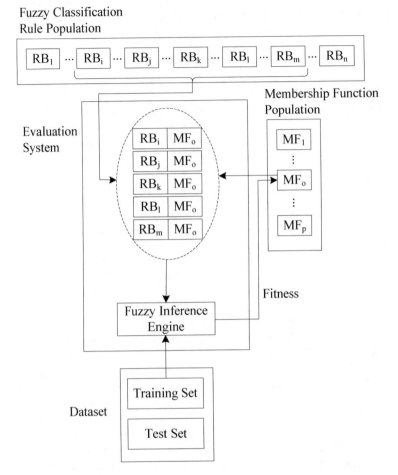

**Fig. 7** Fitness evaluation of membership functions

number of examples that was correctly classified divided by the total number of examples in the test set. We compared our Relational Fuzzy Classification Rule Derivation (RFCRD) system with other evolutionary and machine learning methods over the Iris dataset from UCI machine learning data set repository [35]. The Iris dataset is one of the most well-known datasets in machine learning literature. The dataset contains three classes of Iris plant namely: Setosa, Versicolour and Virginica. There are 50 instances of each plant in the dataset. The algorithm should classify these examples based on four attributes of sepal length, sepal width, petal length and petal width. The Setosa class is linearly separable from the two others, while the remaining two classes are not linearly separable.

**Table 1** Iris Dataset

| Number | Sepal Length | Sepal Width | Petal Length | Petal Width | Species |
|--------|--------------|-------------|--------------|-------------|---------|
| 1 | 5.1 | 3.5 | 1.4 | 0.2 | Iris-Setosa |
| 2 | 7.0 | 3.2 | 4.7 | 1.4 | Iris-Versicolor |
| 3 | 6.3 | 3.3 | 6.0 | 2.5 | Iris-Virginica |
| : | : | : | : | : | : |
| 150 | 4.9 | 3.0 | 1.4 | 0.2 | Iris-setosa |

We evaluated our system using 10-fold cross-validation method. In this method, the original dataset is divided into 10 parts with equal size, and each time one of the 10 parts is selected as the test set and the rest 9 sets form the training set. The overall accuracy of the algorithm is calculated as the average classification accuracy of the algorithm over 10 different training and test sets. In order to compare the classification accuracy of our system, we will also mention two other evolutionary methods and one non-evolutionary method found in the literature. The evolutionary methods are: CEFR-MINER [24], and ESIA [36].CEFR-MINER finds propositional fuzzy classification rules while ESIA builds crisp decision trees. The non-evolutionary method is C4.5 [37] which is a well-known decision tree builder method. We used the results reported in [38] for C4.5 algorithm. All of these systems used 10-fold cross-validation for system evaluation. The parameters that we used in our system are summarized in Table 1 and Table 2. Table 1 shows the parameters of fuzzy classification rule population, and Table 2 shows the parameters of membership function evolution population.

As it can be seen in Table 3, our system got better results compared to other three methods. Recall that our system derives fuzzy rules, so the end rules are more comprehensible for the end user than other crisp rule derivation systems. To show the effectiveness of relational operators we excluded the relational operators from the function set and as a result the final

**Table 2** Fuzzy classification rules population parameters

| Parameter | Value |
|-----------|-------|
| Population Size | 250 |
| Maximum Generations | 100 |
| Initial Number of rules per individual | 2 |
| Initial Number of terminals per rule | 2 |
| Maximum Number of rules per individual | 2 |
| Maximum Number of terminals per rule | 4 |
| Crossover Rate | 0.9 |
| Mutation Rate | 0.05 |
| Reproduction Rate | 0.1 |

**Table 3** Membership function population parameters

| Parameter | Value |
|-----------|-------|
| Population Size | 50 |
| Maximum Generations | 100 |
| Mutation Rate | 0.33 |
| Parent Selection | Truncation Selection |
| Selection Parameter | $\mu = 1, \lambda = 5$ |

**IF**
  *((NOT Sepal-Width = Med) AND (Petal-Length=Low))*
**THEN**
  *Class = Iris-Setosa*

**IF**
  *((NOT Sepal-Width>Petal-Width) AND (NOT Petal-Length>Sepal-Width))*
  *OR*
  *((NOT Petal-Width = Low) AND (NOT Petal-Length>Sepal-Width) AND (NOT Petal-Length>Petal-Width))*
**THEN**
  *Class = Iris-Versicolour*

**IF**
  *((NOT Petal-Length = Low) AND (Petal-Length>Petal-Width))*
  *OR*
  *((Sepal-Width = Low) AND (Petal-Width = Med))*
**THEN**
  *Class = Iris-Virginica*

**Fig. 8** Final fuzzy classification rule set of the system derived from Iris dataset

accuracy of the system decreased. Also, as we derive relational rules we got more understanding of the underlying relation between the attributes of the dataset. A final rule set of the system obtained from the Iris dataset, and the corresponding membership function definitions evolved for $Iris-Versicolour$ flower are shown in Fig. 8, and Fig.9, respectively. Note that the attribute $Petal - Length > Petal - Width$ has appeared in the predicting rule sets of the $Iris - Versicolour$ and $Iris - Virginica$ flowers. So the amount that the $Petal - Length$ of a flower is bigger than its $Petal - Width$ can be used to discriminate its species. The fuzzy membership function definitions

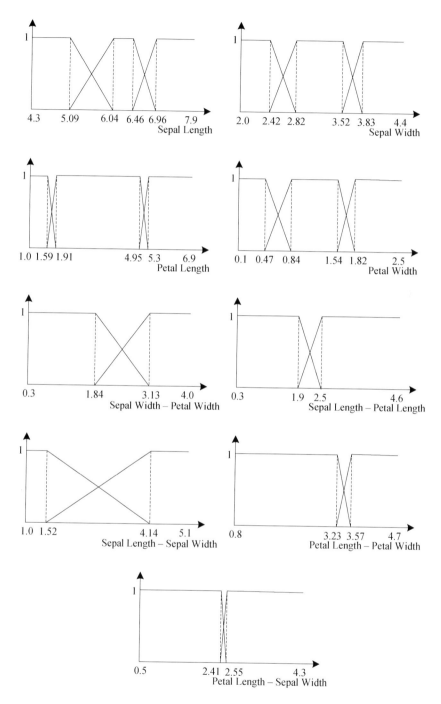

**Fig. 9** The membership functions for $Iris - Versicolour$ flower

support the same idea as the membership function definition for $Petal - Length > Petal - Width$ attribute has been fine tuned to the values that distinguishes $Iris - Versicolour$ and $Iris - Virginica$ flowers.

**Table 4** Comparison of results

| Algorithm | Classification Accuracy | S.D. |
|---|---|---|
| Our System | 98.0% | 4.7 |
| Our System (Excluding Relational Operators) | 96.0% | 9.4 |
| CEFR-MINER | 95.3% | 7.1 |
| ESIA | 95.3% | 0.0 |
| C4.5 | 95.3% | 5.0 |

Compared to our previous work [39], we decreased the maximum number of rules and the maximum number of terminals per rule. This modification did not decrease the final accuracy of our system but dramatically decreased the running time of the system. We ran our system on a dual-processor 2.13 GHz machine with 2.0 GB of memory. Each run of the system for classification of one of the goal attributes took a processing time of 15 seconds.

## 4 Conclusion and Future Research

We proposed a Relational Fuzzy Classification Rule Derivation (RFCRD) system in this paper. Our system uses two populations for discovering classification rules. The first population encodes the fuzzy classification rules, and the second one keeps the membership functions definition for fuzzification of continuous attributes and the relational operators between the attributes that have the same type. The first population uses the second population for defining the rule antecedents evolved by the first population. The GP algorithm evolves the population of fuzzy rules, and a simple mutation based evolutionary algorithm evolves the second population. These two populations co-evolve with each other to better adapt to the underlying dataset. The main advantage of our system that makes it different from previous work is using fuzzy relational operators. We fuzzified the relational operator of "greater than" and "less than" between to attributes that have the same type. More precisely, we fuzzified the degree that one attribute is bigger or less than the other attribute, and used that in the antecedent of our rules. We have evaluated our system over the well-known Iris dataset and compared the end result with three other evolutionary and non-evolutionary classification methods.

- Our system outperformed the other systems with respect to the classification accuracy.
- Our system produces relational fuzzy rules which give the user more understanding of the interaction between the attributes of the dataset.

- Our system produces fuzzy classification rules which are more comprehensible for the user.

There are several directions for the future search. For example, in this paper we only fuzzified the "greater than" and "less than" operator. Fuzzification of other relational operators can be an interesting approach. For example, one can fuzzify how much two attributes are equal to each other. All of these approaches will give the user more understanding of the underlying dataset, and also produces the rules that are comprehensible for the human and can be directly used in knowledge-based systems.

Another important extension of the research presented here is the replacement of the GP algorithm with the Gene Expression Programming [40] (GEP) algorithm. The latter allows for the evolution of multi-genic chromosomes, each gene coding for a different sub-expression. Recall that, in the GP system proposed here, if the problem includes $n$ classes, then we have to run the system $n$ times to obtain $n$ classification rule sets. On the other hand, by using a multi-genic GEP system ( *i.e.*, a GEP system that evolves chromosomes with $n$ genes, each gene coding for a specific classification rule set), any given run would simultaneously evolve $n$ classification rule sets.

# References

1. Freitas, A.A.: A survey of evolutionary algorithms for data mining and knowledge discovery. In: Advances in Evolutionary Computation, pp. 819–845 (2002)
2. Mamdani, E., et al.: Application of fuzzy algorithms for control of simple dynamic plant. Proc. IEE 121(12), 1585–1588 (1974)
3. Takagi, T., Sugeno, M.: Fuzzy identification of systems and its applications to modeling and control. IEEE transactions on systems, man, and cybernetics 15(1), 116–132 (1985)
4. Cordón, O.: Genetic Fuzzy Systems: Evolutionary Tuning and Learning of Fuzzy Knowledge Bases. World Scientific Pub. Co. Inc., Singapore (2001)
5. Goldberg, D.: Genetic Algorithms in Search, Optimization and Machine Learning. Addison-Wesley Longman Publishing Co., Inc., Boston (1989)
6. Holland, J.: Adaption in Natural and Artificial Systems (1975)
7. Pena-Reyes, C., Sipper, M.: Fuzzy CoCo: a cooperative-coevolutionary approach to fuzzy modeling. IEEE Transactions on Fuzzy Systems 9(5), 727–737 (2001)
8. Koza, J.: Genetic Programming IV: Routine Human-Competitive Machine Intelligence. Kluwer Academic Publishers, Dordrecht (2003)
9. Holland, J.: Escaping brittleness: the possibilities of general-purpose learning algorithms applied to parallel rule-based systems. Computation & intelligence: collected readings table of contents, 275–304 (1995)
10. Smith, S.: Flexible learning of problem solving heuristics through adaptive search. In: Proceedings of the Eighth International Joint Conference on Artificial Intelligence, pp. 422–425 (1983)
11. Greene, D., Smith, S.: Competition-based induction of decision models from examples. Machine Learning 13(2), 229–257 (1993)

12. Giordana, A., Neri, F.: Search-Intensive Concept Induction. Evolutionary Computation 3(4), 375–419 (1995)
13. Anglano, C., Botta, M.: NOW G-Net: learning classification programs on networks ofworkstations. IEEE Transactions on Evolutionary Computation 6(5), 463–480 (2002)
14. Freitas, A.: Data Mining and Knowledge Discovery with Evolutionary Algorithms. Springer, Heidelberg (2002)
15. Cordón, O., Herrera, F., del Jesus, M.: Evolutionary approaches to the learning of fuzzy rule-based classification systems. In: The Crc Press International Series On Computational Intelligence, pp. 107–160 (1999)
16. Chien, B., Lin, J., Hong, T.: Learning discriminant functions with fuzzy attributes for classification using genetic programming. Expert Systems With Applications 23(1), 31–37 (2002)
17. Ishibuchi, H., Nakashima, T., Kuroda, T.: A hybrid fuzzy GBML algorithm for designing compact fuzzyrule-based classification systems. In: The Ninth IEEE International Conference on Fuzzy Systems, 2000. FUZZ IEEE 2000, vol. 2 (2000)
18. Walter, D., Mohan, C.: ClaDia: a fuzzy classifier system for disease diagnosis. In: Proceedings of the 2000 Congress on Evolutionary Computation, vol. 2 (2000)
19. Dounias, G., Tsakonas, A., Jantzen, J., Axer, H., Bjerregaard, B., Keyserlingk, D.: Genetic Programming for the Generation of Crisp and Fuzzy Rule Bases in Classification and Diagnosis of Medical Data. In: Proc. 1st Int. NAISO Congr. Neuro Fuzzy Technologies (2002)
20. Kishore, J., Patnaik, L., Mani, V., Agrawal, V.: Application of genetic programming for multicategory patternclassification. IEEE Transactions on Evolutionary Computation 4(3), 242–258 (2000)
21. Fogel, D.B.: Evolving a checkers player without relying on human experience. Intelligence 11(2), 20–27 (2000),
    http://doi.acm.org/10.1145/337897.337996
22. Hillis, W.D.: Co-evolving parasites improve simulated evolution as an optimization procedure. Phys. D 42(1-3), 228–234 (1990), http://dx.doi.org/10.1016/0167-2789(90)90076-2
23. Potter, M.A., Jong, K.A.D.: Cooperative coevolution: An architecture for evolving coadapted subcomponents. Evol. Comput. 8(1), 1–29 (2000), http://dx.doi.org/10.1162/106365600568086
24. Mendes, R., Voznika, F., Freitas, A., Nievola, J.: Discovering Fuzzy Classification Rules with Genetic Programming and Co-evolution. In: Siebes, A., De Raedt, L. (eds.) PKDD 2001. LNCS, vol. 2168, p. 314. Springer, Heidelberg (2001)
25. Regattieri Delgado, M., Von Zuben, F., Gomide, F.: Coevolutionary genetic fuzzy systems: a hierarchical collaborative approach. Fuzzy sets and systems 141(1), 89–106 (2004)
26. Ishibuchi, H., Morisawa, T., Nakashima, T.: Voting schemes for fuzzy-rule-based classification systems. In: Proceedings of the Fifth IEEE International Conference on Fuzzy Systems, vol. 1 (1996)
27. Cordon, O., Herrera, F., Villar, P.: Generating the knowledge base of a fuzzy rule-based system by thegenetic learning of the data base. IEEE Transactions on Fuzzy Systems 9(4), 667–674 (2001)

28. Zadeh, L.: Knowledge representation in fuzzy logic. IEEE Transactions on Knowledge and Data Engineering 1(1), 89–100 (1989)
29. Wang, L.: Adaptive fuzzy systems and control: design and stability analysis. Prentice-Hall, Inc., Upper Saddle River (1994)
30. Hand, D.: Construction and assessment of classification rules. Wiley, New York (1997)
31. Banzhaf, W., Nordin, P., Keller, R., Francone, F.: Genetic Programming: An Introduction. Morgan Kaufmann Publishers, San Francisco (1998)
32. Hoffmann, F., Pfister, G.: Evolutionary design of a fuzzy knowledge base for a mobile robot. International Journal of Approximate Reasoning 17(4), 447–469 (1997)
33. Jang, J., Sun, C.: Neuro-fuzzy and soft computing: a computational approach to learning and machine intelligence. Prentice-Hall, Inc., Upper Saddle River (1996)
34. Schwefel, H.P.: Evolution and Optimum Seeking. John Wiley & Sons, Chichester (1995)
35. Asuncion, A., Newman, D.: UCI machine learning repository (2007), http://www.ics.uci.edu/~mlearn/MLRepository.html
36. Juan Liu, J., Tin-Yau Kwok, J.: An extended genetic rule induction algorithm. In: Proceedings of the 2000 Congress on Evolutionary Computation, vol. 1, pp. 458–463 (2000)
37. Quinlan, J.: C4. 5: Programs for Machine Learning. Morgan Kaufmann, San Francisco (1993)
38. Gama, J.: Oblique Linear Tree. In: Advances in Intelligent Data Analysis: Reasoning About Data: Second International Symposium, Ida 1997, London, Uk, August 4-6, 1997. Proceedings (1997)
39. Akbarzadeh, V., Sadeghian, A., dos Santos, M.: Derivation of relational fuzzy classification rules using evolutionary computation. In: IEEE International Conference on Fuzzy Systems, 2008. FUZZ-IEEE 2008. IEEE World Congress on Computational Intelligence, pp. 1689–1693 (2008)
40. Ferreira, C.: Gene Expression Programming: Mathematical Modeling by an Artificial Intelligence. Studies in Computational Intelligence. Springer, New York (2006)

# Post-processing Evolved Decision Trees

Ulf Johansson, Rikard König, Tuve Löfström, Cecilia Sönströd,
and Lars Niklasson

**Abstract.** Although Genetic Programming (GP) is a very general technique, it is also quite powerful. As a matter of fact, GP has often been shown to outperform more specialized techniques on a variety of tasks. In data mining, GP has successfully been applied to most major tasks; e.g. classification, regression and clustering. In this chapter, we introduce, describe and evaluate a straightforward novel algorithm for post-processing genetically evolved decision trees. The algorithm works by iteratively, one node at a time, search for possible modifications that will result in higher accuracy. More specifically, the algorithm, for each interior test, evaluates every possible split for the current attribute and chooses the best. With this design, the post-processing algorithm can only increase training accuracy, never decrease it. In the experiments, the suggested algorithm is applied to GP decision trees, either induced directly from datasets, or extracted from neural network ensembles. The experimentation, using 22 UCI datasets, shows that the suggested post-processing technique results in higher test set accuracies on a large majority of the datasets. As a matter of fact, the increase in test accuracy is statistically significant for one of the four evaluated setups, and substantial on two out of the other three.

## 1 Introduction

A majority of all data mining projects include *predictive modeling*. The overall purpose of a predictive model is to estimate (or guess) an unknown (often future) value of a specific variable, the *target variable*. If the target variable is restricted to a predefined set of discrete labels (classes), the data mining task is called *classification*. If the target variable is continuous, the task is referred to as *regression*.

Ulf Johansson, Rikard König, Tuve Löfström, and Cecilia Sönströd
School of Business and Informatics, University of Borås, Sweden
e-mail: {ulf.johansson, rikard.konig, tuve.lofstrom,
cecilia.sonstrod}@hb.se

Lars Niklasson
Informatics Research Centre, University of Skövde, Sweden
e-mail: lars.niklasson@his.se

A. Abraham et al. (Eds.): Foundations of Comput. Intel. Vol. 4, SCI 204, pp. 149–164.
springerlink.com                    © Springer-Verlag Berlin Heidelberg 2009

Most often, the predictive model is obtained by using some data mining technique to learn a mapping from a vector input to a scalar output (the target variable) from historical data. More technically, the algorithm uses a set of *training instances*, each consisting of an input vector $x(i)$ and a corresponding target value $y(i)$ to learn the function $y=f(x;\theta)$. When sufficiently trained, the predictive model is able to predict a value $y$, when presented with a novel (test) instance $x$. The process of finding the best estimated parameter values $\theta$, is the core of the data mining technique. Figure 1 below shows a schematic picture of predictive modeling.

**Fig. 1** Predictive modeling

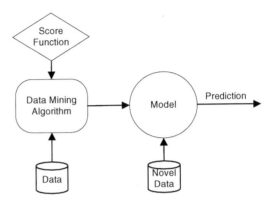

Here, data is fed to the data mining algorithm, which uses a score function to produce an optimized model, which is applied to novel data (a *test set*) to produce the actual predictions.

Most high-accuracy techniques for predictive classification produce opaque models like artificial neural networks (ANNs), ensembles or support vector machines. Opaque predictive models make it impossible for decision-makers to follow and understand the logic behind a prediction, which, in some domains, must be deemed unacceptable. When models need to be interpretable (or even comprehensible) accuracy is often sacrificed by using simpler but transparent models; most typically decision trees. This tradeoff between predictive performance and interpretability is normally called the accuracy vs. comprehensibility tradeoff. With this tradeoff in mind, several researchers have suggested rule extraction algorithms, where opaque models are transformed into comprehensible models, keeping an acceptable accuracy. Most significant, are the many rule extraction algorithms used to extract symbolic rules from trained neural networks; e.g. RX [1] and TREPAN [2]. Several papers have discussed key demands on reliable rule extraction methods; see e.g. [3] or [4]. The most common criteria are: *accuracy* (the ability of extracted representations to make accurate predictions on previously unseen data), *comprehensibility* (the extent to which extracted representations are humanly comprehensible) and *fidelity* (the extent to which extracted representations accurately model the opaque model from which they were extracted).

We have previously suggested a rule extraction algorithm called G-REX (Genetic Rule EXtraction) [5]. G-REX is a *black-box* rule extraction algorithm; i.e. the overall idea is to treat the opaque model as an oracle and view rule extraction as a learning task, where the target concept is the function learnt by the opaque model. Hence, rule sets extracted directly map inputs to outputs. Black-box techniques typically use some symbolic learning algorithm, where the opaque model is used to generate target values for the training examples. The easiest way to understand the process is to regard black-box rule extraction as an instance of predictive modeling, where each input-output pattern consists of the original input vector and the corresponding prediction from the opaque model. From this perspective, black-box rule extraction becomes the task of modeling the function from (original) input attributes to the opaque model predictions; see Figure 2 below.

**Fig. 2** Black-box rule extraction

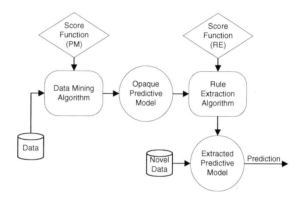

One inherent advantage of black-box approaches, is the ability to extract rules from arbitrary opaque models, including ensembles.

The extraction strategy used by G-REX is based on GP. More specifically, a population of candidate rules is continuously evaluated according to how well the rules mimic the opaque model. The best rules are kept and combined using genetic operators to raise the fitness (performance) over time. After many generations (iterations) the most fit program (rule) is chosen as the extracted rule.

One key property of G-REX is the ability to use a variety of different representation languages, just by choosing suitable function and terminal sets. G-REX has previously been used to extract, for instance, decision trees, regression trees, Boolean rules and fuzzy rules. Another, equally important, feature is the possibility to directly balance accuracy against comprehensibility, by using an appropriate fitness function. Although comprehensibility is a rather complex criterion, the simple choice to evaluate comprehensibility using the size of the model is the most accepted. Consequently, a typical G-REX fitness function includes a positively weighted fidelity term and a penalty for longer rules.

Although initially intended for rule extraction only, G-REX has lately been substantially modified, with the aim of becoming a general data mining framework based on GP; see [6]. For a summary of the G-REX technique and previous studies, see [7].

GP has in many studies proved to be a very efficient search strategy. Often, GP results are comparable to, or even better than, results obtained by more specialized techniques. One particular example is when GP is used for classification, and the performance is compared to, for instance, decision tree algorithms or rule inducers. Specifically, several studies show that decision trees evolved using GP often are more accurate than trees induced by standard techniques like C4.5/C5.0 [8] and CART [9]; see e.g. [10] and [11]. The main reason for this is that GP is a global optimization technique, while decision tree algorithms typically choose splits greedily, working from the root node down. Informally, this means that GP often will make some locally sub-optimal splits, but the overall model will still be more accurate and more general.

On the other hand, GP search is locally much less informed; i.e. each split is only optimized as a part of the entire decision tree. In addition, GP is not able to find all possible splits, simply because the number of constants available to the GP is limited. This problem is actually accentuated in the later stages of evolution, when populations tend to become more homogenous. With this in mind, the main purpose of this study is to investigate whether a straightforward post-processing techniques, where explicit searching is used to find optimal splits, can improve the accuracy of decision trees evolved using GP.

## 2  Background and Related Work

Eggermont, Kok and Kosters in two papers evaluate a refined GP-representation for classification, where the search space is reduced by letting the GP only consider a fixed number of possible splits for each attribute; see [12] and [13]. More specifically, a global set of threshold values for each numerical attribute is determined, and then only these threshold values are used in the evolution. It should be noted that the threshold values are chosen globally instead of at each specific node. In the two papers, a maximum of five thresholds are selected for every attribute, and both *information gain* and *information gain ration* are evaluated as criteria for the selection. Although the results are somewhat inconclusive, the proposed technique is generally more accurate than both C4.5 and standard GP.

As mentioned above, the normal result of rule extraction is another predictive model (the extracted model) which in turn, is used for actual prediction. At the same time, it is important to realize that the opaque model normally is a very accurate model of the relationship between input and target variables. Furthermore; the opaque model could be used to generate predictions for novel instances with unknown target values, as they become available. Naturally, these instances could also be used by the rule extraction algorithm, which is a major difference compared to techniques directly building transparent models from the dataset, where each training instance must have a known target value. Despite this, all rule extraction algorithms that the authors are aware of, use only training data (possibly with the addition of artificially generated instances) when extracting the transparent model. We have previously argued that a data miner often might benefit from also using test data together with predictions from the opaque model when performing rule extraction. Below, test data inputs together with test data predictions from the opaque model is termed *oracle data*, with the motivation that the predictions from the opaque model (the oracle) are regarded as ground truth during rule extraction. Naturally, target values for test data are by definition not available when performing rule extraction, but often input values and predictions from the opaque model could be. With access to a sufficiently sized oracle dataset, the rule extraction algorithm could either use only oracle data, or augment the training data with oracle instances.

The use of oracle data was first suggested in [14], and further evaluated in [15]. The main result was that rules extracted using oracle data were significantly more accurate than both rules extracted by the same rule extraction algorithm (using training data only) and standard decision tree algorithms; i.e. rules extracted using oracle data explained the predictions made on the novel data better than rules extracted using training data only.

Since the use of oracle data means that the same novel data instances used for actual prediction also are used by the rule extraction algorithm, the problem must be one where predictions are made for sets of instances rather than one instance at a time. This is a description matching most data mining problems, but not all. One example, where a sufficiently sized oracle dataset would not be available, is a medical system where diagnosis is based on a predictive model built from historical data. In that situation, test instances (patients), would probably be handled one at a time. On the other hand, if, as an example, a predictive model is used to determine the recipients of a marketing campaign, the oracle dataset could easily contain thousands of instances. Since the use of oracle data has proven to be beneficial for rule extraction using G-REX, we in this study evaluate the suggested post-processing techniques on trees extracted both with and without the use of oracle data. In addition, the technique is also evaluated on trees evolved by G-REX directly from the dataset.

## 3 Method

The overall idea introduced in this chapter is to post-process GP trees; i.e. only the winning tree is potentially modified. In this study, the GP trees are either extracted from opaque models or induced directly from the data, using G-REX. When searching for possible modifications, the structure of the tree is always intact; as a matter of fact, in this study, only the constants in the interior node splits can be modified. More specifically, the post-processing explicitly searches for constant values that would increase the accuracy, one node at a time. The algorithm is presented in pseudo code in Figure 3 below:

```
do
    currentAcc = evaluateTree(tree);
    foreach interior node
        constant=findBestSplit(node, instances);
        modifyNode(&node, constant);
        newAcc = evaluateTree(tree);
while (newAcc > currentAcc)
```

**Fig. 3** Post-processing algorithm

At the heart of the algorithm is the function *findBestSplit*, which takes the current node and the instances reaching that node as arguments. The function should return the best constant value to use in the current split. If, as an example, the node is *If x1 > 7.5* the function will try alternative values for *7.5*, and return the value resulting in the highest accuracy. Which values to consider are determined from the instances reaching the node. In the preceding example, all unique values for *x1*, found in the instances reaching the node, would be tested. It must be noted that the evaluation always starts with the value found by the GP, so post-processing can never decrease accuracy, only increase it. Clearly, different criteria (like information gain) could be used when evaluating the potential splits, but here we use accuracy; i.e. the split is modified if and only if the modified tree classifies more instances correctly.

Changing splits will, of course, affect which instances that reach every node, so when all nodes have been processed, another sweep could very well further increase accuracy. Consequently, the entire procedure is repeated until there is no change in accuracy between two runs.

When modifying the interior nodes, we could start at the root node or at the leaves. In either case, the question in each node is whether it is possible to increase the overall accuracy by finding a split that changes the distribution of instances to the children; i.e. we assume that all nodes further down the tree will remain as they are. The difference is whether the children are optimized before or after their parent. If we start at the root node, we would first optimize that split, and then proceed down the tree. Or, more generally, when we get to a specific node, the parent node is always already modified, but the children are not. When processing the tree in this manner, we are in fact performing a preorder traversal; see Figure 4 below. The numbers in the squares show the order in which the nodes are processed.

**Fig. 4** Preorder traversal

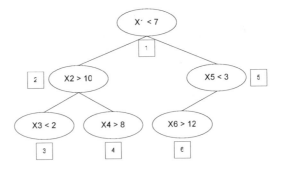

If we instead start at the leaves, no node will be optimized before its descendants. Using this strategy, we perform a *postorder* traversal; see Figure 5 below.

**Fig. 5** Postorder traversal

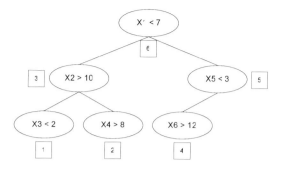

## 3.1 G-REX Settings

The opaque models used to rule extract from are ANN ensembles each consisting of 15 independently trained ANNs. All ANNs are fully connected feed-forward networks where a localist (1-of-C) representation is used. Of the 15 ANNs, seven have one hidden layer and the remaining eight have two hidden layers. The exact number of units in each hidden layer is slightly randomized, but is based on the number of inputs and classes in the current dataset. For an ANN with one hidden layer, the number of hidden units is determined from (1) below.

$$h = \lfloor 2rand \cdot \sqrt{(v \cdot c)} \rfloor \tag{1}$$

Here, $v$ is the number of input variables and $c$ is the number of classes. *rand* is a random number in the interval [0, 1]. For ANNs with two hidden layers, the number of units in the first and second hidden layers are $h_1$ and $h_2$, respectively.

$$h_1 = \lfloor \sqrt{(v \cdot c)}/2 + 4rand \cdot (\sqrt{(v \cdot c)}/c) \rfloor \tag{2}$$

$$h_2 = \lfloor rand \cdot (\sqrt{(v \cdot c)}/c) + c \rfloor \tag{3}$$

Some diversity is introduced by using ANNs with different architectures, and by training each network on slightly different data. More specifically, each ANN uses a "bootstrap" training set; i.e. instances are picked randomly, with replacement, from the available data, until the number of training instances is equal to the size of the available training set. The result is that approximately 63% of the available training instances are actually used by each net. Averaging of the posterior probabilities from each ANN is used to determine ensemble classifications.

When using GP for rule (tree) extraction (induction), the available functions, $F$, and terminals $T$, constitute the literals of the representation language. Functions will typically be logical or relational operators, while the terminals could be, for instance, input variables or constants. Here, the representation language is very similar to basic decision trees. Figure 6 below shows a small but quite accurate (test accuracy is 0.771) sample tree evolved on the Pima Indian diabetes dataset.

```
if (Body_mass_index > 29.132)
 |T: if (plasma_glucose < 127.40)
 |    |T: [Negative] {56/12}
 |    |F: [Positive] {29/21}
 |F: [Negative] {63/11}
```

**Fig. 6** Sample tree evolved on diabetes dataset

The exact grammar used is presented using Backus-Naur form in Figure 7 below.

```
F = {if, ==, <, >}
T = {i₁, i₂, …, iₙ, c₁, c₂, …, cₘ, ℜ}

DTree          :-      (if RExp Dtree Dtree) | Class
Rexp           :-      (ROp ConI ConC) | (== CatI CatC)
Rop            :-      < | >
CatI           :-      Categorical input variable
ConI           :-      Continuous input variable
Class          :-      c₁ | c₂ | … | cₘ
CatC           :-      Categorical attribute value
ConC           :-      ℜ
```

**Fig. 7** G-REX representation language used

The GP parameter settings used in this study are given in Table 1 below. The length penalty used is much smaller than the cost of misclassifying an instance. Nevertheless, it will put some parsimony pressure on the population, resulting in much lower average tree size.

**Table 1** GP parameters

| Parameter | Value | Parameter | Value |
|-----------|-------|-----------|-------|
| Crossover rate | 0.8 | Creation depth | 7 |
| Mutation rate | 0.01 | Creation method | Ramped half-and-half |
| Population size | 1500 | Fitness function | Fidelity - length penalty |
| Generations | 100 | Selection | Roulette wheel |
| Persistence | 25 | Elitism | Yes |

## 3.2 Datasets

The 22 datasets used are all publicly available from the UCI Repository [16]. For a summary of dataset characteristics see Table 2 below. *Ins.* is the total number of instances in the dataset. *Cl.* is the number of output classes in the dataset. *Con.* is the number of continuous input variables and *Cat.* is the number of categorical input variables.

**Table 2** Dataset Characteristics

| Dataset | Ins. | Cl. | Con. | Cat. | Dataset | Ins. | Cl. | Con. | Cat. |
|---------|------|-----|------|------|---------|------|-----|------|------|
| Breast cancer (BC) | 286 | 2 | 0 | 9 | Iris | 150 | 3 | 4 | 0 |
| Liver disorders (Bld) | 345 | 2 | 6 | 0 | Labor | 57 | 2 | 8 | 8 |
| Heart-C (Cleve) | 303 | 2 | 6 | 7 | Diabetes (PID) | 768 | 2 | 8 | 0 |
| Cmc | 1473 | 3 | 2 | 7 | Sick | 2800 | 2 | 7 | 22 |
| Crx | 690 | 2 | 6 | 9 | Sonar | 208 | 2 | 60 | 0 |
| German | 1000 | 2 | 7 | 13 | Tae | 151 | 3 | 1 | 4 |
| Heart Statlog (Heart) | 270 | 2 | 6 | 7 | Tictactoe | 958 | 2 | 0 | 9 |
| Hepatitis (Hepati) | 155 | 2 | 6 | 13 | Vehicle | 846 | 4 | 18 | 0 |
| Horse colic (Horse) | 368 | 2 | 7 | 15 | Breast cancer-W (WBC) | 699 | 2 | 9 | 0 |
| Hypothyroid (Hypo) | 3163 | 2 | 7 | 18 | Wine | 178 | 3 | 13 | 0 |
| Iono | 351 | 2 | 34 | 0 | Zoo | 100 | 7 | 0 | 16 |

It is obvious that the post-processing algorithms suggested will mainly affect splits with continuous attributes. When searching for a categorical split, the alternative values actually represent totally different splits; e.g. $x1 == 3$ instead of $x1 == 5$. It is, therefore, quite unlikely that there are better categorical splits than the ones found by the GP. With this in mind, we first considered using only datasets with mainly continuous attributes. Ultimately, though, we decided to include also datasets with mostly categorical attributes, to confirm our reasoning. Having said that, we did not expect the post-processing algorithms to actually modify any trees from the *Breast cancer, Tictactoe* or *Zoo* datasets.

## 3.3 Experiments

As mentioned above, the post-processing techniques are evaluated on G-REX trees extracted both with and without the use of oracle data. Furthermore, the post-processing techniques are also applied on G-REX trees evolved directly from the original training data; i.e. without the use of the ANN ensemble. Consequently, four experiments are conducted, each comparing three different options for post-processing; *No, Pre* and *Post. No* post-processing is exactly that; i.e. the G-REX tree obtained is evaluated on test data without modifications. *Pre* and *Post* apply the algorithm described in Figure 3 to the tree obtained from G-REX, before the potentially modified tree is evaluated on test data. Naturally, the processing of the nodes is performed in preorder and postorder, respectively. So, to iterate, on each fold, both post-processing algorithms start from the same G-REX tree. This tree is also evaluated to obtain the result for *No*. The difference between the first three experiments is whether oracle data is used or not. When G-REX *original* is used, rules are extracted using training data (but oracle answers) only. G-REX *all* means that both training data and oracle data are used. G-REX *oracle* uses only oracle data. In Experiment 4, G-REX uses only original training data; i.e. training instances with always correct target values. For a summary, see Table 3 below. For the evaluation, stratified 10-fold cross-validation is used.

**Table 3** Experiments

| Experiment | Data used for rule extraction |
|---|---|
| G-REX original | Training (with oracle answers) |
| G-REX all | Training (with oracle answers) and oracle |
| G-REX oracle | Oracle |
| G-REX induction | Original training (with correct answers) |

## 4  Results

Table 4 below shows fidelity results. It should be noted that fidelity values were calculated on the exact data used by G-REX; i.e. for *original* training data only, for *all* both training and oracle data, and for *oracle*, oracle data only. Results where the suggested technique improved the fidelity are given in bold.

From Table 4, it is obvious that the post-processing algorithm was able to find better splits and thereby increasing fidelity. As a matter of fact, when using only training data, the post-processing algorithm increased the fidelity on all datasets but the three "categorical only" mentioned above. This does not mean that the post-processing algorithm always was able to modify the trees; on several datasets there were, in fact, no modifications on a large majority of the folds.

When using oracle data only, the number of instances in the fitness set is quite small, so the fidelity of the starting tree is normally quite high. For five datasets (*Hepati*, *Iris*, *Labor*, *Sonar* and *Wine*) the fidelity is even a perfect 1.0 to start with. Because of this, it is not surprising that relatively few improvements are found.

When using all data, the starting tree has most often, but not always, higher fidelity than the tree extracted using training data only. Still, the post-processing managed to increase fidelity on all datasets but the three "categorical-only" and Horse. Often, however, the increase is quite small, sometimes even the result of a single modification, on just one fold. Overall, however, the picture is clearly that the suggested algorithm was capable of increasing fidelity.

Naturally, fidelity is not vital *per se* and the important question is whether the modifications would also improve test set accuracies. Table 5 below shows test set accuracies. The results obtained by the neural network ensemble used to rule extract from are included for comparison.

**Table 4** Fidelity

| Datasets | Original | | | All | | | Oracle | | |
|---|---|---|---|---|---|---|---|---|---|
| | No | Pre | Post | No | Pre | Post | No | Pre | Post |
| BC | .847 | .847 | .847 | .855 | .855 | .855 | .986 | .986 | .986 |
| Bld | .774 | **.799** | **.797** | .769 | **.781** | **.782** | .918 | **.924** | **.924** |
| Cleve | .861 | **.863** | **.863** | .859 | **.862** | **.862** | .977 | .977 | .977 |
| Cmc | .783 | **.789** | **.790** | .792 | **.803** | **.803** | .817 | **.826** | **.826** |
| Crx | .919 | **.921** | **.922** | .923 | **.924** | **.924** | .958 | .958 | .958 |
| German | .784 | **.787** | **.787** | .789 | **.792** | **.792** | .882 | **.886** | **.886** |
| Heart | .868 | **.871** | **.871** | .869 | **.870** | **.870** | .989 | .989 | .989 |
| Hepati | .883 | **.886** | **.886** | .890 | **.898** | **.897** | 1.00 | 1.00 | 1.00 |
| Horse | .871 | **.873** | **.873** | .879 | .879 | .879 | .972 | .972 | .972 |
| Hypo | .986 | **.988** | **.988** | .988 | **.990** | **.990** | .993 | .993 | .993 |
| Iono | .932 | **.940** | **.940** | .937 | **.946** | **.946** | .986 | .986 | .986 |
| Iris | .979 | **.983** | **.983** | .977 | **.981** | **.981** | 1.00 | 1.00 | 1.00 |
| Labor | .988 | .988 | .988 | .982 | **.984** | **.984** | 1.00 | 1.00 | 1.00 |
| PID | .887 | **.896** | **.896** | .885 | **.894** | **.894** | .941 | **.946** | **.946** |
| Sick | .983 | **.985** | **.985** | .983 | **.984** | **.984** | .987 | **.988** | **.988** |
| Sonar | .827 | **.843** | **.843** | .812 | **.828** | **.826** | 1.00 | 1.00 | 1.00 |
| Tae | .704 | **.719** | **.720** | .693 | **.702** | **.701** | .953 | .953 | .953 |
| Tictactoe | .810 | .810 | .810 | .804 | .804 | .804 | .877 | .877 | .877 |
| Vehicle | .637 | **.654** | **.654** | .628 | **.651** | **.652** | .743 | **.760** | **.760** |
| Wbc | .973 | **.974** | **.974** | .976 | **.977** | **.977** | .996 | .996 | .996 |
| Wine | .961 | **.967** | **.968** | .960 | **.974** | **.974** | 1.00 | 1.00 | 1.00 |
| Zoo | .923 | .923 | .923 | .916 | .916 | .916 | .990 | .990 | .990 |

Table 5 shows that, on a majority of datasets, the increase in fidelity also led to an increase in test set accuracy. Again, it must be noted that these results do not mean that test set accuracies were always increased. As a matter of fact, on several folds, an increase in fidelity led to no change in accuracy. On some folds, the increase in fidelity even led to lower accuracy. When aggregating the results over all ten folds, however, the results are very promising for the suggested algorithm. Table 6 below shows the results for Experiment 4; i.e. when G-REX evolved trees directly from the data.

When performing rule induction instead of rule extraction, the picture is slightly different. The post-processing technique still manages to increase training accuracy on almost all datasets, but this does not always carry over to the test set. On some datasets, a fairly large increase in training accuracy actually leads to clearly worse test accuracy; i.e. *PID*, *Sonar* and *TAE*. The reason is probably that it is easier to overfit when learning a dataset compared to learning an opaque

**Table 5** Test accuracies for the rule extraction

| Datasets | Ensemble | Original | | | All | | | Oracle | | |
|---|---|---|---|---|---|---|---|---|---|---|
| | | No | Pre | Post | No | Pre | Post | No | Pre | Post |
| BC | .729 | .725 | .725 | .725 | .714 | .714 | .714 | .736 | .736 | .736 |
| Bld | .721 | .635 | **.665** | **.659** | .641 | .624 | .626 | .691 | **.697** | **.697** |
| Cleve | .807 | .770 | **.777** | **.777** | .807 | **.810** | **.810** | .810 | .810 | .810 |
| Cmc | .551 | .547 | **.565** | **.565** | .551 | .549 | .551 | .539 | **.540** | **.540** |
| Crx | .859 | .849 | **.854** | **.854** | .848 | **.851** | **.851** | .849 | .849 | .849 |
| German | .749 | .716 | **.720** | **.722** | .726 | **.728** | **.728** | .737 | .735 | .735 |
| Heart | .793 | .785 | **.789** | **.789** | .785 | .785 | .785 | .804 | .804 | .804 |
| Hepati | .860 | .807 | .807 | .807 | .833 | **.840** | **.840** | .860 | .860 | .860 |
| Horse | .817 | .842 | **.844** | **.844** | .847 | .847 | .847 | .806 | .806 | .806 |
| Hypo | .983 | .979 | **.981** | **.981** | .980 | **.982** | **.982** | .980 | .980 | .980 |
| Iono | .934 | .877 | **.889** | **.889** | .914 | **.929** | **.926** | .926 | .926 | .926 |
| Iris | .967 | .953 | .953 | .953 | .947 | **.960** | **.960** | .967 | .967 | .967 |
| Labor | .940 | .840 | .840 | .840 | .920 | .920 | .920 | .940 | .940 | .940 |
| PID | .767 | .746 | **.753** | **.754** | .753 | **.755** | **.757** | .750 | .747 | .747 |
| Sick | .970 | .971 | **.973** | **.973** | .971 | .970 | .970 | .967 | **.968** | **.968** |
| Sonar | .865 | .700 | **.740** | **.740** | .730 | **.750** | **.745** | .865 | .865 | .865 |
| Tae | .567 | .460 | **.480** | **.480** | .480 | .473 | .473 | .547 | .547 | .547 |
| Tictactoe | .898 | .736 | .736 | .736 | .752 | .752 | .752 | .781 | .781 | .781 |
| Vehicle | .848 | .582 | **.586** | **.587** | .605 | **.618** | **.623** | .664 | **.676** | **.676** |
| Wbc | .964 | .957 | **.958** | **.958** | .959 | .959 | .959 | .965 | .965 | .965 |
| Wine | .965 | .918 | .918 | **.929** | .929 | **.953** | **.953** | .965 | .965 | .965 |
| Zoo | .950 | .880 | .880 | .880 | .920 | .920 | .920 | .940 | .940 | .940 |

model. The opaque model actually functions as a filter, reducing the impact of atypical training instances. Still, it is obvious that the post-processing improves test set accuracy much more often than reducing it.

To further analyze the result, the two post-processing algorithms were pair-wise compared to each other and *No*, one experiment at a time. Tables 7 - 10 below show these comparisons for the four experiments. The values tabulated are datasets won, lost and tied, for the row technique when compared to the column technique. Using 22 datasets, a standard sign-test ($\alpha$=0.05) requires 16 wins for statistical significance. When using a sign test, ties are split evenly between the two techniques. Statistically significant differences are shown using bold and underlined values.

First of all, it is interesting to note that neither post-processing algorithm lose a single dataset against *No*. The sign test consequently shows that both algorithms

**Table 6** Results for trees induced directly from the data

| Datasets | Training | | | Test | | |
|---|---|---|---|---|---|---|
| | No | Pre | Post | No | Pre | Post |
| BC | .768 | .768 | .768 | .732 | .732 | .732 |
| Bld | .698 | **.715** | **.714** | .629 | **.644** | **.644** |
| Cleve | .848 | **.851** | **.851** | .780 | **.787** | **.787** |
| Cmc | .552 | **.559** | **.559** | .563 | .560 | .560 |
| Crx | .865 | **.866** | **.866** | .859 | .858 | .858 |
| German | .730 | **.734** | **.734** | .737 | **.740** | **.740** |
| Heart | .841 | **.842** | **.842** | .819 | .811 | .811 |
| Hepati | .873 | **.874** | **.874** | .880 | .880 | .880 |
| Horse | .869 | **.870** | **.870** | .842 | **.844** | **.844** |
| Hypo | .981 | **.984** | **.984** | .980 | **.985** | **.985** |
| Iono | .924 | **.930** | **.930** | .909 | .909 | .909 |
| Iris | .974 | **.976** | **.976** | .953 | **.960** | **.960** |
| Labor | .988 | **.990** | **.990** | .820 | **.840** | **.840** |
| PID | .767 | **.776** | **.776** | .739 | .733 | .734 |
| Sick | .980 | **.982** | **.982** | .980 | **.981** | **.981** |
| Sonar | .812 | **.828** | **.828** | .730 | .705 | .705 |
| Tae | .503 | **.547** | **.547** | .513 | .473 | .487 |
| Tictactoe | .796 | .796 | .796 | .801 | .801 | .801 |
| Vehicle | .660 | **.674** | **.674** | .631 | **.640** | **.646** |
| Wbc | .969 | **.970** | **.970** | .954 | .954 | .954 |
| Wine | .963 | **.972** | **.973** | .912 | **.918** | **.918** |
| Zoo | .908 | .908 | .908 | .960 | .960 | .960 |

**Table 7** Wins, losses and ties using G-REX original

| | No | | | Pre | | | Post | | |
|---|---|---|---|---|---|---|---|---|---|
| | *W* | *L* | *T* | *W* | *L* | *T* | *W* | *L* | *T* |
| No | - | - | - | **0** | **15** | **7** | **0** | **16** | **6** |
| Pre | **15** | **0** | **7** | - | - | - | 1 | 4 | 17 |
| Post | **16** | **0** | **6** | 4 | 1 | 17 | - | - | - |

are significantly more accurate than using no post-processing. The results also show that, on a large majority of the datasets, there is no difference between *Pre* and *Post*.

Here, a sign test reports no statistically significant differences. Still, it is clearly beneficial to use the post-processing algorithm, even for G-REX *all*. As a matter

**Table 8** Wins, losses and ties using G-REX all

|  | No | | | Pre | | | Post | | |
|---|---|---|---|---|---|---|---|---|---|
|  | **W** | **L** | **T** | **W** | **L** | **T** | **W** | **L** | **T** |
| No | - | - | - | 4 | 11 | 7 | 3 | 11 | 8 |
| Pre | 11 | 4 | 7 | - | - | - | 3 | 4 | 15 |
| Post | 11 | 3 | 8 | 4 | 3 | 15 | - | - | - |

**Table 9** Wins, losses and ties using G-REX oracle

|  | No | | | Pre | | | Post | | |
|---|---|---|---|---|---|---|---|---|---|
|  | **W** | **L** | **T** | **W** | **L** | **T** | **W** | **L** | **T** |
| No | - | - | - | 2 | 4 | 16 | 2 | 4 | 16 |
| Pre | 4 | 2 | 16 | - | - | - | 0 | 0 | 22 |
| Post | 4 | 2 | 16 | 0 | 0 | 22 | - | - | - |

of fact, if the three "categorical only" datasets were removed before the analysis, the use of either *Pre* or *Post* would again be significantly more accurate than *No*.

For G-REX *oracle*, the situation is quite different. Here, there is generally very little to gain by using the suggested post-processing. The reason is, of course, the very high fidelity obtained by the rule extraction to start with. Still, it should be noted that the post-processing was able to increase accuracy on five datasets. For two of these (*Bld* and *Vehicle*) the increase was not insignificant; see Table 5.

**Table 10** Wins, losses and ties using G-REX induction

|  | No | | | Pre | | | Post | | |
|---|---|---|---|---|---|---|---|---|---|
|  | **W** | **L** | **T** | **W** | **L** | **T** | **W** | **L** | **T** |
| No | - | - | - | 6 | 10 | 6 | 6 | 10 | 6 |
| Pre | 10 | 6 | 6 | - | - | - | 0 | 3 | 19 |
| Post | 10 | 6 | 6 | 3 | 0 | 19 | - | - | - |

The post-processing manages to improve the test accuracies on almost half of the datasets (10). Unfortunately, it also leads to worse accuracy on six datasets. Still, looking at the overall results, it is obvious that the post-processing was generally successful, even when the model was built directly from the dataset.

## 5 Conclusions

We have in this chapter suggested a novel algorithm for post-processing GP trees. The technique iteratively, one node at a time, searches for possible modifications that would result in higher accuracy. More specifically, for each split, the algorithm evaluates every possible constant value and chooses the best. It should be noted that the algorithm can only increase accuracy (on training data), since modifications are only carried out when the result is a more accurate tree.

In this study, we evaluated the algorithm using the rule extraction technique G-REX. Results show that when the rule extraction uses only training data, the post-processing increased accuracy significantly. When using both training and oracle data, it is still clearly advantageous to apply the post-processing. Even when starting from a tree extracted using oracle data only, the post-processing is sometimes capable of increasing accuracy. Post-processing trees evolved directly from the dataset sometimes results in less accurate models (6 datasets of 22), probably due to overfitting. Still, the post-processing is successful on almost twice as many datasets, improving accuracy on 10 datasets. Overall, the very promising picture is that the post-processing more often than not improves accuracy, and only very rarely decreases it.

## 6 Discussion and Future Work

In this study, we applied the novel technique mainly to rule extraction. One obvious future study is therefore to further evaluate the suggested technique on GP classification, and also on regression trees. Another option is to investigate whether it would be beneficial to search for entirely new splits, not just for constant values. In that study, it would be natural to also evaluate different criteria (like information gain) as alternatives to accuracy. This would, however, be very similar to standard decision tree algorithms. The challenge is to find the right mix between the locally effective greed in decision tree algorithms, and the global optimization in GP.

Finally, we have also performed some initial experiments, with promising results, where the algorithm suggested was instead used as a mutation operator during evolution. Potential benefits are higher accuracy and faster convergence.

**Acknowledgment.** This work was supported by the Information Fusion Research Program (University of Skövde, Sweden) in partnership with the Swedish Knowledge Foundation under grant 2003/0104 (URL: http://www.infofusion.se).

## References

[1] Lu, H., Setino, R., Liu, H.: Neurorule: A connectionist approach to data mining. In: International Very Large Databases Conference, pp. 478–489 (1995)

[2] Craven, M., Shavlik, J.: Extracting Tree-Structured Representations of Trained Networks. In: Advances in Neural Information Processing Systems, vol. 8, pp. 24–30 (1996)

[3] Andrews, R., Diederich, J., Tickle, A.B.: A survey and critique of techniques for extracting rules from trained artificial neural networks. Knowledge-Based Systems 8(6) (1995)

[4] Craven, M., Shavlik, J.: Rule Extraction: Where Do We Go from Here? University of Wisconsin Machine Learning Research Group working Paper, 99-1 (1999)

[5] Johansson, U., König, R., Niklasson, L.: Rule Extraction from Trained Neural Networks using Genetic Programming. In: 13th International Conference on Artificial Neural Networks, Istanbul, Turkey, supplementary proceedings, pp. 13–16 (2003)

[6] König, R., Johansson, U., Niklasson, L.: G-REX: A Versatile Framework for Evolutionary Data Mining. IEEE International Conference on Data Mining (ICDM 2008), Demo paper, Pisa, Italy (in press) (2008)

[7] Johansson, U.: Obtaining accurate and comprehensible data mining models: An evolutionary approach, PhD thesis, Institute of Technology, Linköping University (2007)

[8] Quinlan, J.R.: C4.5: Programs for Machine Learning. Morgan Kaufmann, San Francisco (1993)

[9] Breiman, L., Friedman, J.H., Olshen, R.A., Stone, C.J.: Classification and Regression Trees, Wadsworth International (1984)

[10] Tsakonas, A.: A comparison of classification accuracy of four genetic programming-evolved intelligent structures. Information Sciences 176(6), 691–724 (2006)

[11] Bojarczuk, C.C., Lopes, H.S., Freitas, A.A.: Data Mining with Constrained-syntax Genetic Programming: Applications in Medical Data Sets. In: Intelligent Data Analysis in Medicine and Pharmacology - a workshop at MedInfo 2001 (2001)

[12] Eggermont, J., Kok, J., Kosters, W.A.: Genetic Programming for Data Classification: Refining the Search Space. In: 15th Belgium/Netherlands Conference on Artificial Intelligence, pp. 123–130 (2003)

[13] Eggermont, J., Kok, J., Kosters, W.A.: Genetic Programming for Data Classification: Partitioning the Search Space. In: 19th Annual ACM Symposium on Applied Computing (SAC 2004), pp. 1001–1005 (2004)

[14] Johansson, U., Löfström, T., König, R., Niklasson, L.: Why Not Use an Oracle When You Got One? Neural Information Processing - Letters and Reviews 10(8-9), 227–236 (2006)

[15] Johansson, U., Löfström, T., König, R., Sönströd, C., Niklasson, L.: Rule Extraction from Opaque Models – A Slightly Different Perspective. In: 6th International Conference on Machine Learning and Applications, Orlando, FL, pp. 22–27. IEEE press, Los Alamitos (2006)

[16] Asuncion, A., Newman, D.J.: UCI machine learning repository (2007)

# Part III
# Evolutionary Fuzzy and Swarm in Clustering Problems

# Evolutionary Fuzzy Clustering: An Overview and Efficiency Issues

D. Horta, M. Naldi, R.J.G.B. Campello, E.R. Hruschka, and A.C.P.L.F. de Carvalho

**Abstract.** Clustering algorithms have been successfully applied to several data analysis problems in a wide range of domains, such as image processing, bioinformatics, crude oil analysis, market segmentation, document categorization, and web mining. The need for organizing data into categories of similar objects has made the task of clustering very important to these domains. In this context, there has been an increasingly interest in the study of evolutionary algorithms for clustering, especially those algorithms capable of finding blurred clusters that are not clearly separated from each other. In particular, a number of evolutionary algorithms for fuzzy clustering have been addressed in the literature. This chapter has two main contributions. First, it presents an overview of evolutionary algorithms designed for fuzzy clustering. Second, it describes a fuzzy version of an evolutionary algorithm for clustering, which has shown to be more computationally efficient than systematic (i.e., repetitive) approaches when the number of clusters in a data set is unknown. Illustrative experiments showing the influence of local optimization on the efficiency of the evolutionary search are also presented. These experiments reveal interesting aspects of the effect of an important parameter found in many evolutionary algorithms for clustering, namely, the number of iterations of a given local search procedure to be performed at each generation.

## 1 Introduction

Clustering is a task whose goal is to determine a finite set of categories (clusters) to describe a data set according to similarities among its objects [39][18]. The applicability of clustering is manifold, ranging from market segmentation [12] and image processing [36] through document categorization and web mining [46]. Another application field that has shown to be particularly promising for clustering techniques is bioinformatics [5][6][57].

Clustering techniques can be broadly divided into three main types [36]: overlapping (so-called non-exclusive), partitional, and hierarchical. The last two

D. Horta, M. Naldi, R.J.G.B. Campello, E.R. Hruschka, and A.C.P.L.F. de Carvalho
Department of Computer Sciences of the University of São Paulo at São Carlos (ICMC/USP), São Carlos, SP, Brazil
e-mail: {horta,naldi,campello,erh,andre}@icmc.usp.br

A. Abraham et al. (Eds.): Foundations of Comput. Intel. Vol. 4, SCI 204, pp. 167–195.
springerlink.com                    © Springer-Verlag Berlin Heidelberg 2009

are related to each other in that a hierarchical clustering is a nested sequence of partitional clusterings, each of which represents a hard partition of the data set into a different number of mutually disjoint subsets. A hard partition of a data set $\mathbf{X} = \{\mathbf{x}_1, \mathbf{x}_2, ..., \mathbf{x}_N\}$, where $\mathbf{x}_j$ ($j = 1, ..., N$) stands for an $n$-dimensional feature or attribute vector, is a collection $\mathbf{C} = \{\mathbf{C}_1, \mathbf{C}_2, ..., \mathbf{C}_k\}$ of $k$ non-overlapping data subsets $\mathbf{C}_i \neq \varnothing$ (non-null clusters), such that $\mathbf{C}_1 \cup \mathbf{C}_2 \cup ... \cup \mathbf{C}_k = \mathbf{X}$ and $\mathbf{C}_i \cap \mathbf{C}_j = \varnothing$ for $i \neq j$. If the condition of mutual disjunction ($\mathbf{C}_i \cap \mathbf{C}_j = \varnothing$ for $i \neq j$) is relaxed, then the corresponding data partitions are said to be of overlapping type. Overlapping algorithms produce data partitions that can be *soft* (each object fully belongs to one or more clusters – e.g., see [22], [24], [25], and [55]) or *fuzzy* (each object belongs to one or more clusters to different degrees). The latter type constitutes the focus of the present work.

When a fuzzy clustering algorithm is applied to a data set with $N$ objects, the result is a partition of the data into a certain number $k$ of *fuzzy clusters*, such that:

$$\begin{cases} P = \left[p_{ij}\right]_{k \times N} \\ p_{ij} \in [0,1] \end{cases} \tag{1}$$

where $P$ is a $k \times N$ *fuzzy partition matrix* whose element $p_{ij}$ represents the fuzzy membership of the $j$th object to the $i$th fuzzy cluster. When $p_{ij}$ is limited to the extreme values of its feasibility interval, i.e., $p_{ij} \in \{0,1\}$, then $P$ degenerates to a soft partition. Besides, if the additional constraint $\Sigma_i\, p_{ij} = 1$ is imposed to every column $j$ of the matrix, then $P$ degenerates to a standard hard partition. A fuzzy partition matrix provides additional information about the data that is not available in its soft or hard counterparts. In fact, the fuzzy membership values $p_{ij}$ can help discover more sophisticated relations between the disclosed clusters and the corresponding data objects [59]. In addition, in contrast to their Boolean relatives, the continuous membership values of fuzzy partitions are particularly appropriate to describe boundaries between ambiguous or blurred clusters that are not clearly separated from each other. Owing to these desired properties, the applicability of fuzzy clustering is broad in scope and includes areas such as pattern classification, image segmentation, document categorization, data visualization, and dynamic systems identification, just to mention a few [11][30][4][16].

In spite of the type of algorithm (partitional, hierarchical or overlapping), the main goal of clustering is maximizing both the homogeneity within each cluster and the heterogeneity among different clusters [36][2]. In other words, objects that belong to the same cluster should be more similar to each other than objects that belong to different clusters. The problem of measuring similarity is usually tackled indirectly, i.e., distance measures are used for quantifying the degree of dissimilarity among objects, in such a way that more similar objects have lower dissimilarity values [37]. Several dissimilarity measures can be employed for clustering tasks [36][59]. Each measure has its bias and comes with its own advantages and drawbacks. Therefore, each one may be more or less suitable to a given analysis or application scenario. Indeed, it is well-known that some measures are more suitable for gene clustering in bioinformatics [38], whereas other measures are more appropriate for text clustering and document categorization [51], for instance.

Clustering is deemed one of the most difficult and challenging problems in machine learning, particularly due to its unsupervised nature. The unsupervised nature of the problem implies that its structural characteristics are not known, except if there is some sort of domain knowledge available in advance. Specifically, the spatial distribution of the data in terms of the number, volumes, densities, shapes, and orientations of clusters (if any) are unknown [23]. These difficulties may be intensified even further by an eventual need for dealing with data objects described by attributes of distinct natures (binary, discrete, continuous, and categorical), conditions (complete and partially missing) and scales (ordinal and nominal) [36][37].

From an optimization perspective, clustering can be formally considered as a particular kind of NP-hard grouping problem [19]. This has stimulated the search for efficient approximation algorithms, including not only the use of *ad hoc* heuristics for particular classes or instances of problems, but also the use of general-purpose meta-heuristics (e.g. see [53]). Particularly, evolutionary algorithms are meta-heuristics widely believed to be effective on NP-hard problems, being able to provide near-optimal solutions to such problems in reasonable time. Under this assumption, a large number of evolutionary algorithms for solving clustering problems have been proposed in the literature. These algorithms are based on the optimization of some objective function (i.e., the so-called fitness function) that guides the evolutionary search. We refer the interested reader to [34] for a comprehensive survey on evolutionary algorithms for clustering.

The remaining of this chapter is organized as follows. Section 2 presents a brief overview of evolutionary algorithms for fuzzy clustering. Then, in Section 3, an efficient evolutionary algorithm for fuzzy clustering [13] is described. Original illustrative experiments showing the influence of an important parameter of this algorithm, which is also found in many other evolutionary algorithms for clustering and fuzzy clustering, are reported in Section 4. Finally, Section 5 concludes the chapter and addresses some issues for future research.

## 2  Overview of Evolutionary Fuzzy Clustering

Most of the research on evolutionary algorithms for overlapping clustering has focused on algorithms that evolve fuzzy partitions of data. In this context, many authors have proposed evolutionary algorithms to solve fuzzy clustering problems for which the number of clusters is known or set in advance by the user [27][28][40][8][60][58][17][29][43]. However, the optimal number of clusters is usually unknown in advance. For this reason, more recent papers have proposed to optimize both the number of clusters and the corresponding fuzzy partitions by some form of evolutionary search [52][42][45][49][32][1][13][20]. Regardless of the fixed or variable nature of the number of clusters, the evolutionary algorithms for fuzzy clustering are mostly based on extensions – to the fuzzy domain – of the fundamental ideas developed in the context of hard partitions [34][1]. This is in conformity with the fact that most fuzzy clustering algorithms are based on generalizations of traditional algorithms for hard clustering, as it is the case of the well-known Fuzzy $c$-Means (FCM) algorithm and its variants [11][30][4], which are essentially generalizations of the classic $k$-means algorithm. Only a few exceptions (e.g. see [56]) try to develop operators that could act directly on fuzzy partitions of data. Contrarily, most authors have chosen to adapt the existing evolutionary clustering techniques to the fuzzy domain, not only for convenience, but mainly because, to date, there is no strong evidence that the more complex fully fuzzy formulation of the problem can be counterbalanced by efficiency and/or efficacy gains. This is an interesting open question still to be tackled.

Essentially, the only differences between evolutionary algorithms for fuzzy clustering and their hard counterparts thoroughly discussed in [34] are: (i) they have to compute the fuzzy partition corresponding to every genotype; and (ii) they use, as fitness functions, clustering validity criteria that are capable of assessing fuzzy partitions. In what concerns fuzzy clustering validity criteria, we refer the interested reader to [30][14][50] and references therein.

Roughly speaking, the evolutionary algorithms for fuzzy clustering can be broadly divided into two main categories. The first (and most representative) one is composed of algorithms that encode and evolve prototypes for the FCM algorithm or for one of its variants [17][3][43][52][40][8][27][28][29][45] [49][42]. The second category is composed of algorithms that use some variant of FCM as a local search operator to speed up their convergence by refining rough partitions explored by the evolutionary search, while providing the necessary computations to get the fuzzy partition [1][13][32]. In [13] it is shown that an evolutionary algorithm is able to outperform, in terms of computational efficiency, traditional approaches to determine satisfactory estimates of the unknown number of fuzzy clusters and their structures, under both the theoretical (asymptotic time complexity analyses) and experimental (statistical analyses) perspectives. In Section 3, we describe such an evolutionary fuzzy clustering algorithm in details. Since it uses FCM as a local search operator, FCM is briefly reviewed in the sequel.

Fuzzy $c$-Means (FCM) [11] is essentially a fuzzy extension of the well-known $k$-means algorithm [44]. Like $k$-means, there are several variants of the basic algorithm, as discussed in [13]. The basic algorithm, which will be considered here for the sake of simplicity, is an iterative procedure shown to be able to find local solutions to the following problem:

---

[1] Reference [34] comprises an extensive and detailed description of evolutionary approaches for hard partitional clustering.

$$\min_{\mathbf{v}_i, p_{ij}} \quad J = \sum_{j=1}^{N} \sum_{i=1}^{c} \left( p_{ij} \left\| \mathbf{x}_j - \mathbf{v}_i \right\| \right)^2$$

$$\text{s. to} \quad \begin{cases} 0 \le p_{ij} \le 1 \\ \sum_{i=1}^{c} p_{ij} = 1 \quad \forall j \in \{1,...,N\} \\ 0 < \sum_{j=1}^{N} p_{ij} < N \quad \forall i \in \{1,...,c\} \end{cases} \tag{2}$$

where $\mathbf{x}_j \in \mathfrak{R}^n$ ($j = 1, ..., N$) are the data objects to be clustered into $c \ge 2$ clusters, $\mathbf{v}_i \in \mathfrak{R}^n$ ($i = 1, ..., c$) are the cluster prototypes, $p_{ij}$ stands for the membership of the $j$th object to the $i$th fuzzy cluster, and $\| \cdot \|$ denotes an inner-product norm (e.g. Euclidean norm). The second constraint, so-called "probabilistic" constraint, prevents the algorithm to find the trivial solution $p_{ij} = 0$ ($i = 1, ..., c$ ; $j = 1, ..., N$) and forces the minimum feasible value for $J$ to be such that objects $\mathbf{x}_j$ in the near vicinity of a given cluster prototype $\mathbf{v}_i$ have a high degree of membership ($p_{ij}$) to the $i$th fuzzy cluster, whereas objects far from $\mathbf{v}_i$ have a low membership degree to that cluster. The third constraint, in turn, prevents the algorithm to find degenerated fuzzy partitions containing null fuzzy clusters. The iterative algorithm to solve the problem formulated in (2) is summarized next:

1. Select the number of fuzzy clusters $c$.
2. Select initial cluster prototypes $\mathbf{v}_1, \mathbf{v}_2, ..., \mathbf{v}_c$.
3. Compute the distances $\| \mathbf{x}_j - \mathbf{v}_i \|$ between objects and prototypes.
4. Compute the elements of the fuzzy partition matrix:

$$p_{ij} = \left[ \sum_{l=1}^{c} \left( \frac{\left\| \mathbf{x}_j - \mathbf{v}_i \right\|}{\left\| \mathbf{x}_j - \mathbf{v}_l \right\|} \right)^2 \right]^{-1} \tag{3}$$

5. Compute the cluster prototypes:

$$\mathbf{v}_i = \frac{\sum_{j=1}^{N} p_{ij}^2 \mathbf{x}_j}{\sum_{j=1}^{N} p_{ij}^2} \tag{4}$$

6. Stop if either converged attained or the number of iterations $t$ exceeded a given limit. Otherwise, go back to Step 3.

*Remark 1*: Equation (4) requires that $\| \mathbf{x}_j - \mathbf{v}_i \| > 0$ for all $j \in \{1, ..., N\}$ and $i \in \{1, ..., c\}$. For every $j$, if $\| \mathbf{x}_j - \mathbf{v}_i \| = 0$ for $i \in I \subseteq \{1, ..., c\}$, define $p_{ij}$ in such a way that: *a)* $p_{ij} = 0$ for $i \in \bar{I}$; and *b)* $\sum_{i \in I} p_{ij} = 1$.

*Remark 2*: Convergence properties of alternating optimization schemes, like the one just described, have been well studied [9][10]. Given that the algorithm converges, an effective stopping criterion is that the maximum absolute difference between elements of the partition matrix in two consecutive iterations be lower than a given positive threshold $\varepsilon$. A usual setting is $\varepsilon=10^{-3}$ [4][29][54]. An alternative stopping criterion (possibly additional – see step 6 in the previous algorithm) involves setting up a fixed maximum number of iterations, $t$. This alternative criterion can be particularly suitable when FCM is embedded into an evolutionary search scheme, as it will be discussed and illustrated by means of examples in Section 4.

Leaving out minor variants of the previous algorithm – concerned simply with the use of a slightly different stopping rule and/or initialization procedure (e.g. starting from an initial partition matrix, instead of initial prototypes) – there is an impressive number of different algorithms based on modifications and/or extensions of the ordinary FCM formulation that have been developed during the past 35 years (see [13] for details). Probably, the most popular one regards the use of a different exponent (other than two), named fuzzifier $m$, to control the weight of the partition matrix elements in the optimization functional $J$ in (2) and, as a consequence, the fuzziness of the resulting clusters [11]. Whether such a fuzzifier is adopted or not, the use of a fixed inner-product norm in the FCM algorithm induces fuzzy clusters of a particular shape. For instance, hyperspherical clusters are induced when the Euclidean norm is adopted. This is precisely the case of the algorithm used in the illustrative experiments to be described in Section 4 of this chapter.

## 3 An Evolutionary Algorithm for Fuzzy Clustering

The Evolutionary Algorithm for Clustering (EAC) [33][31][35] is capable of optimizing the number of clusters and initial prototypes (centroids) for the popular $k$-means. It has been shown – from a statistical perspective – to be more efficient than multiple runs of $k$-means [33]. Since FCM can be viewed as an extension of $k$-means for fuzzy applications, EAC naturally becomes an interesting choice to optimize the number of clusters and initial prototypes for FCM. This approach, for which the resulting algorithm is named EAC-FCM [13], is here described.

EAC-FCM is based on a simple encoding scheme. In order to explain it, let us consider a data set composed of $N$ objects. A partition is encoded as an integer string (genotype) of $N$ positions. Each string position corresponds to an object, i.e., the $i$th position represents the $i$th object of the data set. Thus, each string component has a value over the possible cluster labels $\{1, 2, 3, ..., c\}$. Fig. 1 illustrates the EAC-FCM encoding scheme assuming a data set containing 20 objects. For illustration purposes, right below the encoded partition ($c=5$) are the corresponding object numbers. In this case, four objects $\{1, 7, 8, 12\}$ form the cluster whose label is 2. The cluster whose label is 1 has three objects $\{2, 6, 13\}$, and so on.

| 2 | 1 | 3 | 5 | 5 | 1 | 2 | 2 | 4 | 5 | 3 | 2 | 1 | 4 | 5 | 4 | 4 | 5 | 5 | 3 |
|---|---|---|---|---|---|---|---|---|---|---|---|---|---|---|---|---|---|---|---|
| 1 | 2 | 3 | 4 | 5 | 6 | 7 | 8 | 9 | 10 | 11 | 12 | 13 | 14 | 15 | 16 | 17 | 18 | 19 | 20 |

**Fig. 1** EAC-FCM encoding scheme

Standard genetic operators are usually not suitable for clustering problems for several reasons, as detailed in [34][19]. In brief, such operators often just manipulate objects by means of their corresponding cluster labels, without taking into account their connections with other clusters. Indeed, the interconnections among string positions (given by cluster labels) constitute the genuine optimization goal in clustering problems. For this reason, the development of evolutionary operators specially designed for clustering tasks has been investigated (e.g., see [34]). The cluster-oriented (i.e. context-sensitive) EAC operators [33], which have been adapted here for fuzzy clustering via FCM, are detailed in the following.

### 3.1 Mutation Operators

Two operators for mutation are used by EAC-FCM [13]. The first operator can only be applied to genotypes that encode more than two clusters. It eliminates one or more randomly selected clusters, placing its objects into the nearest remaining clusters (according to their prototypes). Fig. 2 provides the pseudo code for Mutation Operator 1 ($MO_1$).

---

Let **g** be a given genotype that represents $c_g$ clusters;
1. If $c_g > 2$ then:
    1.1 Randomly generate a number $k \in \{1,..., c_g-2\}$; // at least two clusters must remain in **g** //
    1.2 For $i = (1,...,k)$ do:
        1.2.1 Randomly choose a cluster $\mathbf{C}_s$ encoded into **g**;
        1.2.2 Place the objects that belong to $\mathbf{C}_s$ into the nearest remaining clusters $\mathbf{C}_j \in \mathbf{g}$, $j \neq s$, according to the distances between objects and cluster prototypes;
    End For;
Else **g** is not mutated.

---

**Fig. 2** Pseudo code for Mutation Operator 1 ($MO_1$)

The second mutation operator can only be applied to clusters formed by at least two objects. It splits one or more randomly selected clusters $(1,...,c_g)$, each of which into two new clusters. Let us assume that cluster $\mathbf{C}_s$ has been selected for mutation. Initially, an object $s_1$ of $\mathbf{C}_s$ is randomly chosen. This object will be used as a *seed* to generate a new cluster, $\mathbf{C}_s'$. The farthest object ($s_2 \in \mathbf{C}_s$) from $s_1$ is selected as the seed to generate another new cluster, $\mathbf{C}_s''$. Then, the objects of $\mathbf{C}_s$

closer to $s_1$ are allocated to $\mathbf{C}_s^{'}$, whereas those objects closer to $s_2$ form cluster $\mathbf{C}_s^{''}$. Fig. 3 provides the pseudo code for *Mutation Operator 2* (MO$_2$).

In summary, some genotypes (candidate solutions) may be slightly changed (with just a few clusters being affected, perhaps one or two), whereas others may be strongly modified (with several clusters being affected, which means a large jump in the search space). This is a desirable property for an evolutionary algorithm, since it contributes to keep diversity of the population, thus causing the algorithm to be less susceptible to get stuck at local optima.

---

Let **g** be a given genotype that represents $c_g$ clusters;
1. Randomly generate a number $k \in \{1,..., c_g\}$;
2. For $i = (1,...,k)$ do:
    2.1 Randomly choose a cluster $\mathbf{C}_s \in \mathbf{g}$;
    2.2 If $|\mathbf{C}_s| > 2$ then: // $\mathbf{C}_s$ must have more than two objects to be eligible for this operator //
        2.2.1 Randomly choose an object $s_1 \in \mathbf{C}_s$;
        2.2.2 Determine the farthest object $s_2 \in \mathbf{C}_s$ from $s_1$ ;
        2.2.3 Generate two new clusters $\mathbf{C}_s^{'}$ and $\mathbf{C}_s^{''}$, placing the object(s) of $\mathbf{C}_s$ closer to $s_1$ into $\mathbf{C}_s^{'}$ and the object(s) closer to $s_2$ into $\mathbf{C}_s^{''}$ ;
    Else do not split $\mathbf{C}_s$;
    End For.

---

**Fig. 3** Pseudo code for Mutation Operator 2 (MO$_2$)

## 3.2 FCM as a Local Search Engine

EAC [33] originally makes use of the popular $k$-means algorithm as a local search procedure, which performs a fine-tuning of partitions obtained by the evolutionary search, thus speeding up EAC convergence. Since FCM is an extension of $k$-means to the fuzzy domain, it can be used to adapt EAC to fuzzy clustering, as detailed in [13]. To do so, the local search is performed by FCM, which now substitutes $k$-means in the EAC adapted for fuzzy clustering. There is a synergy between FCM and the EAC operators. On the one hand, the EAC operators can lessen the two main drawbacks of FCM, namely: (i) it may get stuck at suboptimal prototypes; and (ii) the user has to specify the number of prototypes. Since the EAC operators can eliminate, split, and merge clusters through an evolutionary search, they are able to evolve better partitions in terms of both the number of clusters and prototypes. These partitions may provide better initial prototypes for FCM, thus reducing the probability of getting stuck at suboptimal solutions. On the other hand, FCM minimizes the variances of the clusters achieved by the EAC operators, thus yielding to more compact clusters. Clearly, FCM also provides the additional information about the spatial distribution of the data contained in the fuzzy partition matrix – equations (1) and (3). Finally, since that the evolutionary search performed by EAC-FCM favors a cumulative refinement of the fuzzy partitions, a few FCM iterations are expected to suffice at each local search. This hypothesis – which was only conjectured in [13] – will be illustrated by means of experiments in Section 4 of the present chapter.

*Remark 3*: In the original EAC, $k$-means plays another peripheral yet important role besides the local search procedure previously described. By construction, it automatically discards useless (non-representative) prototypes that have been placed farther from every object in the data set when compared to the other prototypes. In other words, if no object belongs to a certain (null) cluster in the initial iteration of $k$-means, the corresponding prototype will no longer exist in the subsequent iterations. This feature helps the evolutionary search to decrease an overestimated initial number of clusters much more quickly [33] and is also helpful to keep such a number under control when the fitness function adopted exhibits a secondary peak for very large numbers of clusters. In principle, EAC-FCM is not endowed with this capability, since every object always belongs to every fuzzy cluster in FCM (membership being just a matter of degree). A very simple, yet useful, heuristic can however be introduced into EAC-FCM in order to mimic $k$-means' ability to eliminate non-representative clusters. It starts by checking for non-representative fuzzy clusters after FCM has been run as a local search procedure. A fuzzy cluster is deemed *non-representative* if no object in the data set belongs more to it than to the other fuzzy clusters. Then, the corresponding prototypes are ruled out and FCM is run again with the remaining ones only.

The heuristic described in Remark 3 has shown to be very effective in previous studies [1][13]. However, its use can clearly affect the number of local search (FCM) iterations that are performed at each stage of the evolutionary search. Once the amount of local search is the main subject of investigation here in this chapter, this heuristic will not be adopted in the experiments reported in Section 4.

## 3.3 *Fitness Function*

In principle, any relative clustering validity criterion (e.g. see [48][26]) that is not monotonic with the number of clusters can be potentially used as a fitness function for an evolutionary algorithm designed to optimize the number of clusters. The fitness function (objective function) originally employed by EAC is a simplified version of the Average Silhouette Width Criterion (*ASWC*) proposed by Kaufman and Rousseeuw [39]. In order to define this very simple, intuitive and effective criterion, consider a data object $j \in \{1,2,...,N\}$ *belonging* to cluster $r \in \{1,...,c\}$. In the context of hard partitions produced by a prototype-based clustering algorithm (e.g., $k$-means), this means that object $j$ is closer to the prototype of cluster $r$ than to any other prototype. In the more general context of fuzzy partitions, on the other hand, this means that the membership of the $j$th object to the $r$th fuzzy cluster, $p_{rj}$, is higher than the membership of this object to any other fuzzy cluster, i.e., $p_{rj} > p_{qj}$ for every $q \in \{1,...,c\}, q \neq r$.

Let the average distance of object $j$ to all other objects *belonging* to cluster $r$ be denoted by $a_{rj}$. Also, let the average distance of this object to all objects *belonging* to another cluster $q$, $q \neq r$, be called $d_{qj}$. Finally, let $b_{rj}$ be the minimum $d_{qj}$ computed over $q = 1, ..., c, q \neq r$, which represents the dissimilarity of object $j$ to its closest neighboring cluster. Then, the *silhouette* of object $j$ is defined as:

$$s_j = \frac{b_{rj} - a_{rj}}{\max\{a_{rj}, b_{rj}\}} \qquad (5)$$

where the denominator is used just as a normalization term. Clearly, the larger $s_j$, the better the assignment of object $j$ to cluster $r$. In case $r$ is a singleton, i.e., if it is constituted uniquely by object $j$, then the silhouette of this object is defined as $s_j = 0$ [39]. This prevents $ASWC$, defined as the average of $s_j$ over $j = 1, 2, ..., N$, i.e.

$$ASWC = \frac{1}{N} \sum_{j=1}^{N} s_j \qquad (6)$$

from finding the trivial solution $c = N$, with each object of the data set forming a cluster on its own. Therefore, the best partition is achieved when $ASWC$ in (6) is maximized, which implies minimizing the intra-cluster distance $a_{rj}$ while maximizing the inter-cluster distance $b_{rj}$.

A problem with the $ASWC$ measure is that it depends on the highly intensive computation of all distances among all data objects. In order to get around this problem, Hruschka et al. [33] proposed to replace the terms $a_{rj}$ and $b_{rj}$ in equation (5) with simplified versions of them based on the distances among the objects and the prototypes of the corresponding clusters. This modification has shown not to degrade accuracy while being able to significantly reduce the computational burden from $O(N^2)$ to $O(N)$. Moreover, it does not change the dependency of $ASWC$ on average distances (in this case represented by the prototypes), which is a desirable property concerning robustness to noise in the data [7].

The original and simplified (prototype-based) versions of $ASWC$ – equation (6) – can be used to evaluate fuzzy partitions [18][39]. To do so, however, they do not make explicit use of the fuzzy partition matrix in their calculations. Instead, the fuzzy partition matrix $P = [p_{ij}]_{c \times N}$ is used only to impose to the data set a hard partition matrix $H = [h_{ij}]_{c \times N}$ to which the $ASWC$ measure can be applied. Specifically, partition matrix $H$ is such that $h_{ij} = 1$ if $i = \arg \max_l p_{lj}$ ($l \in \{1, ..., c\}$) and $h_{ij} = 0$ otherwise. Consequently, $ASWC$ may not be able to discriminate between overlapped data clusters – even if these clusters have each their own (distinct) regions with higher data densities – since it neglects information contained in the fuzzy partition matrix $P$ on degrees to which clusters overlap one another. This information can be used to reveal those regions with high data densities (if any) by stressing importance of data objects concentrated in the vicinity of the cluster prototypes while reducing importance of objects lying in overlapping areas. To do so, a generalized silhouette criterion, named Fuzzy Silhouette ($FS$), has been defined as [14]:

$$FS = \frac{\sum_{j=1}^{N} \left(p_{rj} - p_{qj}\right)^{\alpha} s_j}{\sum_{j=1}^{N} \left(p_{rj} - p_{qj}\right)^{\alpha}} \qquad (7)$$

where $p_{rj}$ and $p_{qj}$ are the first and second largest elements of the $j$th column of the fuzzy partition matrix, respectively, $\alpha \geq 0$ is a weighting coefficient, and $s_j$ is the silhouette of object $j$ according to equation (5) – possibly in its (faster) prototype-based version. The prototype-based version of $FS$ in (7) will be hereafter referred to as Simplified Fuzzy Silhouette ($SFS$).

The influence of the exponent $\alpha$ on the behavior of the fuzzy silhouette in (7) has been conceptually discussed in [14]. Note that if $\alpha$ is set equal to zero, than the original (crisp) silhouette is recovered. The existence of this exponent gives rise to an interesting (and possibly exclusive) feature of $FS$: it can be adjusted to different degrees of fuzziness, e.g. induced by different values of the fuzzifier $m$ of FCM (see Section 2). However, an appropriate range of values for $\alpha$ (possibly as a function of $m$) has not yet been determined and is an interesting open question still to be addressed. The unit value ($\alpha = 1$) was arbitrarily adopted as a blind choice in the original study involving the EAC-FCM algorithm [13]. Such an arbitrary choice was made because that study was essentially focused on computational efficiency issues, rather than on efficacy issues. The experiments in the present chapter are mainly intended to illustrating some additional efficiency aspects not explored in [13], so the same value is adopted here.

Regardless the possible benefits and drawbacks associated with the presence of exponent $\alpha$ in $FS$, there are many different fuzzy clustering validity criteria whose particular features may make each of them capable of outperforming the others in specific classes (or instances) of problems [50][14]. For this reason, practical applications should in principle rely on multiple validity criteria (e.g. a voting scheme), rather than on just one [50][14]. Despite this, some of these criteria have shown satisfactory individual results in a wide variety of different application scenarios [34]. For the illustrative purposes of the present chapter, only the prototype-based fuzzy silhouette ($SFS$) with $\alpha = 1$ will be adopted.

## 3.4  Initial Population, Selection, Main Steps, and Asymptotic Computational Time Complexity

The population of the EAC-FCM can be initialized by randomly drawing both the number ($c$) and the locations of the initial prototypes. Then, each string component takes a value from the set $\{1, 2, ..., c\}$ according to the distance of the corresponding object to the closest prototype. This approach favors diversity in the initial population, by creating initial partitions representing different numbers of clusters. Overestimated and underestimated values for $c$ can be dealt with by EAC-FCM, which is able to decrease or increase this initial number of clusters towards a better estimate – according to the information provided by the fitness function. In what concerns selection, in principle, any selection operator (e.g. roulette wheel [15][21]) can be used. In the original study involving the EAC-FCM algorithm [13], the well-known ($\mu+\lambda$) deterministic selection strategy [21] was adopted. To keep consistency with the original study, the same strategy will be adopted in this chapter. However, it is worth warning that the use of a deterministic selection operator makes it theoretically possible (though expectedly rare in practice) that the evolutionary search gets stuck at a local optimum. Indeed, the algorithm might get stuck if all the candidate solutions (population of genotypes) converge to a local optimum in the space of possible solutions from where no solution with better fitness value can be obtained from the application of the (possibly very aggressive) mutation operators $MO_1$ and $MO_2$ in Figures 2 and 3, respectively. Clearly, a non-deterministic selection operator could easily rule out such a possibility.

In summary, the main steps of the algorithm used here to evolve fuzzy clusters are listed in Fig. 4.

---

1. Initialize a population of random genotypes;

2. Apply Fuzzy $c$-Means (FCM) to each genotype;

3. Evaluate each genotype according to the objective function;

4. Apply a linear normalization [15];

5. Select genotypes;

6. Apply the mutation operators;

7. Replace the old genotypes by the ones formed in step 6;

8. If convergence is attained, stop; else, go to step 2.

---

**Fig. 4** Main steps of the EAC-FCM algorithm

The evolutionary algorithm summarized in Fig. 4 basically runs FCM for partitions that evolve through mutation and selection operators. Such operators use probabilistic rules to process partitions sampled from the search space. Roughly speaking, more fitted partitions (i.e., those with higher fitness function values) have higher probability of being sampled. In other words, the evolutionary search is biased towards more promising clustering solutions. Therefore, the evolutionary search tends to perform a more computationally efficient search than traditional approaches based on running the clustering algorithm (e.g., FCM) multiple times for different number of clusters (e.g. see [13] for a detailed discussion on this topic). In brief, the overall computational cost of the evolutionary algorithm summarized in Fig. 4 is (over) estimated as $O(G \cdot P_s \cdot t \cdot N \cdot \hat{c}_{max}^2 \cdot n)$ in the worst-case, where $G$ is the number of generations, $P_s$ is the population size (number of genotypes), $t$ is the number of FCM iterations, $N$ is the number of data set objects, $\hat{c}_{max}$ is the maximum number of clusters ever encoded by a genotype during the evolutionary search, and $n$ is the number of attributes. Quantities $G$ and $P_s$ are inversely interdependent on each other, since increasing the size of the evolutionary population typically reduces the number of generations needed for convergence and vice versa. Their product, $G \cdot P_s$, does not depend either on $N$ or on $\hat{c}_{max}$. Accordingly, the overall computational cost of EAC-FCM can be estimated in terms of the potentially critical quantities of the problem as $O(N \cdot \hat{c}_{max}^2)$ in the worst-case. If the linear time FCM implementation [41] that smartly runs in $O(t \cdot N \cdot c \cdot n)$ time were adopted, then the overall computational cost of EAC-FCM would become $O(N \cdot \hat{c}_{max})$ [13].

## 4 Illustrative Examples

Campello et al. [13] showed that the EAC-FCM described in Section 3 could be more computationally efficient than systematic (i.e., repetitive) approaches when the number of clusters in a data set is unknown. In this context, sensitivity analyses of configuration parameters of the repetitive and evolutionary approaches were reported in that reference. The experimental results reported in the present chapter illustrate a complementary topic that has not been empirically addressed in [13]. In particular, we here illustrate how the evolutionary search performed by EAC-FCM favors a cumulative refinement of the fuzzy partitions by showing that a very few FCM iterations (fewer than usual) suffice as an efficient local search procedure at each stage of the evolutionary process. The cumulative refinement is favored by the design of the evolutionary algorithm, which evolves data partitions in terms of both the number of clusters and their structures.

It is intuitive that the number of FCM iterations ($t$), as a local search procedure, has some influence on the performance of the evolutionary algorithm. Roughly speaking, the more accurate the estimates of the quality of the partitions from a given region of the search space (here defined in terms of both the number of clusters and positions of prototypes), the better the information provided to the (biased) sampling process performed by the evolutionary search, which tends to converge in fewer generations. The accuracy of the estimate of the quality of partitions from a given region of the search space depends on the number of FCM iterations ($t$). The larger the value of $t$, the better is such an estimate (to be provided to the next stage of the evolutionary process). However, larger values of $t$ imply more computationally demanding generations of the evolutionary search. Under this perspective, and aimed at clarifying the influence of the parameter $t$ in the overall performance of the evolutionary algorithm, let us first elaborate on the likely consequences of setting $t$ to overestimated or underestimated values by considering two different stages of the evolutionary process. On the one hand, it is expected that, if the value of $t$ is overestimated, then the local search engine (i.e., FCM) may spend significant computational time refining unpromising data partitions that should be promptly discarded by the selection operator. This scenario may be particularly critical in early generations. On the other hand, if the number of FCM iterations is underestimated, then, in principle, it is reasonable to expect that potentially good clustering solutions may be neglected by the evolutionary search due to a lack of refinement of such solutions before they are evaluated by the objective function. This scenario may be particularly critical in late generations. In summary, overestimating the value of $t$ may waste a lot of computational time, whereas underestimating the value of $t$ may mislead the evolutionary search when it is about to converge to near-optimal solutions.

The underestimation of $t$ has two possible consequences. The first one is merely the need for a larger number of generations to find an arbitrarily accurate solution. The second (and more serious one) is the impossibility of finding such a solution. Indeed, it is theoretically possible that the algorithm will no longer be able to find a given arbitrarily accurate solution, which could be found for some sufficiently "high" value of $t$, if $t$ is underestimated. The reason is that, depending on how the conceptual algorithm described in Section 3 is actually implemented in practice, the cumulative refinement of the local search may not take place, e.g., due to the approximation of prototypes by cluster centroids at each new generation. In brief, if $t$ is too small and all the resulting FCM prototypes at a given generation are discarded and replaced with cluster centroids computed directly from the crisp genotypes in the next generation, no matter whether the corresponding clusters have been affected by mutation or not, then the non-mutated centroids may retreat precisely to the same values they had before FCM was applied. In this case, the local search is useless and the crisp mutation operators alone may not be able to meet the requirements in terms of the expected fuzzy partitions. This problem will be discussed in Sections 4.3 and 4.4.

In summary, one may hypothesize that, ideally, the value of $t$ should be gradually increased from early to late generations of the evolutionary process. However, it is not clear how this idea could be put into practice in a principled way, particularly because speed of convergence of evolutionary algorithms is problem dependent. If care is taken when implementing the algorithm, in such a way that the eventual critical effect of the underestimation of $t$ mentioned in the previous paragraph is eliminated (see Section 4.4), then one may conjecture that underestimating the optimal value for $t$ is likely better than overestimating it. The reason is that additional computations associated with the greater number of generations that will be required for convergence may be at least partially counterbalanced by the lower processing time of each individual generation. Taking into account such a conjecture and having in mind that EAC-FCM favors a cumulative refinement of the fuzzy partitions throughout generations, we feel that the use of small values for the number of FCM iterations ($t$) should be preferred, as illustrated in Section 4.3. Before proceeding with the results, however, some methodological aspects of the empirical evaluations conducted in this chapter are discussed in the sequel.

## 4.1 Experimental Methodology

The experimental methodology adopted here is based on the use of a baseline exhaustive-like algorithm, called *Ordered Multiple Runs of FCM* (OMR-FCM), to derive reference partitions with high objective function values as goals to be achieved by EAC-FCM. OMR-FCM runs FCM repeatedly for an increasing number of clusters ($c$). For each value of $c$, a number of partitions achieved by FCM are assessed by means of some validity index (here the Simplified Fuzzy Silhouette – *SFS*), for which the best obtained value is kept for further reference. After running FCM multiple times for every value of $c$ in a given range, the best obtained partition (according to the validity index) is chosen. In the present context, this means a peak of the *SFS* criterion. This procedure has been widely used in practice to estimate the number of clusters in data[2] and is summarized by the algorithmic framework depicted in Fig. 5.

Considering the OMR-FCM algorithm depicted in Fig. 5, the methodology here adopted to assess the influence of the number of local search iterations ($t$) on the computational efficiency of the evolutionary algorithm can be summarized by means of the following steps:

1. Run OMR-FCM and store the value $V_{VC}*$ of the clustering validity criterion (*SFS*) for the best partition found by the algorithm (Fig. 5).
2. Run EAC-FCM (with a given value of $t$) until the reference value $V_{VC}*$ obtained in Step 1 (or greater) is found and store the corresponding computing time. For better confidence of the results, repeat this procedure a number of times with the same value of $t$ and, then, estimate the average computing time.
3. Repeat step 2 for different values of $t$ and compare the resulting average computing times.

---

Let $c_{max}$ be the maximum acceptable number of clusters, $c^*$ be the number of clusters to be estimated by the algorithm, $S_C$ be the stopping criterion for a single run of FCM, $V_{VC}$ be the value of the validity criterion after a single run of FCM, $V_{VC}^*$ be the value of the validity criterion for the best partition found by the algorithm, and $n_p$ be the number of different partitions generated for each number $2 \leq c \leq c_{max}$ of clusters. Let us also assume that the smallest possible value for $V_{VC}$ is $V_{min}$ (zero in case of a prototype-based silhouette criterion). Then, OMR-FCM can be summarized as follows:

Step 1: Choose $c_{max}$ and $S_C$;
Step 2: $V_{VC}^* \leftarrow V_{min}$;
Step 3: For $c=2, ..., c_{max}$ do:
      {
      For $i=1, ..., n_p$ do:
         {
         3.1 Generate a random partition with $c$ clusters;
         3.2 Run FCM until $S_C$ is met;
         3.3 Compute $V_{VC}$ for the resulting partition;
         3.4 If ($V_{VC} > V_{VC}^*$) then
            {
            $V_{VC}^* \leftarrow V_{VC}$;
            $c^* \leftarrow c$;
            Hold the resulting partition $S$ for $c^*$;
            }
         } // End loop for $i$ //
      } // End loop for $c$ //
Step 4: Return {$V_{VC}^*, c^*$, and the corresponding partition $S$ for $c^*$}

**Fig. 5** *Ordered Multiple Runs of FCM (OMR-FCM)*

---

[2] A well-known, analogous approach involves recognizing a knee in a chart that depicts the number of clusters versus some measure of their compactness.

These steps provide a way of making EAC-FCM find partitions of equivalent quality in relation to those found by OMR-FCM. As far as the value for the maximum number of acceptable clusters in OMR-FCM is concerned, a commonly used rule of thumb is to set $c_{max} = N^{1/2}$ [50][49]. This rule is adopted here. The stopping criterion for a single run of FCM in OMR-FCM, $S_C$, is that the maximum absolute difference between elements of the partition matrix in two consecutive iterations be lower than a given positive threshold $\varepsilon=10^{-3}$. The number of different partitions generated for each number $2 \leq c \leq c_{max}$ of clusters in OMR-FCM is $n_p = 50$. For better confidence of the results (see step 3), 30 runs of EAC-FCM have been performed for each value of $t \in \{1, 3, 5, 7, 10, 15\}$ on each data set to be described in Section 4.2.

## 4.2 Data Sets

The first data set used in the experiments is formed by 900 objects distributed in nine balanced, overlapping clusters, which were randomly generated from bidimensional Gaussian distributions with standard deviations equal to 0.5 and centers (x, y) given in the left column of Table 1. This data set, now referenced as *Gauss9*, is depicted in Fig. 6. The nine clusters of *Gauss9* can be recognized by visual inspection.

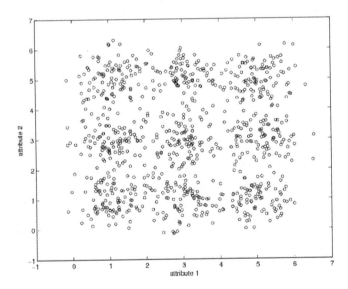

**Fig. 6** *Gauss9* data set [32][13][14]

The second data set has five clusters with 100 objects each. The clusters were generated according to a multivariate Gaussian distribution with standard deviation $\sigma = 0.1$. The cluster centers were generated according to the following steps: (i) Randomly generate the first cluster center; (ii) The remaining cluster centers are also randomly generated, but with the constraint that every new center must have Euclidean distance equal to or greater than 4·σ in relation to the centers previously generated. In addition, at least one of such distances must be equal to or less than 5·σ. The data set generated according to this procedure will be named here *fbase_5c_5d*. Each center has five dimensions and their coordinates are presented in the left column of Table 1.

Finally, the third data set adopted here has been generated as described in the study by Milligan and Cooper [47]. This data set, from now referred to as *mc_10c_5d*, contains 1,000 objects (described by 5 attributes) equally distributed into ten clusters, whose centers are also presented in the left column of Table 1.

**Table 1** Actual cluster centers and FCM prototypes resulting from OMR-FCM (three data sets)

| Gauss9 | |
|---|---|
| Original Centers | OMR-FCM Prototypes |
| (1, 1) (1, 3) (1, 5) (3, 1) (3, 3) (3, 5) (5, 1) (5, 3) (5,5) | (1.05, 0.96) (1.07, 2.93) (1.08, 5.05) (3.04, 1) (3.04, 2.9) (3.05, 5.13) (4.98, 1) (5.16, 3.07) (5.06, 5.02) |
| fbase_5c_5d | |
| Original Centers | OMR-FCM Prototypes |
| (0.0568, 0.2194, 0.8310, 0.0195, 0.8646) | (0.0399, 0.2377, 0.8176, 0.0317, 0.8969) |
| (-0.1888, 0.4724, 0.6568, 0.1152, 1.0053) | (-0.1694, 0.4517, 0.6581, 0.1147, 0.9964) |
| (0.1109, 0.0390, 0.8220, 0.2589, 1.1902) | (0.1126, 0.0694, 0.8258, 0.2488, 1.1773) |
| (0.2901, 0.3493, 0.7650, 0.1267, 1.3551) | (0.2716, 0.3314, 0.7735, 0.1312, 1.3265) |
| (-0.1647, 0.4022, 0.8906, -0.1037, 0.5818) | (-0.1532, 0.3998, 0.8929, -0.0964, 0.6102) |
| mc_10c_5d | |
| Original Centers | OMR-FCM Prototypes |
| (0.0454, 0.3007, 0.7534, 0.1249, 0.5322) | (0.0454, 0.3001, 0.7533, 0.1243, 0.5326) |
| (0.1392, 0.8499, 0.9159, 0.8988, 0.5774) | (0.1393, 0.8497, 0.9159, 0.8992, 0.5773) |
| (0.1992, 0.2062, 0.0812, 0.1058, 0.8553) | (0.1996, 0.2061, 0.0801, 0.1060, 0.8552) |
| (0.2625, 0.3114, 0.8227, 0.1035, 0.9346) | (0.2625, 0.3114, 0.8228, 0.1038, 0.9353) |
| (0.3498, 0.4036, 0.1285, 0.1280, 0.6419) | (0.3495, 0.4032, 0.1285, 0.1277, 0.6420) |
| (0.4719, 0.5865, 0.2777, 0.7566, 0.6020) | (0.4717, 0.5865, 0.2776, 0.7566, 0.6019) |
| (0.6032, 0.6564, 0.9561, 0.1076, 0.7757) | (0.6031, 0.6562, 0.9560, 0.1077, 0.7757) |
| (0.7470, 0.9204, 0.3086, 0.4693, 0.0842) | (0.7469, 0.9206, 0.3087, 0.4697, 0.0843) |
| (0.8656, 0.0590, 0.5029, 0.7005, 0.9510) | (0.8656, 0.0588, 0.5027, 0.7003, 0.9509) |
| (0.9679, 0.1588, 0.5417, 0.3571, 0.0757). | (0.9678, 0.1587, 0.5418, 0.3573, 0.0757) |

## 4.3 *Experimental Results*

Except for the number of local search iterations, $t$, which is under investigation here, the remaining parameters of EAC-FCM are set in accordance to satisfactory values observed in the study performed in [13]. In particular, the population size is set to $P_s = 10$ and the proportions of the $(\mu+\lambda)$ selection procedure are set to $\mu = 4$ and $\lambda = 6$. Following the methodology addressed in Section 4.1, OMR-FCM found the values of $V_{VC}^*$ and $c^*$ reported in Table 2. As described in Section 4.2, three data sets for which the right data partitions are known have been used in the experiments. OMR-FCM recovered clustering structures that fit the data very well with regard to the known data partitions. For the sake of illustration, the prototypes of the best solution provided by the OMR-FCM algorithm are shown in the right column of Table 1.

EAC-FCM also recovered the expected clusters in most cases. Only a few runs of the algorithm led to partitions with the target value $V_{VC}$* or larger, but with too many clusters ($c$ much larger than $c$*). Such anomalous results have not been observed in the original (crisp) EAC algorithm [35][33][31] and the reasons behind them are manifold: (i) The important heuristic described in *Remark 3* has not been adopted in the experiments reported here, for the reasons already explained in the end of Section 3.2[3]; (ii) EAC-FCM persistently succeeded in achieving the goal ($V_{VC}$*), but the goal here relies on a single validity criterion (*SFS*), rather than on a pool of criteria. Unlike conjectured in [14], the authors realized further that *SFS* might actually exhibit a secondary peak at partitions with too many clusters, where "too many" depends both on the data set and on the value of the exponent $\alpha$ in (7). For instance, "too many" means over 100 clusters for the *Gauss9* data set in Figure 6 when $\alpha=1$. It is worth remarking that *SFS* has been originally designed to detect clusters as regions of high data densities. So, if such regions do not exist or are not very clear – e.g. for the data have relatively many dimensions with respect to the number of objects – then the criterion may tend to point out a large number of small sub-clusters as a good solution. Historically, such a behavior has not been observed in the original, crisp silhouette criterion ($\alpha = 0$) and deserves further investigations; (iii) The implementation of the algorithm for the experiments reported here in this section discards, at each new generation, the FCM prototypes of the clusters not affected by mutation, replacing them with the corresponding centroids (computed directly from the recovered genotypes). As discussed in the beginning of Section 4, this implementation strategy may lose information by approximating prototypes with centroids, thus being subject to mislead the evolutionary search; and (iv) Unlike OMR-FCM, EAC-FCM was executed without any constraint imposed to the maximum acceptable number of clusters;

All the previously mentioned drawbacks could be properly circumvented if desired, as already discussed in different points of this text. However, since none of them affects the analyses under interest here, they will be just ignored in the remaining of this section.

**Table 2** Optimal values for *SFS* ($V_{VC}$*) and number of clusters ($c$*) found by OMR-FCM

| Data Set | $V_{VC}$* | $c$* |
|:---:|:---:|:---:|
| *Gauss9* | 0.712 | 9 |
| *fbase_5c_5d* | 0.536 | 5 |
| *mc_10c_5d* | 0.893 | 10 |

---

[3] Using such a heuristic to discard non-representative fuzzy clusters has made EAC-FCM able to find the expected numbers of clusters in all the experiments reported in [1][13].

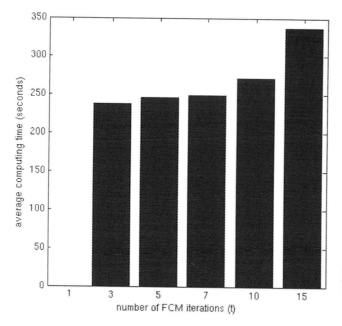

**Fig. 7** Average computing times for *Gauss9*

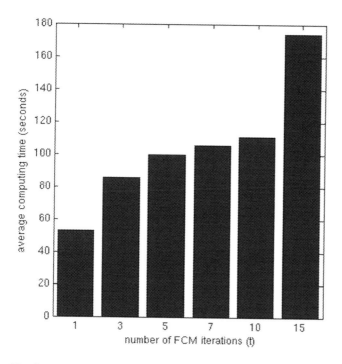

**Fig. 8** Average computing times for *fbase_5c_5d*

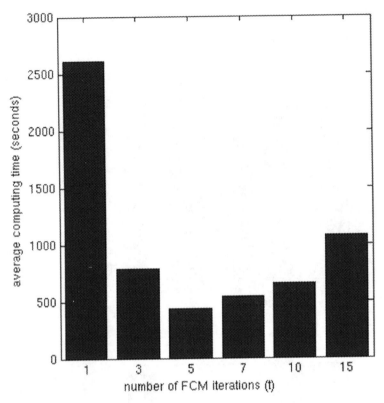

**Fig. 9** Average computing times for *mc_10c_5d*

In order to illustrate the influence of the number of FCM iterations ($t$) on EAC-FCM, this algorithm has been executed 30 times for each $t \in \{1, 3, 5, 7, 10, 15\}$ over each of the employed data sets. All the experiments involving a given data set have been  run in a single computer, so that they can be compared to each other, but experiments involving different data sets have been run in different computers. During the experiments, the computers ran exclusively the operational system and the Matlab environment. Figures 7, 8, and 9 illustrate the average computing times achieved.

Due to the "naïve" implementation of the algorithm used here, in which FCM prototypes are replaced with centroids at each new generation, EAC-FCM was unable to find the referential objective value $V_{VC}^{*}$ in the experiments with the *Gauss9* data set and $t = 1$, at least not in less than a maximum limit of 500 generations. For this reason, the corresponding average computing time has been omitted in Figure 7. Note that this is a practical example of the critical problem that can take place with this sort of implementation of EAC-FCM when the value of $t$ is too underestimated, as previously envisaged in the introductory part of Section 4. A similar problem occurred with the *mc_10c_5d* data set for some experiments

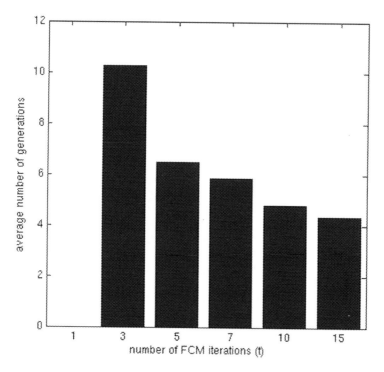

**Fig. 10** Average numbers of generations for *Gauss9*

with $t = 1$, but these represented less than 15% of the amount of experiments with that data and, so, the corresponding result has not been omitted in Figure 9.

Generally speaking, the results in Figures 7, 8, and 9 show that fewer FCM iterations may lead to better (or at least comparable) performances of the evolutionary algorithm, except for $t = 1$ in *Gauss9* and *mc_10c_5d*, for the reasons previously explained. In particular, three FCM iterations have shown to favor a more efficient search for the referential objective value established by OMR-FCM in the case of the *Gauss9* data set. More precisely, this result suggests that, for this data set, sufficiently accurate estimates of the quality of the partitions from a given region of the search space (in terms of both the number of clusters and positions of prototypes) can be obtained by setting $t = 3$, minimizing as much as possible the FCM computational burden and, thus, favoring the minimization of the overall computational cost of EAC-FCM.

The experiments performed with *fbase_5c_5d* and *mc_10c_5d* also illustrate how EAC-FCM can indeed favor a cumulative refinement of the fuzzy partitions throughout generations. For *fbase_5c_5d*, a single FCM iteration ($t = 1$) has shown to provide more computationally efficient evolutionary searches. For *mc_10c_5d*, in its turn, $t = 5$ has shown to be the best value, possibly due to the higher complexity of this data set in terms of both the number of clusters (10) and the number of attributes (5). It is intuitive that more complex data sets tend to demand more accurate refinements of the solutions.

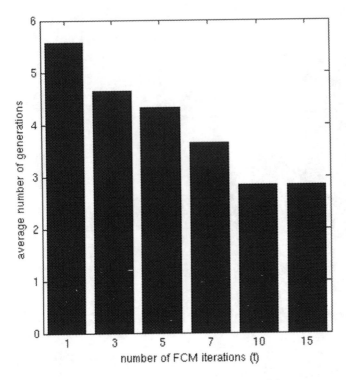

**Fig. 11** Average numbers of generations for *fbase_5c_5d*

By looking at Figure 8, one may wonder if such data set is not too simple, so that the referential objective value $V_{VC}^{*}$ can be found from a single iteration of FCM ($t = 1$) assuming that the correct number of clusters is given. To investigate such a possibility, we have performed 500 independent runs of FCM (with $c = c^{*}$ = 5 and $t = 1$) for *fbase_5c_5d*. *SFS* values equal to or larger than those found by EAC-FCM were not obtained in such 500 FCM runs. This result suggests that it is unlikely that the partitions found by EAC-FCM are outcomes of pure chance, especially when noticing that the average number of FCM iterations performed by EAC-FCM over 30 runs with this data set for $t = 1$ is about 55 (11% of the 500 independent runs that were not able to reach $V_{VC}^{*}$). In other words, this result strongly suggests that EAC-FCM does favor a cumulative refinement of the fuzzy partitions throughout generations of the evolutionary search. Note that this can be true even using the naïve implementation of the algorithm adopted here, provided that $t$ is "large" enough, so that the improvements achieved by local search are at least partially transferred to the subsequent generations of the evolutionary process. Preliminary experiments with a little more sophisticate implementation of the algorithm have shown that this can be true for any value of $t$ (see Section 4.4).

Recalling the discussions in the introductory part of Section 4, there is a trade-off between the number of generations $G$ of EAC-FCM and the number of FCM iterations, $t$. In order to illustrate this relationship, Figures 10, 11, and 12 present

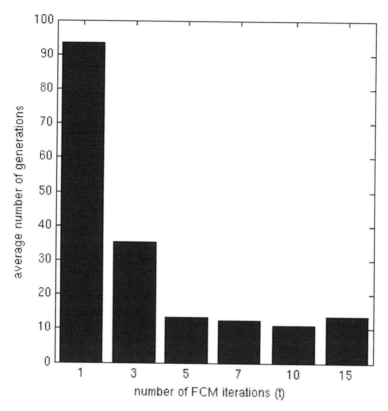

**Fig. 12** Average numbers of generations for *mc_10c_5d*

the average numbers of generations for distinct values of *t*. As expected, smaller values of *t* demand more generations to improve partitions toward the referential *SFS* values. Note, however, that the computing times in Figures 7, 8, and 9 are not just linear functions of *G* and *t*. Among other reasons, it follows that the FCM implementation adopted here has squared time complexity with respect to the number of clusters *c*. Accordingly, the computing times of EAC-FCM are non-linearly related to the numbers of clusters encoded into the population of geno-types throughout the evolutionary process.

To conclude, Figures 13, 14, and 15 show typical convergence curves for EAC-FCM. These figures show that the algorithm converges very quickly to the right number of clusters and to the referential (aimed) objective function value.

## 4.4 Complementary Results

Figure 7 has shown that, due to the implementation of the algorithm used up to this point, in which FCM prototypes are replaced with centroids at each new generation of the evolutionary search, EAC-FCM was unable to find the referen-tial objective value $V_{VC}^*$ in the experiments with the *Gauss9* data set and *t* = 1, at

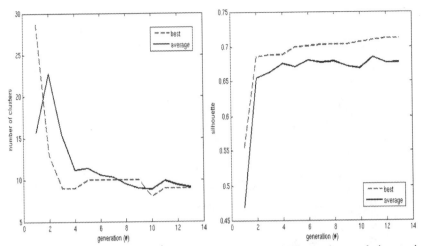

**Fig. 13** Average and best values for the no. of clusters and *SFS* over the population: typical curves for *Gauss9* and *t* =3

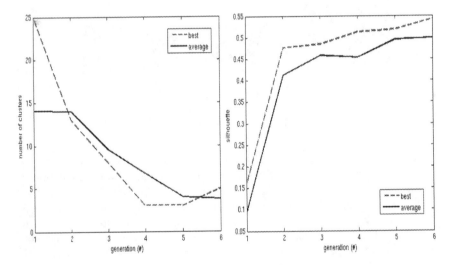

**Fig. 14** Average and best values for the no. of clusters and *SFS* over the population: typical curves for *fbase_5c_5d* and *t* =1

least not in less than a maximum limit of 500 generations. For this reason, a complementary experiment over the same data set is performed here using a little more sophisticate implementation of the algorithm that uses some auxiliary data structures to keep the information on the prototypes between two consecutive generations. The results are shown in Figure 16. Comparing Figures 7 and 16 makes it clear the overall improvement in performance, especially for smaller values of *t*. In particular, the algorithm with *t* = 1 not only became able to achieve the goal, but became the optimal choice in terms of computational efficiency.

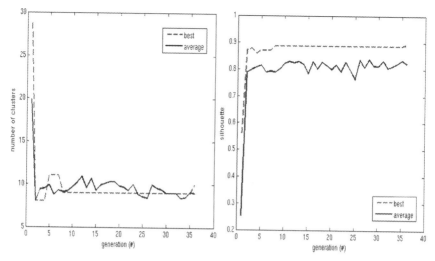

**Fig. 15** Average and best values for the no. of clusters and *SFS* over the population: typical curves for *mc_10c_5d* and *t* =3

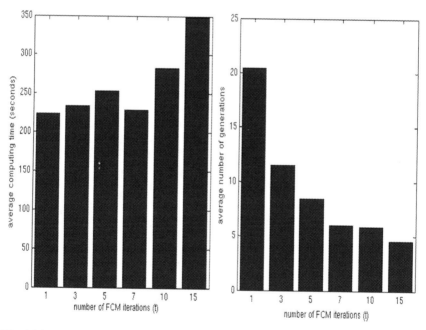

**Fig. 16** Average computing times for *Gauss9*: alternative implementation of EAC-FCM

# 5  Conclusions and Future Trends

This chapter has presented a brief overview of evolutionary algorithms designed for fuzzy clustering and has described an algorithm, called EAC-FCM, that uses the popular Fuzzy c-Means (FCM) method as a local search engine. In a previous study, EAC-FCM has shown to be more computationally efficient than systematic (i.e., repetitive) approaches when the number of clusters in a data set is unknown. This chapter has investigated a complementary aspect of the algorithm that has not been empirically addressed in [13]. In particular, a preliminary study on the influence of an important parameter of EAC-FCM, which is also found in many other evolutionary algorithms for clustering and fuzzy clustering, has been presented. Such a parameter is the number of iterations of a given local search procedure to be performed at each generation of the evolutionary search. In the case of EAC-FCM, this parameter is the number of iterations of FCM ($t$). In [13], the authors have only conjectured (but not shown) that EAC-FCM favors a cumulative refinement of the fuzzy partitions throughout generations of the evolutionary search. Taking into account such a conjecture, a set of experiments have been performed on three data sets with EAC-FCM for several values of $t$. As expected, the evolutionary search performed by EAC-FCM favored a cumulative refinement of the fuzzy partitions and was able to find high quality referential fuzzy clustering solutions with an unusually small numbers of iterations of FCM (e.g. $t = 1$).

It has also been shown by preliminary experiments that the medium level implementation of the high level conceptual algorithm may have a significant impact on the performance of EAC-FCM. More specifically, it has been shown that, if the FCM prototypes at a given generation are discarded and replaced with cluster centroids computed directly from the crisp genotypes in the next generation, no matter whether the corresponding clusters have been affected by mutation or not, the local search may become useless for very small values of $t$ and the algorithm may become much less effective and/or computationally efficient. Preliminary experiments with a more sophisticated implementation that uses some auxiliary data structures to keep the information on the prototypes between two consecutive generations have shown promising results. In particular, this new implementation made the algorithm very efficient and effective with $t = 1$ in a scenario in which the original implementation had not been able to achieve the referential goal with such a value of $t$. This result suggests that the medium level implementation of the algorithm deserves further attention. A more extensive study on this topic, including alternative data structures to implement better interfaces between the encoding scheme, the mutation operators, and the local search operator, without losing important information on the fuzzy partitions and prototypes, is an interesting subject for future research. Other potential topic for future research concerns the fitness function. In particular, appropriate values or interval of values for the exponent $\alpha$ of the Fuzzy Silhouette must be determined. In addition, schemes to compose more robust fitness functions, based on the combination of multiple clustering validity criteria, also deserve further investigation.

**Acknowledgements.** The authors acknowledge the Brazilian Research Agencies CNPq and FAPESP for their financial support to this work.

# References

[1] Alves, V.S., Campello, R.J.G.B., Hruschka, E.R.: A Fuzzy Variant of an Evolutionary Algorithm for Clustering. In: Proc. IEEE Int. Conference on Fuzzy Systems, pp. 375–380 (2007)

[2] Arabie, L.J., Hubert, G., DeSoete, P.: Clustering and Classification. World Scientific, Singapore (1999)

[3] Babu, G.P., Murty, M.N.: Clustering with Evolution Strategies. Pattern Recognition 27, 321–329 (1994)

[4] Babuška, R.: Fuzzy Modeling for Control. Kluwer, Dordrecht (1998)

[5] Baldi, P., Brunak, S.: Bioinformatics - The Machine Learning Approach, 2nd edn. MIT Press, Cambridge (2001)

[6] Bertone, P., Gerstein, M.: Integrative Data Mining: The New Direction in Bioinformatics – Machine Learning for Analyzing Genome-Wide Expression Profiles. IEEE Engineering in Medicine and Biology 20, 33–40 (2001)

[7] Bezdek, J.C., Pal, N.R.: Some new indexes of cluster validity. IEEE Trans. on Systems, Man and Cybernetics – B 28, 301–315 (1998)

[8] Bezdek, J.C., Hathaway, R.J.: Optimization of Fuzzy Clustering Criteria using Genetic Algorithms. In: Proc. IEEE World Congress on Computational Intelligence, pp. 589–594 (1994)

[9] Bezdek, J.C., Hathaway, R.J., Howard, R.E., Wilson, C.A., Windham, M.P.: Local Convergence Analysis of a Grouped Variable Version of Coordinate Descent. Journal of Optimization Theory and Applications 54, 471–477 (1987)

[10] Bezdek, J.C., Hathaway, R.J., Sabin, M.J., Tucker, H.T.: Convergence Theory for Fuzzy C-Means: Counterexamples and Repairs. IEEE Trans. on Systems, Man and Cybernetics SMC-17, 873–877 (1987)

[11] Bezdek, J.C.: Pattern Recognition with Fuzzy Objective Function Algorithm. Plenum Press (1981)

[12] Bigus, J.P.: Data Mining with Neural Networks. McGraw-Hill, New York (1996)

[13] Campello, R.J.G.B., Alves, V.S., Hruschka, E.R.: On the Efficiency of Evolutionary Fuzzy Clustering. Journal of Heuristics, doi:10.1007/s10732-007-9059-6

[14] Campello, R.J.G.B., Hruschka, E.R.: A Fuzzy Extension of the Silhouette Width Criterion for Cluster Analysis. Fuzzy Sets and Systems 157(21), 2858–2875 (2006)

[15] Davis, L.: Handbook of Genetic Algorithms. International Thomson Computer Press (1996)

[16] de Oliveira, J.V., Pedrycz, W.: Advances in Fuzzy Clustering and its Applications. Wiley, Chichester (2007)

[17] Egan, M.A., Krishnamoorthy, M., Rajan, K.: Comparative Study of a Genetic Fuzzy C-Means Algorithm and a Validity Guided Fuzzy C-Means Algorithm for Locating Clusters in Noisy Data. In: Proc. IEEE World Congress on Computational Intelligence, pp. 440–445 (1998)

[18] Everitt, B.S., Landau, S., Leese, M.: Cluster Analysis. Arnold Publishers (2001)

[19] Falkenauer, E.: Genetic Algorithms and Grouping Problems. John Wiley & Sons, Chichester (1998)

[20] Fazendeiro, P., Valente de Oliveira, J.: A Semantic Driven Evolutive Fuzzy Clustering Algorithm. In: Proc. IEEE Int. Conference on Fuzzy Systems, pp. 1–6 (2007)

[21] Fogel, D.B.: Evolutionary Computation: Toward a New Philosophy of Machine Intelligence. IEEE Press, New York (1995)

[22] Fogel, D.B., Simpson, P.K.: Evolving Fuzzy Clusters. In: Proc. IEEE Int. Conference on Neural Networks, pp. 1829–1834 (1993)

[23] Fralley, C., Raftery, A.E.: How Many Clusters? Which Clustering Method? Answer via Model-Based Cluster Analysis. The Computer Journal 41, 578–588 (1998)

[24] Freitas, A.: A Review of Evolutionary Algorithms for Data Mining. In: Maimon, O., Rokach, L. (eds.) Soft Computing for Knowledge Discovery and Data Mining, pp. 61–93. Springer, Heidelberg (2007)

[25] Ghozeil, A., Fogel, D.B.: Discovering Patterns in Spatial Data using Evolutionary Programming. In: Proc. 1st Annual Conference on Genetic Programming, pp. 521–527 (1996)

[26] Halkidi, M., Batistakis, Y., Vazirgiannis, M.: On Clustering Validation Techniques. Journal of Intelligent Information Systems 17, 107–145 (2001)

[27] Hall, L.O., Bezdek, J.C., Boggavarpu, S., Bensaid, A.: Genetic Fuzzy Clustering. In: Proc. Annual Conference of the North American Fuzzy Information Processing Society (NAFIPS), pp. 411–415 (1994)

[28] Hall, L.O., Özyurt, B.: Scaling Genetically Guided Fuzzy Clustering. In: Proc. Int. Symposium on Uncertainty Modeling and Analysis & Annual Conference of the North American Fuzzy Information Processing Society (ISUMA-NAFIPS), pp. 328–332 (1995)

[29] Hall, L.O., Özyurt, I.B., Bezdek, J.C.: Clustering with a Genetically Optimized Approach. IEEE Trans. on Evolutionary Computation 3, 103–112 (1999)

[30] Höppner, F., Klawonn, F., Kruse, R., Runkler, T.: Fuzzy Cluster Analysis: Methods for Classification. In: Data Analysis and Image Recognition. Wiley, Chichester (1999)

[31] Hruschka, E.R., Campello, R.J.G.B., de Castro, L.N.: Clustering Gene-Expression Data: A Hybrid Approach that Iterates between k-Means and Evolutionary Search. In: Grosan, C., Abraham, A., Ishibuchi, H. (eds.) Hybrid Evolutionary Algorithms, pp. 313–335. Springer, Heidelberg (2007)

[32] Hruschka, E.R., Campello, R.J.G.B., de Castro, L.N.: Evolutionary Search for Optimal Fuzzy C-Means Clustering. In: Proc. Int. Conference on Fuzzy Systems, pp. 685–690 (2004)

[33] Hruschka, E.R., Campello, R.J.G.B., de Castro, L.N.: Evolving Clusters in Gene-Expression Data. Information Sciences 176, 1898–1927 (2006)

[34] Hruschka, E.R., Campello, R.J.G.B., Freitas, A.A., Carvalho, A.C.P.L.F.: A Survey of Evolutionary Algorithms for Clustering. IEEE Transactions on Systems, Man, and Cybernetics - Part C: Applications and Reviews (to appear)

[35] Hruschka, E.R., de Castro, L.N., Campello, R.J.G.B.: Evolutionary Algorithms for Clustering Gene-Expression Data. In: Proc. 4th IEEE Int. Conference on Data Mining, pp. 403–406 (2004)

[36] Jain, A.K., Dubes, R.C.: Algorithms for Clustering Data. Prentice-Hall, Englewood Cliffs (1988)

[37] Jain, A.K., Murty, M.N., Flynn, P.J.: Data Clustering: A Review. ACM Computing Surveys 31, 264–323 (1999)

[38] Jiang, D., Tang, C., Zhang, A.: Cluster Analysis for Gene Expression Data: A Survey. IEEE Trans. on Knowledge and Data Engineering 16, 1370–1386 (2004)

[39] Kaufman, L., Rousseeuw, P.J.: Finding Groups in Data – An Introduction to Cluster Analysis. Wiley Series in Probability and Mathematical Statistics (1990)

[40] Klawonn, F.: Fuzzy Clustering with Evolutionary Algorithms. In: Proc. of 7th Int. Fuzzy Systems Association (IFSA) World Congress, pp. 312–323 (1997)

[41] Kolen, J.F., Hutcheson, T.: Reducing the Time Complexity of the Fuzzy C-Means Algorithm. IEEE Trans. on Fuzzy Systems 10, 263–267 (2002)

[42] Liu, H., Li, J., Chapman, M.A.: Automated Road Extraction from Satellite Imagery using Hybrid Genetic Algorithms and Cluster Analysis. Journal of Environmental Informatics 1(2), 40–47 (2003)

[43] Liu, J., Xie, W.: A Genetics-Based Approach to Fuzzy Clustering. In: Proc. Int. Conference on Fuzzy Systems, pp. 2233–2240 (1995)

[44] MacQueen, J.B.: Some Methods of Classification and Analysis of Multivariate Observations. In: Proc. 5th Berkeley Symposium on Mathematical Statistics and Probability, pp. 281–297 (1967)

[45] Maulik, U., Bandyopadhyay, S.: Fuzzy Partitioning Using Real Coded Variable Length Genetic Algorithm for Pixel Classification. IEEE Trans. on Geosciences and Remote Sensing 41(5), 1075–1081 (2003)

[46] Mecca, G., Raunich, S., Pappalardo, A.: A New Algorithm for Clustering Search Results. Data and Knowledge Engineering 62, 504–522 (2007)

[47] Milligan, G.: A Monte Carlo Study of Thirty Internal Criterion Measures for Cluster Analysis. Psychometrika 46(2), 187–199 (1981)

[48] Milligan, G.W., Cooper, M.C.: An Examination of Procedures for Determining the Number of Clusters in a Data Set. Psychometrika 50, 159–179 (1985)

[49] Pakhira, M.K., Bandyopadhyay, S., Maulik, U.: A Study of some Fuzzy Cluster Validity Indices, Genetic Clustering and Application to Pixel Classification. Fuzzy Sets and Systems 155, 191–214 (2005)

[50] Pal, N.R., Bezdek, J.C.: On Cluster Validity for the Fuzzy c-Means Model. IEEE Transactions on Fuzzy Systems 3(3) (1995)

[51] Pantel, P.A.: Clustering by Commitee, PhD Thesis, Department of Computer Sciences of the University of Alberta, Canada (2003)

[52] Park, H.-S., Yoo, S.-H., Cho, S.-B.: Evolutionary Fuzzy Clustering Algorithm with Knowledge-Based Evaluation and Applications for Gene Expression Profiling. Journal of Computational and Theoretical Nanoscience 2, 1–10 (2005)

[53] Rayward-Smith, V.J.: Metaheuristics for Clustering in KDD. In: Proc. IEEE Congress on Evolutionary Computation, pp. 2380–2387 (2005)

[54] Rezaee, M.R., Lelieveldt, B.P.F., Reiber, J.H.C.: A New Cluster Validity Index for the Fuzzy c-Mean. Pattern Recognition Letters 19, 237–246 (1998)

[55] Srikanth, R., George, R., Warsi, N., Prabhu, D., Petry, F.E., Buckles, B.P.: A Variable-Length Genetic Algorithm for Clustering and Classification. Pattern Recognition Letters 16, 789–800 (1995)

[56] Sun, H., Wang, S., Jiang, Q.: FCM-Based Model Selection Algorithms for Determining the Number of Clusters. Pattern Recognition Letters 37, 2027–2037 (2004)

[57] Valafar, F.: Pattern Recognition Techniques in Microarray Data Analysis: A Survey. Annals of New York Academy of Sciences 980, 41–64 (2002)

[58] Van Le, T.: Evolutionary Fuzzy Clustering. In: Proc. IEEE Congress on Evolutionary Computation, pp. 753–758 (1995)

[59] Xu, R., Wunsch II, D.: Survey of Clustering Algorithms. IEEE Trans. on Neural Networks 16, 645–678 (2005)

[60] Yuan, B., Klir, G.J., Swan-Stone, J.F.: Evolutionary Fuzzy C-Means Clustering Algorithm. In: Proc. Int. Conference on Fuzzy Systems, pp. 2221–2226 (1995)

# Stability-Based Model Order Selection for Clustering Using Multiple Cooperative Particle Swarms

Abbas Ahmadi, Fakhri Karray, and Mohamed S. Kamel

**Summary.** Data clustering is the organization of a set of unlabelled data into similar groups. In this chapter, stability analysis is proposed to determine the model order of the underlying data using multiple cooperative swarms clustering. The mathematical explanations demonstrating why multiple cooperative swarms clustering leads to more stable and robust results than those of single swarm clustering are also provided. The proposed approach is evaluated using different data sets and its performance is compared with that of other clustering techniques.

## 1 Introduction

The goal of data clustering is to discover the natural groups or structures in the given unlabelled data. Many clustering techniques are available to cluster unlabelled data based on different assumptions about distribution, shape and size of the data.

Particle swarm optimization (PSO), inspired by mimicking the social behavior of bird flocks [8], has been applied for data clustering tasks and its performance promoted the quality of the clustering solutions significantly [3], [6], [10], [19], [23]. Some of the recent swarm clustering techniques use a single swarm approach to find a final clustering solution [19], [20], [23]. Recently, clustering using multiple cooperative swarms (MCS) approach has also been introduced [3]. The MCS approach distributes the search task among multiple swarms, each of which explores its associated division while cooperating with others. MCS approach is effective to deal with the problems of higher dimensions and large number of clusters [3].

A challenging problem in most of the clustering methods, especially partitional techniques including MCS, is to select the proper number of clusters also known

Abbas Ahmadi, Fakhri Karray, and Mohamed S. Kamel
Pattern Analysis and Machine Intelligence Lab, Department of Electrical and
Computer Engineering, University of Waterloo, Canada
e-mail: {aahmadi,karray,mkamel}@uwaterloo.ca

A. Abraham et al. (Eds.): Foundations of Comput. Intel. Vol. 4, SCI 204, pp. 197–218.
springerlink.com          © Springer-Verlag Berlin Heidelberg 2009

as model order selection. Normally, this information is provided by the user and sometimes is chosen arbitrarily.

In this chapter, we focus on the model order selection for MCS approach. We employ stability analysis to extract the number of clusters for the underlying data [11]. We show that the model order selection using MCS is more robust as compared to single swarm clustering approach and other clustering approaches.

In the following section, an introduction to data clustering is given. Particle swarm optimization and particle swarm clustering approaches are next explained. In section 4, model order selection using stability analysis is outlined. Finally, experimental results using four different data sets and concluding remarks are provided.

## 2   Data Clustering

Determining underlying groups in a given data is a difficult problem as apposed to the classification task. In other words, the labels of data are known in classification, but hidden in data clustering. Data clustering is a critical task for many applications such as bioinformatics, document retrieval, image segmentation and speech recognition [3], [5], [19], [20], [25], [28].

Let $Y$ denote a set of unlabelled data that is required to be clustered into different groups $Y^1, Y^2, \cdots, Y^K$, where $K$ shows the number of clusters. Subsequently, each group $Y^k$ consists of a set of similar data points given by $Y^k = \{y_j^k\}_{j=1}^{n_k}$, where $n_k$ indicates the number of data points in cluster $k$. Furthermore, let $A_K(Y)$ denote a clustering algorithm aiming to cluster data set $Y$ into $K$ distinct clusters. Moreover, assume the solution of the clustering algorithm $A_K(Y)$ for the given data points $Y$ of size $N$ is presented by $T := A_K(Y)$ which is a vector of labels $T = \{\mathbf{t}_i\}_{i=1}^N$, where $t_i \in \mathscr{L} := \{1, ..., K\}$.

To solve the clustering problem, there are two main approaches which are hierarchical and partitional clustering [28]. Hierarchical clustering approaches provide a hierarchy of clusters known as a dendrogram. Fig. 1 presents an example of the dendrogram for the given data set.

To construct the dendrogram, agglomerative and divisive approaches are available. A divisive approach starts with a single cluster consisting of all data points. Then, it splits this cluster into two distinct clusters. This procedure proceeds until each cluster includes a single data point. In contrast to the divisive approach, an agglomerative approach assumes that each data point is a cluster initially. Then, two close clusters merge together and make a new cluster. Merging close clusters is continued until all points build a single cluster.

Partitional clustering approaches cluster the data set into a predefined number of clusters by optimizing a certain criterion [19]. There are different partitional clustering approaches such as $K$-means, $K$-harmonic means and fuzzy $c$-means. Moreover, PSO-based clustering approaches belong to the class of partitional clustering. An introduction to PSO-based clustering approaches will be given in next section. Here, the other partitional clustering approaches are briefly outlined.

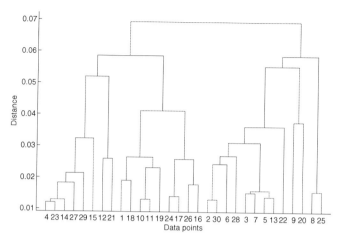

**Fig. 1** A typical dendrogram

$K$-means clustering is the most popular partitional clustering algorithm. It commences with $K$ arbitrary points as initial centers. Next, each data point is assigned to the closest center. Then, new centers are estimated by calculating the mean of all data points associated to each cluster $k$. This procedure is repeated until the convergence is observed. $K$-mean clustering has many characteristics that make it an attractive approach. It converges to a final solution very quickly and it is easy to comprehend and implement. However, it inherits serious limitations as well. $K$-means algorithm is highly dependent on the effect of initial solution and it may converge to local optimal solutions.

An alternative approach for data clustering is known as $K$-harmonic means (KHM) [17]. KHM employs harmonic averages of distances from every data point to the centers. It is claimed that KHM clustering is less sensitive to initial solutions empirically [17]. As compared to $K$-means algorithm, it improves the quality of clustering results in certain cases [17].

Bezdeck has extended $K$-means using fuzzy logic and has suggested a new clustering approach referred to as fuzzy $c$-means (FCM) clustering [15]. Each data point in FCM is associated to each cluster with some degree of belongness. In other words, it has a membership in all clusters [15].

Now, we want to explain how the quality of the obtained clustering solutions is measured. Evaluating the quality of clustering solution is not an easy task since the labels of data are unknown. There are two widely used measures, namely within-cluster distance and between-cluster distance. Within-cluster distance reflects the compactness of the obtained clusters. In other words, it indicates how compact the obtained clusters are. Between-cluster distance, however, represents the separation of the obtained clusters. Clustering algorithms are expected to minimize the within-cluster distance and to maximize the between-cluster distance [14].

Compactness measure is given by

$$\mathcal{F}_c(\mathbf{M}) = \frac{1}{K} \sum_{k=1}^{K} \frac{1}{n_k} \sum_{j=1}^{n_k} d(\mathbf{m}^k, \mathbf{y}_j^k), \tag{1}$$

where $\mathbf{M} = (\mathbf{m}^1, ..., \mathbf{m}^K)$, $\mathbf{m}^k$ denotes the center of the cluster $k$ and $d(\cdot)$ stands for Euclidean distance.

Further, the separation measure is defined as

$$\mathcal{F}_s(\mathbf{M}) = \frac{1}{K(K-1)} \sum_{j=1}^{K} \sum_{k=j+1}^{K} d(\mathbf{m}^j, \mathbf{m}^k). \tag{2}$$

It is aimed to maximize this measure, or equivalently minimize $-\mathcal{F}_s(\mathbf{M})$. There are some other measures, also known as cluster validity measures, which are mainly based on the combination of the compactness and separation measures as presented next.

## 2.1  Combined Measure

Combined measure is a linear combination of the compactness and separation measures given by

$$\mathcal{F}_{Combined}(\mathbf{M}) = w_1 \mathcal{F}_c(\mathbf{M}) - w_2 \mathcal{F}_s(\mathbf{M}), \tag{3}$$

where $w_1$ and $w_2$ are weighting parameters such that $\sum_{i=1}^{2} w_i = 1$ [12].

## 2.2  Turi's Validity Index

Turi's validity index [18] is computed as

$$\mathcal{F}_{Turi}(\mathbf{M}) = (c \times \mathcal{N}(2,1) + 1) \times \frac{intra}{inter}, \tag{4}$$

where $c$ is a user-specified parameter and $\mathcal{N}$ is a Gaussian distribution with $\mu = 2$ and $\sigma = 1$. The $intra$ denotes the within-cluster distance provided in equation (1). Besides, the $inter$ term is the minimum distance between the cluster centers given by

$$\begin{aligned} inter = min\{\|\mathbf{m}^k - \mathbf{m}^{kk}\|\}, \quad & k = 1, 2, ..., K-1, \\ & kk = k+1, ..., K. \end{aligned} \tag{5}$$

Different clustering approaches are required to minimize Turi's index [18].

## 2.3  Dunn's Index

Let us define $\alpha(C^k, C^{kk})$ and $\beta(C^k)$ as follows

$$\alpha(C^k, C^{kk}) = \min_{\mathbf{x} \in C^k, \mathbf{z} \in C^{kk}} d(\mathbf{x}, \mathbf{z}),$$

$$\beta(C^k) = \max_{\mathbf{x}, \mathbf{z} \in C^k} d(\mathbf{x}, \mathbf{z}).$$

Now, Dunn's index [22] can be computed as

$$\mathscr{F}_{Dunn}(\mathbf{M}) = \min_{1 \le k \le K} \left\{ \min_{k+1 \le kk \le K} \left( \frac{\alpha(C^k, C^{kk})}{\max\limits_{1 \le \tilde{k} \le K} \beta(C^{\tilde{k}})} \right) \right\}. \tag{6}$$

The goal of the different clustering approaches is to maximize Dunn's index.

## 2.4 S_Dbw Index

Let us consider the average scattering of the clusters as a measure of compactness, defined as

$$Scatt = K^{-1} \sum_{k=1}^{K} \frac{\|\sigma(C^k)\|}{\|\sigma(Y)\|}, \tag{7}$$

where $\sigma(\cdot)$ shows the variance of the data and $\|\mathbf{x}\|$ is given by $\|\mathbf{x}\| = \sqrt{\mathbf{x}^T \mathbf{x}}$. Also, the separation measure is formulated as

$$Den\_bw = \frac{1}{K(K-1)} \sum_{k=1}^{K} \sum_{kk=1, kk \ne k}^{K} \frac{\mathscr{D}(\mathbf{z}_{k,kk})}{max\{D(\mathbf{m}^k), \mathscr{D}(\mathbf{m}^{kk})\}}, \text{ where } \mathbf{z}_{k,kk} \text{ is the mid-}$$

dle point of the line segment characterized by cluster centers $\mathbf{m}^k$ and $\mathbf{m}^{kk}$. Moreover, $\mathscr{D}(\mathbf{m}^k)$ represents a density function around point $\mathbf{m}^k$ which is estimated by

$$\mathscr{D}(\mathbf{m}^k) = \sum_{j=1}^{n_k} f(\mathbf{m}^k, y_j^k), \text{ and}$$

$$f(\mathbf{m}^k, y_j^k) = \begin{cases} 1 & \text{if } d(\mathbf{m}^k, y_j^k) > \tilde{\sigma} \\ 0 & \text{Otherwise,} \end{cases} \tag{8}$$

where $\tilde{\sigma} = K^{-1} \sqrt{\sum_{k=1}^{K} \|\sigma(C^k)\|}$.

Eventually, S_Dbw index [24, 9] is defined as

$$\mathscr{F}_{S\_Dbw}(\mathbf{M}) = Scatt + Den\_bw. \tag{9}$$

The aim of the different clustering approaches is to minimize S_Dbw index.

## 3 Particle Swarm Optimization for Data Clustering

Particle swarm optimization (PSO) was introduced to tackle optimization problems [12, 13]. PSO begins with an initial swarm of particles and explores a search space through a number of iterations to find an optimum solution for a predefined objective function $\mathscr{F}$. Each particle $i$ is represented by its position $\mathbf{x}_i$ and velocity $\mathbf{v}_i$. More-over, each particle follows its corresponding best position and the best position among the swarm obtained so far. These two positions are also known as personal best and global best, denoted by $\mathbf{x}_i^{pb}$ and $\mathbf{x}^*$, respectively. A new position for each particle is obtained by

$$\mathbf{x}_i(t+1) = \mathbf{x}_i(t) + \mathbf{v}_i(t+1), \tag{10}$$

and the velocity is updated as

$$\mathbf{v}_i(t+1) = w\mathbf{v}_i(t) + c_1 r_1 (\mathbf{x}_i^{pb}(t) - \mathbf{x}_i(t)) + c_2 r_2 (\mathbf{x}^*(t) - \mathbf{x}_i(t)), \tag{11}$$

where $w$ denotes the impact of the previous history of velocities on the current ve-locity, $c_1$ and $c_2$ are cognitive and social components, respectively, and $r_1$ and $r_2$ are generated randomly using a uniform distribution in interval $[0,1]$. When the mini-mization of the objective function is of interest, the personal best position of particle $i$ at iteration $t$ is determined by

$$\mathbf{x}_i^{pb}(t+1) = \begin{cases} \mathbf{x}_i^{pb}(t) & \text{if } \mathscr{F}(\mathbf{x}_i(t+1)) \geq \mathscr{F}(\mathbf{x}_i^{pb}(t)), \\ \mathbf{x}_i(t+1) & \text{otherwise.} \end{cases} \tag{12}$$

Further, the global best solution is updated as

$$\mathbf{x}^*(t+1) = arg \min_{\mathbf{x}_i^{pb}(t)} f(\mathbf{x}_i^{pb}(t)), \ i \in [1,...,n]. \tag{13}$$

There are several methods to terminate a PSO procedure such as reaching the maximum number of iterations, having a number of iterations with no improve-ment, and reaching minimum objective function criterion [1]. In this chapter, the first strategy is considered.

Particle swarm optimization has a number of features that makes it a suitable alternative for clustering tasks. PSO procedure is less sensitive to the effect of the initial conditions due to its population-based mechanism. It also performs a global search of the solution space. Accordingly, it is more likely to provide a near-optimal solution. Further, PSO can manipulate multiple objectives simultaneously. As a re-sult, it is an excellent tool for solving clustering problems where optimizing different objectives is necessary.

## 3.1 Single Swarm Clustering

In single swarm clustering, the particle $i$ is represented by $\mathbf{x}_i = (\mathbf{m}^1,...,\mathbf{m}^K)_i$. To model the clustering problem as an optimization problem, it is required to formulate an objective function. Cluster validity measures described earlier are considered as the objective function. By setting $\mathscr{F}(\mathbf{m}^1,...,\mathbf{m}^K)$ or $\mathscr{F}(\mathbf{M})$ as the required objec-tive function, PSO procedure can be used to find the cluster centers. The required procedure for single swarm clustering is provided in Algorithm **1**.

**Algorithm 1.** Single swarm clustering

---

initialize a swarm of size $n$
**repeat**
   **for** each particle $i \in [1...n]$ **do**
      update position and velocity
      **if** $\mathscr{F}((\mathbf{m}^1,...,\mathbf{m}^K)_i(t+1)) < \mathscr{F}((\mathbf{m}^1,...,\mathbf{m}^K)_i^{pb}(t))$ **then**
         $(\mathbf{m}^1,...,\mathbf{m}^K)_i^{pb}(t+1) \leftarrow (\mathbf{m}^1,...,\mathbf{m}^K)_i(t+1)$
      **end if**
   **end for**
   $(\mathbf{m}^1,...,\mathbf{m}^K)^*(t+1) \leftarrow argmin\{\mathscr{F}((\mathbf{m}^1,...,\mathbf{m}^K)_i^{pb}(t))|i \in [1,...,n]\}$
**until** termination criterion is met

---

There are several situations in which a single swarm is not able to search all of the space sufficiently, and so fails to return satisfactory results. In the cases where the dimensionality of data is high or there is a considerable number of clusters, multiple cooperative swarms can perform better than a single swarm, due to the exponential increase in the volume of the search space as the dimension of the problem increases [3].

## 3.2 Multiple Cooperative Swarms Clustering

Multiple cooperative swarms clustering approach assumes that the number of swarms is equal to that of clusters and particles of each swarm are candidates for the corresponding cluster's center.

The core idea is to divide the search space into different divisions. Each division $s^k$ is characterized by its center $\mathbf{z}^k$ and width $R^k$; i.e., $s^k = f(\mathbf{z}^k, R^k)$. To distribute the search space into different divisions, a super-swarm is used. The super-swarm which is a population of particles aims to find the center of divisions $\mathbf{z}^k, k = 1,...K$. Each particle of super-swarm is defined as $(\mathbf{z}^1,...,\mathbf{z}^K)$, where $\mathbf{z}^k$ denotes the center of division $k$. By repeating PSO procedure using one of the mentioned cluster validity measures as the objective function, the centers of different divisions are obtained. Then, the width of divisions are computed by

$$R^k = \alpha\lambda_{max}^k , \quad k = 1,...,K, \tag{14}$$

where $\alpha$ is a positive constant and $\lambda_{max}^k$ is the square root of the biggest eigen value of data points belonging to division $k$. The $\alpha$ is selected such that an appropriate coverage of the search space is attained [3].

After distributing search space, one swarm is assigned to each division. That is, the number of swarms is equal to the number of divisions, or clusters, and particles of each swarm are potential candidates for the associated cluster's center. In this stage, there is information exchange among swarms and each swarm knows the global best, the best cluster center, of the other swarms obtained so far. Therefore, there is a cooperative search scheme where each swarm explores its related division

**Fig. 2** Schematic illustration of multiple cooperative swarms. First, the cooperation between multiple swarms initiates and each swarm investigates its associated division (a). When the particles of each swarm converge (b), the final solution for cluster centers is revealed

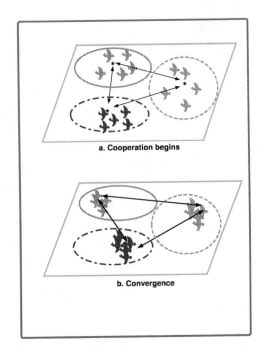

a. Cooperation begins

b. Convergence

to find the best solution for the associated cluster center interacting with other swarms. The schematic presentation of the multiple cooperative swarms is depicted in Fig. 2.

In this cooperative search scheme, particles of each swarm tend to optimize the following problem:

$$\min \mathscr{F}(\mathbf{x}_i^1, \cdots, \mathbf{x}_i^K)$$
$$s.t. : \mathbf{x}_i^k \in s^k,$$
$$k = 1, \cdots, K,$$
$$i = 1, \cdots, n,$$

(15)

where $\mathscr{F}(\cdot)$ denotes one of the introduced cluster validity measures and $\mathbf{x}_i^k$ is particle $i$ of swarm $k$.

The search procedure using multiple swarms is performed in parallel scheme. First, a new candidate for the cluster centers, $\mathbf{M}' = (\mathbf{m}'^1, ..., \mathbf{m}'^K)$, is obtained using equation (15). To update the cluster centers, the following rule is used:

$$\mathbf{M}^{(new)} = \begin{cases} \mathbf{M}' & \text{if } \mathscr{F}(\mathbf{M}') \leq \mathscr{F}(\mathbf{M}^{(old)}), \\ \mathbf{M}^{(old)} & \text{otherwise.} \end{cases}$$

(16)

In other words, if the fitness value of the new candidate for cluster centers is smaller than that of the former candidate, the new solution is accepted; otherwise, it is rejected. The overall algorithm of multiple cooperative swarms clustering is provided in Algorithm **2**.

---

**Algorithm 2.** Multiple cooperative swarms clustering

---

Stage 1: Distributing search space into various divisions $s^1, ..., s^K$

- Obtaining divisions' center $z^1, ..., z^K$
- Obtaining divisions' width $R^1, ..., R^K$

Stage 2: Cooperating till convergence

- Exploring within each swarm

  – 1.1. Compute new positions of all particles of swarms.
  – 1.2. Obtain the fitness value of all particles using the corresponding cluster validity measure.
  – 1.3. Select the set of positions which minimize the optimization problem using equation (15) and denote it as a new candidate for cluster centers $(m'^1, ..., m'^K)$.

- Updating rule

  – 2.1. If the fitness value of the new candidates for centers of clusters $(m'^1, ..., m'^K)$ is smaller than that of previous iteration, accept the new solution; otherwise, reject it.
  – 2.2. If termination criterion is achieved, stop; otherwise, continue this stage.

---

**Fig. 3** Search space and optimal solution region

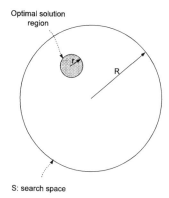

S: search space

## 3.3 Single Swarm vs. Multiple Swarm Clustering

Suppose PSO algorithm tends to find the solution of the following optimization problem

$$\mathcal{L} = \min \ \mathcal{F}(x)$$
$$\text{s.t.:} \quad x \in S, \tag{17}$$

where $S$ denotes the search space. Assume that $S$ is a $d$-dimensional hyper-sphere of radius $R$ and the optimal solution is located in a smaller $d$-dimensional hyper-sphere of radius $r$ (Fig. 3 ). The volume of the $d$-dimensional hyper-sphere of radius $R$ is defined as

$$V(R,d) = \frac{\pi^{\frac{d}{2}}}{\Gamma(\frac{d}{2}+1)} R^d,$$

(18)

where $\Gamma(.)$ stands for Gamma function. The probability of finding a solution in the optimal region using any search technique is as follows:

$$P_r(\text{converge to an optimal solution}) = \frac{V(r,d)}{V(R,d)} = (\frac{r}{R})^d$$

(19)

In other words, the probability of finding an optimal solution decreases by increasing the dimensionality of data if $r$ and $R$ remain constant.

Now suppose single swarm clustering approach is used to find cluster centers. Assume the optimal solutions for centers of the clusters are located in $K$ different $d$-dimensional hyper-spheres of radius $r_1, r_2, ..., r_K$. To achieve an optimal solution, cluster centers should be selected from optimal solution regions. In single swarm clustering, we denote $P_r(\text{converge to optimal solution})$ by $P_r^1$ defined as

$$P_r^1 = \prod_{k=1}^{K} P_r(\mathbf{m}^k \in C^k),$$

(20)

where $P_r(\mathbf{m}^k \in C^k)$ which stands for the probability of selecting the center of cluster $k$ from its corresponding optimal region calculated by

$$P_r(\mathbf{m}^k \in C^k) = (\frac{r}{R})^d.$$

(21)

Using this expression, equation (20) can be rewritten as

$$P_r^1 = \prod_{k=1}^{K} (\frac{r_k}{R})^d = \frac{(r_1 r_2 \cdots r_K)^d}{R^{d.K}}.$$

(22)

However in the case of multiple swarms, each swarm explores a part of the search space characterized by a $d$-dimensional hyper-sphere of radius $R_k$. As $R_k < R$ for all $k$, the following inequality is valid:

$$R_1...R_K < R^K.$$

(23)

Since $d \geq 1$, inequality (23) can be modified as

$$(R_1...R_K)^d < R^{d.K}.$$

(24)

Assume the optimal solution for each swarm $k$ in multiple swarms approach is situated in a $d$-dimensional hyper-sphere of radius $r_k$. Accordingly, the probability of getting an optimal solution using multiple swarms at each iteration is calculated as

$$P_r^M = \prod_{k=1}^{K} P_r(\mathbf{m}^k \in C^k)$$

(25)

and it can be simplified as

$$P_r^M = \prod_{k=1}^{K}(\frac{r_k}{R_k})^d = \frac{(r_1 r_2 ... r_K)^d}{(R_1 . R_2 ... R_K)^d}. \tag{26}$$

According to equations (22) and (26), we have

$$\frac{P_r^M}{P_r^I} = \frac{R^{d.K}}{(R_1 . R_2 ... R_K)^d}. \tag{27}$$

Considering equation (24), we have

$$\frac{P_r^M}{P_r^I} > 1. \tag{28}$$

In other words, $P_r^M > P_r^I$. Therefore, the probability of obtaining the optimal solution using multiple cooperative swarms clustering is absolutely greater than that of single swarm clustering.

Similar to most of the partitional clustering approaches, the multiple cooperative swarms approach needs to be provided a priori the number of clusters. This has always been a challenging task in the area of partitional clustering. In the following section, it is shown how stability analysis can be used to estimate the number of clusters for the underlying data using multiple cooperative swarms approach.

## 4  Model Order Selection Using Stability Approach

Finding the number of clusters in data clustering is known as a model order selection problem. There are two main steps in model order selection. First, a clustering algorithm should be selected. Then, the model order needs to be estimated for the underlying data [29].

Many clustering approaches assume that the model order is known in advance. Here, stability analysis is considered to obtain the number of clusters using multiple cooperative swarms clustering approach. An explanation of stability analysis is outlined before describing the core algorithm.

Stability concept is used to evaluate the robustness of a clustering algorithm. In other words, stability measure indicates how much the results of the clustering algorithm is reproducible on other data drawn from the same source. One of the issues that affects the stability of the given algorithm is the model order. For instance, by considering a large number of clusters the algorithm generates random groups of data influenced by the changes observed in different samples. On the other hand, by choosing a very small number of clusters, algorithm may combine separated structures together and return unstable clusters [29]. As a result, one can utilize the stability measure for estimating the model order of the unlabeled data.

The multiple cooperative swarms clustering data requires a priori knowledge of model order in advance. To enable this approach for estimating the number of clusters, the stability approach is taken into consideration. This chapter uses the stability method introduced by Lange et al. [29] due to the following reasons:

- it requires no information about the data being processed,
- it can be applied on any clustering algorithm,
- it returns the correct model order using the notion of maximal stability.

The required procedure for model order selection using stability analysis is provided in Algorithm **3**. The core idea of this algorithm is depicted in Fig. 4.

---

**Algorithm 3.** Model order selection using stability analysis

---

  **for** $k \in [2...K]$ **do**
    **for** $r \in [1...rmax]$ **do**
      -Randomly split the given data $Y$ into two halves $Y_1, Y_2$.
      -Cluster $Y_1, Y_2$ independently using an **appropriate clustering approach**; i.e., $T_1 := A_k(Y_1)$, $T_2 := A_k(Y_2)$.
      -Use $(Y_1, T_1)$ to train classifier $\phi(Y_1)$ and compute $T_2' = \phi(Y_2)$.
      -Calculate the distance of the two solutions $T_2$ and $T_2'$ for $Y_2$; i.e., $d_r = d(T_2, T_2')$.
      -Again cluster $Y_1, Y_2$ by assigning random labels to points.
      -Extend *random clustering* as above, and obtain the distance of the solutions; i.e., $dn_r$.
    **end for**
    -Compute the stability $stab(k) = mean_r(d)$.
    -Compute the stability of random clustering $stab_{rand}(k) = mean_r(dn)$.
    - $s(k) = \frac{stab(k)}{stab_{rand}(k)}$
  **end for**
  -Select the model order $k^*$ such that $k^* = \arg\min_k \{s(k)\}$.

---

As can be seen in Fig. 4, the algorithm first splits the given data set randomly into two subsets $Y_1, Y_2$. Then, these subsets are clustered using the proposed multiple cooperative swarms clustering. Let $T_1$ and $T_2$ denote the clustering solutions (estimated labels) for subsets $Y_1$ and $Y_2$, respectively. Now, one can use subset $Y_1$ and its estimated labels to train a classifier $\phi(\cdot)$. The trained classifier is then used to generate a set of labels for subset $Y_2$ denoted by $T_2'$. As a result, there are two labels for subset $Y_2$ which are $T_2$ and $T_2'$. First label, $T_2$, is estimated by the multiple cooperative swarms clustering, whereas the other one , $T_2'$, is obtained by the trained classifier. The ideal case is that the proposed clustering approach and the trained classifier produce the same labels for the subset $Y_1$. However, there is a gap between these two in reality. This distance is used to evaluate the stability of the proposed clustering approach.

Regarding stability-based model order selection algorithm, some issues require more clarifications.

**Fig. 4** The core idea of
the model order selection
algorithm

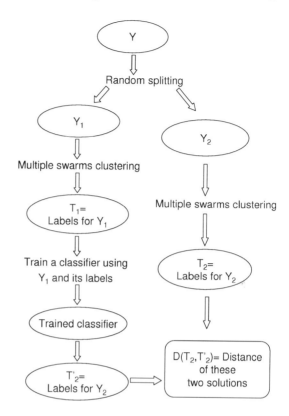

- *Classifier $\phi(\cdot)$*
  As can be seen in the algorithm, there is a need for a classifier. There are a
  vast range of classifiers can be used for classification. Here, $k$-nearest neighbor
  classifier was chosen as it requires no assumption on the distribution of data.
- *Distance of solutions provided by clustering and classifier*
  Having a set of training data, the classifier can be tested using a test data $Y_2$.
  Its solution is represented by $T_2' = \phi(Y_2)$. But, there exists another solution
  for the same data obtained from multiple cooperative swarms clustering technique, i.e., $T_2 := A_k(Y_2)$. The distance of these two solutions is calculated by
  $$d(T_2, T_2') = \arg\min_{\omega \in \rho_k} \sum_{i=1}^{N} \vartheta\{\omega(t_{2,i}) \neq t_{2,i}'\}, \text{ where } \vartheta\{t_{2,i} \neq t_{2,i}'\} = 1 \text{ if } t_{2,i} \neq t_{2,i}' \text{ and}$$
  zero, otherwise. Also, $\rho_k$ contains all permutations of $k$ labels and $\omega$ is the op-
  timal permutation in $\rho_k$ which produces the maximum agreement between two
  solutions [29].
- *Random clustering*
  The stability rate depends on the number of classes or clusters. For instance, the
  accuracy rate of 50% for binary classification is more or less the same as that of a
  random guess. However, this rate for $k = 10$ is much better than a random predictor. In other words, if a clustering approach returns the same accuracy for model

**Fig. 5** The effect of the random clustering on the selection of the model order

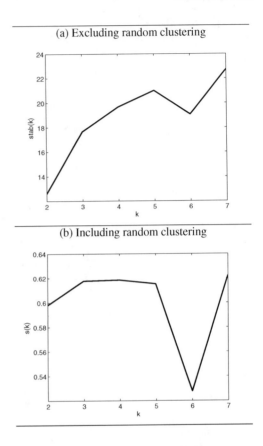

(a) Excluding random clustering

(b) Including random clustering

orders $k_1$ and $k_2$, where $k_1 < k_2$, the clustering solution for $k_2$ is more reliable than the other solution. Hence, the primary stability measure obtained for a certain value $k$, $stab(k)$ in Algorithm **3**, should be normalized using the stability rate of a random clustering, $stab_{rand}(k)$ in Algorithm **3**, [29]. The random clustering simply means to assign a label between one and $K$ to each data point randomly. The final stability measure for the model order $k$ is obtained as follows:

$$s(k) = \{ \frac{stab(k)}{stab_{rand}(k)} \}. \tag{29}$$

The effect of the random clustering is studied on the performance of the Zoo data set provided in section 5 to determine the model order of the data using $K$-means algorithm. The stability measure for different number of clusters with and without using random clustering is shown in Fig. 5.

As depicted in Fig. 5, the model order of the zoo data using $K$-means clustering is recognized as two without considering random clustering, while it becomes six, which is near to the true model order, by normalizing the primary stability measure to the stability of the random clustering.

For a given data set, the algorithm does not provide the same result for multiple runs. Moreover, the model order is highly dependent on the type of **appropriate clustering approach** that is used in this algorithm (see Algorithm 3), and there is no specific emphasis in [29] on the type of clustering algorithm that should be used. $K$-means, $K$-harmonic means and fuzzy $c$-means clustering algorithms are either sensitive to the initial conditions or to the type of data. In other words, they can not capture true underlying patterns of the data, and consequently the estimated model order is not robust. However, PSO-based clustering approaches such as single swarm or multiple cooperative swarms clustering do not rely on initial conditions and they are the search schemes which can explore the search space more effectively and can escape from local optimums. Moreover as demonstrated earlier, multiple cooperative swarms clustering is more probable to get the optimal solution than single swarm clustering.

Furthermore, multiple cooperative swarms approach distributes the search space among multiple swarms and enables cooperation between swarms leading to an effective search strategy. Accordingly, we propose to use multiple cooperative swarms clustering in stability analysis-based approach to find the model order of the given data.

## 5 Experimental Results

The performance of the proposed approach is evaluated and compared with other approaches such as single swarm clustering, $K$-means and $K$-harmonic means clustering using four different data sets selected from the UCI machine learning repository [30]. The name of data sets, their associated number of classes, samples and dimensions are provided in Table 1.

**Table 1** Data sets selected from UCI machine learning repository

| Data set | classes | samples | dimensionality |
|---|---|---|---|
| Iris | 3 | 150 | 4 |
| Wine | 3 | 178 | 13 |
| Zoo | 7 | 101 | 17 |
| Glass identification | 7 | 214 | 9 |

The performance of the multiple cooperative swarms clustering approach is compared with $K$-means and single swarm clustering techniques in terms of Turi's validity index over 80 iterations (Fig. 6). The results are obtained by repeating the algorithms 30 independent times. For these experiments, the parameters are set as $w = 1.2$ (decreasing gradually), $c_1 = 1.49$, $c_2 = 1.49$, $n = 30$ (for all swarms). In addition, the model order is considered to be equal to the number of classes.

As illustrated in Fig. 6, multiple cooperative swarms clustering provides better results as compared with $K$-means, as well as single swarm clustering approaches, in terms of Turi's index for majority of the data sets.

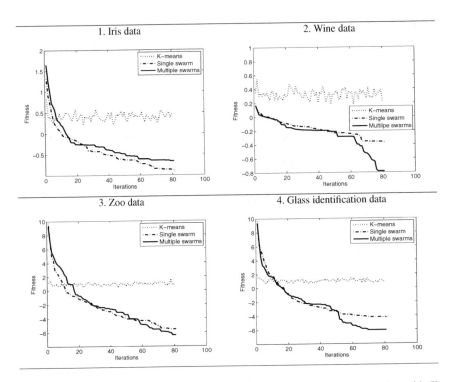

**Fig. 6** Comparing the performance of the multiple cooperative swarms clustering with *K*-means and single swarm clustering in terms of Turi's index

**Table 2** Average and standard deviation comparison of different measures for iris data

| Method | Turi's index[$\sigma$] | Dunn's index[$\sigma$] | S_Dbw[$\sigma$] |
|---|---|---|---|
| *K*-means | 0.4942[0.3227] | 0.1008[0.0138] | 3.0714[0.2383] |
| *K*-harmonic means | 0.82e05[0.95e05] | 0.0921[0.0214] | 3.0993[0.0001] |
| Single swarm | **−0.8802**[0.4415] | **0.3979**[0.0001] | 1.4902[0.0148] |
| Cooperative swarms | −0.7[1.0164] | **0.3979**[0.0001] | **1.48**[0.008] |

**Table 3** Average and standard deviation comparison of different measures for wine data

| Method | Turi's index[$\sigma$] | Dunn's index[$\sigma$] | S_Dbw[$\sigma$] |
|---|---|---|---|
| *K*-means | 0.2101[0.3565] | 0.016[0.006] | 3.1239[0.4139] |
| *K*-harmonic means | 2.83e07[2.82e07] | **190.2**[320.75] | 2.1401[0.0149] |
| Single swarm | −0.3669[0.4735] | 0.1122[0.0213] | 1.3843[0.0026] |
| Cooperative swarms | **−0.7832**[0.8564] | 0.0848[0.009] | **1.3829**[0.0044] |

**Table 4** Average and standard deviation comparison of different measures for zoo data

| Method | Turi's index[$\sigma$] | Dunn's index[$\sigma$] | S_Dbw[$\sigma$] |
|---|---|---|---|
| $K$-means | 0.8513[1.0624] | 0.2228[0.0581] | 2.5181[0.2848] |
| $K$-harmonic means | 1.239[1.5692] | 0.3168[0.0938] | 2.3048[0.1174] |
| Single swarm | $-5.5567$[3.6787] | **0.5427**[0.0165] | **2.0528**[0.0142] |
| Cooperative swarms | $-\mathbf{6.385}$[4.6226] | 0.5207[0.0407] | 2.0767[0.025] |

**Table 5** Average and standard deviation comparison of different measures for glass identification data

| Method | Turi's index | Dunn's index | S_Dbw |
|---|---|---|---|
| $K$-means | 0.7572[0.9624] | 0.0286[0.001] | 2.599[0.2571] |
| $K$-harmonic means | 0.89$e$05[1.01$e$05] | 0.0455[0.0012] | **2.0941**[0.0981] |
| Single swarm | $-4.214$[3.0376] | 0.1877[0.0363] | 2.6797[0.3372] |
| Cooperative swarms | $-\mathbf{6.0543}$[4.5113] | **0.225**[0.1034] | 2.484[0.1911] |

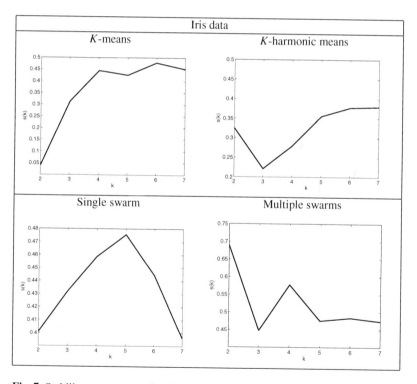

**Fig. 7** Stability measure as a function of model order: iris data

In Tables 2-5, the multiple cooperative swarms clustering is compared with other clustering approaches using different cluster validity measures over 30 independent runs. The results presented for different data sets are in terms of average and standard deviation ($[\sigma]$) values.

As observed in Tables 2-5, multiple swarms clustering is able to provide better results in terms of the different cluster validity measures for most of the data sets. This is because it is capable of manipulating multiple-objective problems, in contrast to $K$-means (KM) and $K$-harmonic means (KHM) clustering, and it distributes the search space between multiple swarms and solves the problem more effectively.

Now, the stability-based approach for model order selection in multiple cooperative swarms clustering is studied. The PSO parameters are kept the same as before, and $rmax = 30$ and $k$ is considered to be 25 for KNN classifier. The stability measures of different model orders for the multiple cooperative swarms and other clustering approaches using different data sets are presented in Fig. 7 - Fig. 10. In these figures, $k$ and $s(k)$ indicate model order and stability measure for the given model order $k$, respectively. The corresponding curves for single swarm and multiple swarms clustering approaches are obtained using Turi's validity index.

According to Fig. 7 - Fig. 10, the proposed approach using multiple cooperative swarms clustering is able to identify the correct model order for most of the data sets. Moreover, the best model order for different data sets can be obtained as provided

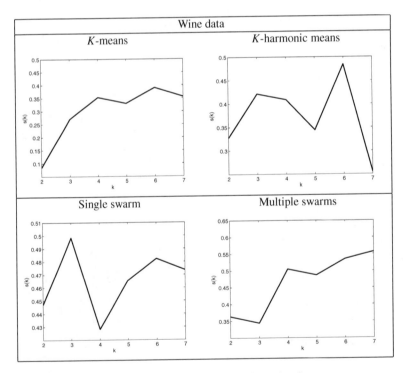

**Fig. 8** Stability measure as a function of model order: wine data

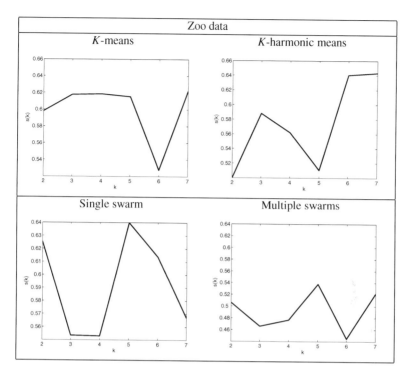

**Fig. 9** Stability measure as a function of model order: zoo data

in Table 6. The minimum value for stability measure given any clustering approach corresponds to the best model order ($k^*$); i.e.,

$$k^* = arg \min_{k}\{s(k)\}. \qquad (30)$$

As presented in Table 6, $K$-means and $K$-harmonic means clustering approaches do not converge to the true model order using the stability-based approach for the most of the data sets. The performance of the single swarm clustering is partially better than that of $K$-means and $K$-harmonic means clustering because it does not depend on initial conditions and can escape trapping in local optimal solutions. Moreover, the multiple cooperative swarms approach using Turi's index provides

**Table 6** The best model order ($k^*$) for different data sets

| Data set | KM | KHM | Single swarm | | | Multiple swarms | | |
|---|---|---|---|---|---|---|---|---|
| | | | Turi | Dunn | S_Dbw | Turi | Dunn | S_Dbw |
| Iris | 2 | **3** | 7 | 2 | 2 | **3** | 4 | 2 |
| Wine | 2 | 7 | 4 | 4 | 2 | **3** | 5 | **3** |
| Zoo | **6** | 2 | 4 | 2 | 2 | **6** | 2 | 2 |
| Glass identification | 2 | 3 | 4 | 2 | 2 | **7** | 2 | 2 |

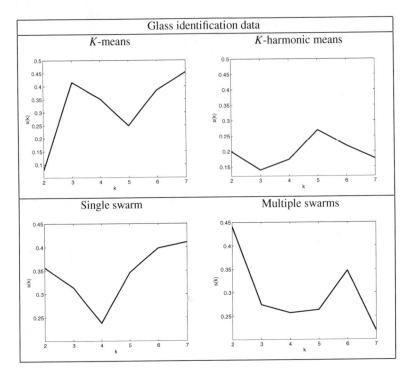

**Fig. 10** Stability measure as a function of model order: glass identification data

the true model order for majority of the data sets. As a result, Turi's validity index is appropriate for model order selection using the proposed clustering approach. Its performance, based on Dunn's index and S_Dbw index, is also considerable as compared to the other clustering approaches. Consequently, the proposed multiple cooperative swarms can provide better estimates for model order, as well as stable clustering results as compared to the other clustering techniques by using the introduced stability-based approach.

## 6   Conclusions

In this chapter, the stability analysis-based approach was introduced to estimate the model order of the given data using multiple cooperative swarms clustering. We proposed to use multiple cooperative swarms clustering to find the model order of the data, due to its robustness and stable solutions. Moreover, it has been shown that the probability of providing an optimal solution by multiple cooperative swarms clustering is higher than that of a single swarm scenario. To demonstrate the scalability of the proposed algorithm, it has been evaluated using four different data sets. Its performance has also been compared with other clustering approaches.

# References

1. Abraham, A., Guo, H., Liu, H.: Swarm Intelligence: Foundations. In: Nedjah, N., Mourelle, L. (eds.) Perspectives and Applications, Swarm Intelligent Systems. Studies in Computational Intelligence. Springer, Germany (2006)
2. Kazemian, M., Ramezani, Y., Lucas, C., Moshiri, B.: Swarm Clustering Based on Flowers Pollination by Artificial Bees. In: Abraham, A., Grosan, C., Ramos, V. (eds.) Swarm Intelligence in Data Mining, pp. 191–202. Springer, Heidelberg (2006)
3. Ahmadi, A., Karray, F., Kamel, M.: Multiple Cooperating Swarms for Data Clustering. In: Proceeding of the IEEE Swarm Intelligence Symposium (SIS 2007), pp. 206–212 (2007)
4. Ahmadi, A., Karray, F., Kamel, M.: Cooperative Swarms for Clustering Phoneme Data. In: Proceeding of the IEEE Workshop on Statistical Signal Processing (SSP 2007), pp. 606–610 (2007)
5. Cui, X., Potok, T.E., Palathingal, P.: Document Clustering Using Particle Swarm Optimization. In: Proceeding of the IEEE Swarm Intelligence Symposium (SIS 2005), pp. 185–191 (2005)
6. Xiao, X., Dow, E.R., Eberhart, R., Miled, Z.B., Oppelt, R.: Gene Clustering Using Self-Organizing Maps and Particle Swarm Optimization. In: Proceeding of International Parallel Processing Symposium (IPDPS 2003), 10 p. (2003)
7. Eberhart, R., Kennedy, J.: A New Optimizer Using Particle Swarm Theory. In: Proceeding of 6th Int. Symp. Micro Machine and Human Scince, pp. 39–43 (1995)
8. Eberhart, R., Kennedy, J.: Particle Swarm Optimization. In: Proceeding of the IEEE International Conference on Neural Networks, vol. 4, pp. 1942–1948 (1995)
9. Halkidi, M., Vazirgiannis, M.: Clustering Validity Assessment: Finding the Optimal Partitioning of a Data Set. In: Proceeding of the IEEE International Conference on Data Mining (ICDM 2001), pp. 187–194 (2001)
10. Van der Merwe, D.W., Engelbrecht, P.: Data Clustering Using Particle Swarm Optimization. In: Proceeding of the IEEE Congress on Evolutionary Computation, pp. 215–220 (2003)
11. Ahmadi, A., Karray, F., Kamel, M.S.: Model order selection for multiple cooperative swarms clustering using stability analysis. In: Proceeding of the IEEE Congress on Evolutionary Computation (IEEE CEC 2006), pp. 3387–3394 (2008)
12. Engelbrecht, A.P.: Fundamentals of Computational Swarm Intelligence. John Wiley and Sons, Chichester (2005)
13. Kennedy, J., Eberhart, R., Shi, Y.: Swarm Intelligence. Morgan Kaufmann, San Francisco (2001)
14. Duda, R., Hart, P., Stork, D.: Pattern Classification. John Wiley & Sons, Chichester (2000)
15. Bezdek, J.: Pattern Recognition with Fuzzy Objective Function Algoritms. Plenum Press, New York (1981)
16. Sharkey, A.: Combining Artificial Neural Networks. Springer, Heidelberg (1999)
17. Zhang, Z., Hsu, M.: K-Harmonic Means – A Data Clustering Algorithm, Technical Report in Hewlett-Packard Labs, HPL-1999-124
18. Turi, R.H.: Clustering-Based Colour Image Segmentation, PhD Thesis in Monash University (2001)
19. Omran, M., Salman, A., Engelbrecht, E.P.: Dynamic Clustering Using Particle Swarm Optimization with Application in Image Segmentation. Pattern Analysis and Applications 6, 332–344 (2006)

20. Omran, M., Engelbrecht, E.P., Salman, A.: Particle Swarm Optimization Method for Image Clustering. International Journal of Pattern Recognition and Artificial Intelligence 19(3), 297–321 (2005)
21. Van den Bergh, F., Engelbrecht, E.P.: A Cooperative Approach to Particle Swarm Optimization. IEEE Transactions on Evolutionary Computing 8(3), 225–239 (2004)
22. Dunn, J.C.: A Fuzzy Relative of the ISODATA Process and Its Use in Detecting Compact Well-Separated Clusters. Cybernetics 3, 32–57 (1973)
23. Cui, X., Gao, J., Potok, T.E.: A Flocking Based Algorithm for Document Clustering Analysis. Journal of Systems Architecture 52(8-9), 505–515 (2006)
24. Halkidi, M., Batistakis, Y., Vazirgiannis, M.: On Clustering Validation Techniques. Intelligent Information Systems 17(2-3), 107–145 (2001)
25. Xiao, X., Dow, E., Eberhart, R., Miled, Z., Oppelt, R.: A Hybrid Self-Organizing Maps and Particle Swarm Optimization Approach. Concurrency and Computation: Practice and Experience 16(9), 895–915 (2004)
26. Waibel, A., Hanazawa, T., Hinton, G., Shikano, K., Shikano, K., Lang, L.: Phoneme Recognition Using Time-Delay Neural Networks. IEEE Transactions on Acoustics, Speech and Signal Processing 37(3), 328–339 (1989)
27. Auda, G., Kamel, M.S.: Modular Neural Networks: A Survey. International Journal of Neural Systems 9(2), 129–151 (1999)
28. Jain, A.K., Murty, M.N., Flynn, P.J.: Data Clustering: A Review. IACM Computing Surveys 31(3), 264–323 (1999)
29. Lange, T., Roth, V., Braun, M.L., Buhmann, J.M.: Stability-Based Validation of Clustering Solutions. Neural Computing 16, 1299–1323 (2004)
30. Blake, C.L., Merz, C.J.: UCI Repository of Machine Learning Databases (1998), http://www.ics.uci.edu/~mlearn/MLRepository.html

# Part IV
# Genetic and Evolutionary Algorithms in Bioinformatics

# Data-Mining Protein Structure by Clustering, Segmentation and Evolutionary Algorithms

Matej Lexa, Václav Snášel, and Ivan Zelinka

**Summary.** In this participation are discussed principles of bioinformatics, datamining, evolutionary computing and its mutual intersection. Datamining by means of selected evolutionary techniques are discussed with attention on protein structure, segmentation and state of art on the field of evolutionary algorithms use. Basic principles and terminology of evolutionary algorithms, as well as two evolutionary algorithms are mentioned here - differential evolution and self-organizing migrating algorithm.

## 1 Introduction

Biology has traditionally been an observational rather than a deductive science. Modern developments with the onset of modeling, simulation and the emergence of systems biology as a discipline have altered this orientation. Expansion of molecular biology, genomics and proteomics in the last decades has radically changed the nature of biological data as well. Data have become not only much more quantitative and precise, but, in the case of nucleotide and amino acid sequences, they have also become discrete. Analysis of this data gave birth to the new discipline of bioinformatics.

Clustering and segmentation techniques find a wide range of applications in bioinformatics, from hierarchical clustering of genes and proteins by

Matej Lexa
Masaryk University, Brno, Czech Republic
e-mail: `lexa@fi.muni.cz`

Václav Snášel
VSB - Technical University of Ostrava, Ostrava, Czech Republic
e-mail: `vaclav.snasel@vsb.cz`

Ivan Zelinka
Tomáš Baťa University, Zlín, Czech Republic
e-mail: `zelinka@fai.utb.cz`

A. Abraham et al. (Eds.): Foundations of Comput. Intel. Vol. 4, SCI 204, pp. 221–248.
springerlink.com                                        © Springer-Verlag Berlin Heidelberg 2009

structural and functional characteristics to splitting DNA and protein sequences into modules that can be assigned specific molecular functions. In this chapter we focus on protein aminoacid sequences and how they relate to local and overall structure of protein molecules. We review clustering and segmentation methods that have been instrumental in understanding protein sequence-structure relationships and discuss various similarity measures used in such studies. We discuss how these problems relate to (or can possibly benefit from) current and former advances in fields like computational linguistics, discrete data-mining or softcomputing and optimization. We highlight the most important applications of clustering and segmentation, including references to our own work. In our opinion, the most promising direction of research is the approach where clustering and segmentation methods are adapted to the hierarchical nature of protein sequences. Exploring all possible hierarchies that can be constructed above a protein sequence (or a set of sequences) is a complex task and may well require the use of heuristics and techniques from the domain of softcomputing. Evolutionary algorithms are one class of approaches that have been particularly succesfull in this respect.

## 1.1   Protein Sequence and Structure Primer

Proteins are an important group of molecules supporting life as we know it. With the exception of a small subgroup of viruses (which depend on proteins of their hosts anyway), all life forms on Earth are made of an array of various proteins, each with its specific structure and function. It is impossible to know how organisms function at the molecular level without knowing how proteins interact with each other, with the rest of the living matter and with the environment.

Detailed description of proteins and their molecular and biological functions can be found in numerous molecular biology and biochemistry textbooks [8, 20, 37, 63].

By evolutionary measures, proteins are perhaps younger than other "life molecules", such as DNA and RNA. However, they are the key component of all cells. They have important role as the building blocks of cells, tissues and are important in executing and regulating most biological processes. Enzymes, hormones, transcription factors, pumps and antibodies are all examples of the diverse functions fulfilled by proteins in a living organism.

Chemically, proteins can be described as macromolecules made of a linear chain of aminoacids connected by peptide bonds (a type of covalent bond between the carboxyl -COOH and amino -NH2 groups present in all aminoacids). Generally speaking, there are 20 aminoacids available for building proteins in every cellular organism. Mature proteins, however, can differ from this elementary building plan, by having some parts of the linear chain chemically cross-linked or further modified by addition of chemical groups, such as phosphates or saccharides.

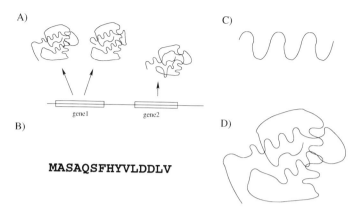

A)

B)

**MASAQSFHYVLDDLV**

C)

D)

**Fig. 1** Illustration of gene expression, primary, secondary and tertiary structures

For the purpose of this text, we will mostly work with symbolic representations of the molecules. This is routinely done by identifying the individual aminoacids that proteins are made of and representing them by a three-letter or a one-letter code. At this level, proteins can be viewed as sequences on the alphabet {A,C,D,E,F,G,H,I,K,L,M,N,P,Q,R,S,T,V,W,Y}. This so-called *primary structure* (Fig.1B) is quite important because only information about the primary structure is directly passed from one generation of cells to the next. Physically, this happens at the level of DNA molecules, which carry genes for individual proteins. Each gene specifies the primary structure of one protein, or at most a group of related proteins (Fig 1C). The function of proteins, however, depends on their ability to adopt a precise three-dimensional structure (*tertiary structure*, the process of aquiring the final shape is termed folding) (Fig.1D). Many proteins adopt their native structure automatically, without specific interactions with other cell components [17]. Other proteins only arrive at their functional structure upon interaction with other molecules, often proteins themselves [12, 64]. Different proteins may interact among themselves or with other proteins to form a temporary or permanent complex. These may be either homomeric or heteromeric (dimers, tetramers, this level is also termed *quarternary structure*), combining their partial functions into higher-level functions.

It is the process of folding, which still troubles many biologists and chemists today. We are quite able to determine the nucleotide sequence of DNA molecules, as has been done in numerous genome-sequencing projects, culminating a few years ago by sequencing the human genome [1]. Today, focus is moving to personal genomics. Having the genome available and using the genetic code we can identify genes in the sequence and determine the most likely primary structure of proteins that will be present in cells at some point of their life cycle. However, we are still unable to formulate the exact laws by which the proteins adopt their final structure based on their aminoacid sequence. This "knowledge" is apparently contained in the genes and in the resulting protein sequences [17], but this so-called "second code" or "folding code" has not been discovered yet [21]. Many of the methods described later in this text are aimed at deciphering parts of this second code.

## 1.2 Determination and Classification of Protein Structure

Tertiary structure of proteins can be determined experimentally. In the past, this has been done solely by protein crystallization, X-ray diffractometry, electron density calculation and subsequent model atom placement to satisfy the obtained electron density [10]. More than 42000 structures of proteins have been solved by X-ray to this date [2]. This approach has its fallbacks in differential ability of various proteins to crystallize. Protein chemists have even identified features associated with proteins that are difficult to crystallize [50]. Nuclear Magnetic Resonance (NMR) is a high-resolution alternative to X-ray that does not require crystallization. Molecules are probed in their native or near-native environment. It, however, requires labeling the protein with suitable isotopes for detailed analysis, restricting its use mostly to proteins that can be expressed from recombinant DNA [45],[31]. More than 7000 structures in PDB are due to NMR to date [2].

The moderate number of available structures in the PDB database contrasts sharply with the large number of protein-coding sequences identified individually or in one of the genome-sequencing projects (cca 2 mil., [3]). Obviously, our ability to solve structures of newly discovered proteins lags behind our ability to determine their primary structure. This is one of the reasons why methods that predict protein structure from sequence are sought. Currently, four main approaches to structure prediction exist (Table 1).

**Table 1** Different approaches used in protein structure prediction from sequence

| Approach | Requirements |
|---|---|
| homology-based modelling | similar sequence exists in PDB |
| threading | belongs to a known fold |
| fragment-based modelling | has segments with predictable structure |
| ab initio modelling | no similarity data used |

The discrepancy between the number of structures and sequences becomes less dramatic when we realize, that most of the known proteins fall into large protein families which follow a common structural plan (fold). It is estimated, that the PDB database currently covers most of the existing folds [4]. As more and more structures are solved experimentally, the percentage of unrepresented families may approach zero. In that case, knowing the folding code may become an academic issue, since empirical methods, such as threading may be able to assign the best fold based on protein primary structure.

Once the fold is known, computational refinement procedures are available to arrive at a probable tertiary structure.

Regardless of our ability to predict structure from sequence, the available structures have been classified by clustering based on their pairwise similarity. The two most respected databases created by such approach are CATH (http://www.cathdb. info/, [27, 9]) and SCOP (http://scop.mrc-lmb.cam.ac.uk/scop/, [6]).

## 1.3  Classification and Assignment of Protein Function

In theory, it should be possible to deduce the function of a protein from inspecting its structure in context of other structures present in the cell (or even by inspecting its sequence). In practice, all functional information is currently derived (directly or indirectly) from experimentally determined biological function. Molecular biologists are interested in several aspects of protein function. They strive to understand regulatory control of gene activation and the subsequent accumulation and degradation of proteins. If the protein is known to be present, we would like to know where in the organism or the cell it accumulates, which molecules it interacts with and how this reflects on the overall functioning of the organism or the cell. This kind of information has been present in scientific literature for many years, but in this form, it cannot be readily accessed by computational methods. One of the important milestones in bioinformatics was the creation of a curated

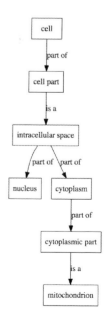

**Fig. 2** Gene Ontology is a directed acyclic graph of terms connected by "is a" and "is part of" relationships

database of protein (aka. gene product) functions, Gene Ontology [5]. The need for computational treatment of functional data dictates the way data is organized. Possible molecular and biological functions and cellular locations are organized into an ontology. Ontologies are directed acyclic graphs of terms, in this case connected by "is a" and "is part of" relationships (Fig.2).

## 1.4 Selected Techniques of Protein Clustering and Classification

It is a well-known fact that proteins are the building blocks of all organisms. Knowledge of their structure and their classification in general is one of vitally important tasks in bioinformatics today. Many methods have been developed within the few last years of research. For example methods measuring sequence similarity are Smith-Waterman method [72], which searches for local alignment. It compares substrings of all possible lengths. The Smith-Waterman algorithm is based on dynamic programming. This is an algorithmic technique where a problem is solved step by step by caching the solutions of sub-problems, and these are used in later stages of the computation. Another well-known method, FASTA [74], is a heuristic method, which provides an approximation of the local alignment score or BLAST [70] which is another heuristic method for sequence alignment. The oldest motif-based classification of proteins is PROSITE (http://www.expasy.ch/prosite, [26]).

Methods for classification based on full protein sequence are ProtoMap, which provides a fully automated hierarchical clustering of the protein space (http://protom ap.cs.cornell.edu, [76]). ProtoMap is based on a graph approach. Sequence space is represented by a directed graph where vertices are protein sequences and edges represent similarity between proteins. The weight associated with an edge measures the similarity, or the significance of the relationship. This method uses a combination of Smith-Waterman, FASTA, and BLAST to determine similarity. Another method is ProtoNet (http://www.protonet.cs.huji.ac.il, [69]). Hierarchical clustering of the SwissProt proteins is provides here. ProtoNet implements an average-link hierarchical agglomerative clustering algorithm. The novelty of ProtoNet is the use of several averaging methods. The averaging methods provided are arithmetic mean, geometric mean, and harmonic mean. Classification of the protein space based on single linkage agglomerative clustering is Systers (http://systers.molgen.mpg.de, [46]). Here so called single-linkage clustering (compare with average-linkage clustering used in ProtoMap and ProtoNet) defined the similarity between two clusters as the highest similarity between pairs from the two clusters.

Other clustering methods are ProClust (http://promoter.mi.uni-koeln. de/-procl ust, [60]). It is a protein clustering derived using a graph-based approach. A graph of proteins is constructed where edge-weights are based on Smith-Waterman similarity scores scaled with respect to self-similarity. Method using SwissProt/TREMBL proteins is CluSTr (http://www.ebi.ac.uk/clustr, [47]). The clustering is based on single-linkage hierarchical clustering; underlying scores are based on Smith-Waterman. Some of those methods are incorporated into software packages like Tribe-MCL (http://www.ebi.ac.uk/research/cgg/tribe, [25], which is a protein clustering software package. Results of BLAST are used as input, producing output in the form of proteins clustered into families.

Full-sequence analysis is important in protein classification with known sequences. Typical perception today is such that protein function stems from protein structure and not so much from sequence. Thus the idea of classifying proteins based on structure. One of the significant drawbacks of structure-based clustering is that the number of available sequences is relatively small, due to the complexity of obtaining high-resolution structural information. Several different algorithms and various software packages are available to measure the similarity between two protein structures. Examples are VAST [38], CE [35], PrISM [13], Dali [49] and STRUCTAL [59].

A full description of all of these methods is of course beyond the scope of this chapter. Methods mentioned here are only a fraction of today's set of various methods used for protein clustering and classification. However, the problem of protein structure prediction can also be stated as an optimization problem. In that case, optimal protein alignment or/and classification can be represented by multimodal functions or/and N dimensional hyperlanes with many local optimas and usually one global optimum. This global optimum may represent an optimal alignment or a preferrred structure. Traditional search techniques usually fall into local optimas. This is not the case of a special class of search algorithms - evolutionary algorithms.

# 2   Evolutionary Algorithms and Bioinformatics

Evolutionary algorithms (EAs) are widely used in bioinformatics today for tasks like sequence and structure alignment, protein structure alignment, protein folding problem, clustering of microarray data, identifying metabolic pathways and gene regulation, etc. The most representative literature sources demonstrating the use of EAs in the bioinformatics field are [22] and [29].

Evolutionary computation can be easily found in many applications in bioinformatics. For example in [32] is proposed a Chaotic Genetic Algorithm to cluster protein interaction data to find protein complexes. Compared with other selected computation methods can be stated that the most significant advantage of this method is that it can (according to author's report in this article) find as many potential protein complexes as possible. This method has been tested in application on the Yeast genomic data. In [61] has been investigated a new search strategy in combination with the simple genetic algorithm on a two-dimensional lattice model to improve protein folding simulations. In proposed strategy (called by author also systematic crossover) are coupled the best individuals, tests every possible crossover point, and takes the two best individuals for the next generation. The standard genetic algorithm with and without this new implementation for various chain lengths has been tested. Another paper [56] also uses evolutionary algorithms, especially novel evolutionary algorithm for regulatory motif discovery in DNA promoter sequences. The algorithm uses data clustering to logically distribute the evolving population across the search space. Mating then takes place within local regions of the population, promoting overall solution diversity and encouraging discovery of multiple solutions. For the experiments were used synthetic data sets to find position frequency matrix models of known regulatory motifs in relatively long promoter sequences. Another paper [19] develops an evolutionary method that learns inductively to recognize the makeup and the position of very short consensus sequences, cis-acting sites, which are a typical feature of promoters in genomes. The method combines a Finite State Automata and Genetic Programming to discover candidate promoter sequences in primary sequence data. An experiment measures the success of the method for promoter prediction in the human genome. Similar methods from genetic programming field is [75]. Genetic Programming has again been used for biopattern discovery. In [77] has been adopted evolutionary computation, which combines a clustering method and genetic algorithms to produce the schemata for the visible natures of protein secondary structures. There are two major roles of clustering algorithm, one is to generate part of the initial chromosomes in the genetic algorithms and the other is to assist schemata in predicting secondary protein structures. By performing the new approach, the accuracy of Q3 can be improved 12% more than the previous algorithm.

When confronted with a difficult optimization task, the method of first choice will usually be a problem-specific heuristic. By incorporating expert knowledge into their design, such techniques invariably achieve superior performance when compared to general methods like EAs. For example, if the objective function is a linear polynomial, then efficient methods exist, like the simplex method and Karmakar's algorithm, which can provide globally optimal solutions for problems containing hundreds, even thousands, of variables. If the objective function is not a linear polynomial and if problem-specific methods are not available, then an appeal is usually made to the Levenberg-Marquardt algorithm, branch-and-bound methods, or to real-valued, simplex derivatives like that of Nelder and Mead. Unfortunately, functional pathologies like optima), epistasis (parameter interaction), constraints, flatness and non-differentiability, either alone or in combination, can easily render these deterministic methods impotent. As stochastic algorithms, EA's have the potential to overcome many of these limitations, but if they do not incorporate problem-specific knowledge into their design, their performance will likely be sub-par. Fortunately, knowledge that the optimization process will be restricted to the domain of totally ordered spaces and in particular to the space of continuous variation, is itself problem-specific information that can be used to enhance an EA's performance without jeopardizing its status as a general-purpose numerical optimizer.

While the value of evolutionary algorithms as general-purpose numerical optimizers has often been questioned, much of this criticism has (justifiably) been based on results obtained with combinatorial algorithms that were not originally designed for the task. Like a chain, a successful numerical optimization algorithm will only be as effective as its weakest link, and within the context of real parameter global optimization, integer encoding, non-isomorphic search strategies, non-scaling mutation operators and susceptibility to coordinate rotation are all *weak links* that can significantly degrade performance. Even once these design challenges have been adequately addressed, an effective global optimizer must still find a way to accommodate the peculiarities posed by each objective function without becoming victimized by them. To this end, differential evolution (DE) introduces a novel, self-referential mutation scheme that is not only elegantly simple, but also remarkably effective.

Unlike EAs that rely on the output of a predefined probability distribution function, differential evolution algorithms drive mutation with the differences of randomly sampled pairs of object vectors: $\overrightarrow{x}_{r1} - \overrightarrow{x}_{r2}$. Obviously, the distribution of these object vector differences is determined by the distribution of the object vectors themselves. Moreover, the way in which object vectors are distributed depends primarily on their response to the objective function's topography. EAs are also:

fast
simple
easy to use (*tune*) and modify
effective; superior global optimization ability
inherently parallel
does not require that the objective function be differentiable
operates on flat surfaces
works with noisy, epistatic and time-dependent objective functions
can provide multiple solutions in a single run
can provide co-evolutionary solutions for games and simulations
effective in nonlinear constraint optimization in conjunction with penalty functions

## 2.1   Evolutionary Algorithms - A Brief Survey

Evolutionary algorithms (EAs), in general, are based in principle on Darwinian idea of evolution, i.e. on survival of the fittest. As a father of an evolutionary process implemented in a computer can be accepted british mathematician A.M.Turing which says in essay "Intelligent Machines" (1948): ...*if the untrained infant's mind is to become an intelligent one, it must acquire both discipline and initiative ... discipline is certainly not enough in itself to produce intelligence. That it is required in addition we call initiative ... our task is to discover the nature of this residue as it occurs in man, and to try and copy it in machines.* Continuing he made the connection between searches and AI by saying: *...further research into intelligence of machinery will be probably very greatly concerned with "searches".* These ideas have continued in "Computering Machinery and Intelligence" (1950): *...we cannot expect to find a good child-machine at the first attempt. One must experiment with teaching one such machine and see how well it learns. Then try to another and see if it is better or worse. There is an obvious connection between this process and evolution, by the identifications:*

*structure of child machine = hereditary material*
*changes of the child machine = mutations*
*natural selection = judgement of the experimenter*

Simply put, EA is a numerical process, during which several (or many) possible solutions of a given problem are processed together. These solutions are called *individuals* and usually are represented by a vector consisting of arguments of a predefined cost function. A set of individuals is called a *population* and each individual is associated with a *fitness*, which can be in the simplest case the value of the cost function indicating its suitability. For example, with a cost function $F_{cost}(x_1, x_2, x_3, x_4, x_5)$, all individuals constitute a set of x, e.g., $I = \{x_1, x_2, x_3, x_4, x_5\}$, a population consist of N individuals with numerical values instead of x, like $I_1 = \{2, 44, 51, -3.24, -22, 2\}, I_2, I_3, ..., I_N = \{0.22, 3.4, 44, 1, 0.001, 0\}$, as illustrated by Fig.3. Parameter values in the individuals are assigned randomly at the beginning of the evolutionary process,

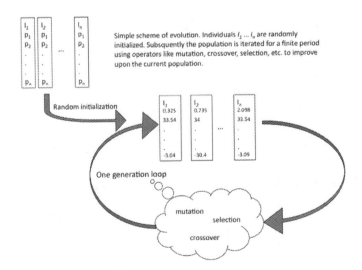

Simple scheme of evolution. Individuals $I_1$ ... $I_n$ are randomly initialized. Subsquently the population is iterated for a finite period using operators like mutation, crossover, selection, etc. to improve upon the current population.

**Fig. 3** Population in the evolutionary algorithm

i.e., the whole population is randomly generated. The population is then used to create the so-called offsprings' new individuals, by means of selecting some individuals from the population (also called parents). This is done by operations like cross-over, mutation, etc. A number of variants of such operations are available thanks to the existence of a large family of evolutionary algorithms. In principle, these operations are only arithmetical (or geometrical) operations, which combine selected individuals from the parental population.

When an offspring is created, it is evaluated for its fitness by using the cost function. The individual with the best fitness (parents or offspring) survives, i.e., taking a place in the new population. When the new population is filled by successful individuals, the old one is removed and the offspring population becomes the new population of parents for the next cycle of offspring creation. This loop is usually called *generation*, or *migration*, etc.

Evolutionary process is thus an iterative process with selection and survival of the temporarily best solutions, which are used in the next generation to create better solutions. In the end, the best individual (i.e., problem solution) is selected from the last population and is regarded as the best solution from the evolution just ended. The cost function, used in the population, should be defined such that its minimization or maximization could lead to an optimal solution. From this point of view, evolution can also be regarded as a mutually parallel search on an N-dimensional, nonlinear and complicated surface, where each point represents a possible solution. Typical example of the *cost function landscape* of the individuals with two parameters, I = {x, y}, is depicted on Fig.4, in which the cost value is marked on the $z$-axis.

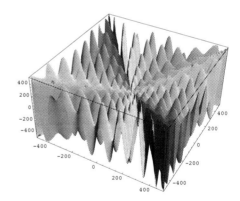

**Fig. 4** Surface representation of cost function, the test function called Rana's function

Today, a large set of variants of evolutionary algorithms exists. They differ each other by the mathematical principles in driving them and by the fundament philosophy employed. Another major difference is that individual representation may consist of integers and/or real numbers, like I = {2, 44, 51, -3.24, -22, 2}, or of binary strings, like I = {0010101101010101}, which is typical for a genetic algorithm in its canonical form. For a more detailed study of EAs, which is normally quite time consuming, it is recommended to consult with the literature, for example, [39], [48] on genetic algorithms, [42]; [43]; [44] on differential evolution, [67]; [73] on simulated annealing, [62] on particle swarm, [34] on SOMA, [33] on evolutionary strategies, [51] on AntColony Optimization, and [71] on general issues.

## 2.2 A Brief Review of Selected Evolutionary Algorithms

For both numerical and symbolic experiments described below, stochastic optimization algorithms such as Differential Evolution (DE) [44], SelfOrganizing Migrating Algorithm (SOMA) [34] and Genetic Algorithms [39] are selected to use.

DE (Fig.5) is a population-based optimization method that works on real-number-coded individuals. For each individual $\vec{x}_{i,G}$ in the current generation G, DE generates a new trial individual $\vec{x}'_{i,G}$ by adding the weighted difference between two randomly selected individuals $\vec{x}_{r1,G}$ and $\vec{x}_{r2,G}$ to a randomly selected third individual $\vec{x}_{r3,G}$. The resulting individual $\vec{x}'_{i,G}$ is crossed-over with the original individual $\vec{x}_{i,G}$. The fitness of the resulting individual, referred to as a perturbed vector $\vec{u}_{i,G+1}$, is then compared with the fitness of $\vec{x}_{i,G}$. If the fitness of $\vec{u}_{i,G+1}$ is greater than the fitness of $\vec{x}_{i,G}$, then $\vec{x}_{i,G}$ is replaced with $\vec{u}_{i,G+1}$, otherwise, $\vec{x}_{i,G}$ remains in the population as $\vec{x}_{i,G+1}$. DE is quite robust, fast, and effective, with a global optimization ability. It does not require the objective function be differentiable, and it works well even with noisy, epistatic and time-dependent objective functions.

**Table 2** DE pseudocode

1. Input: $D, G_{max}, NP \geq 4, F \in (0, 1+), CR \in [0, 1]$, and initial bounds : $\vec{x}^{(lo)}, \vec{x}^{(hi)}$.

2. Initialize: $\begin{cases} \forall i \leq NP \wedge \forall j \in D : x_{i,j,G=0} = x_j^{(lo)} + rand_j[0,1] \bullet (x_j^{(hi)} - x_j^{(lo)}) \\ i = \{1, 2, \ldots, NP\}, j = \{1, 2, \ldots, D\}, G = 0, rand_j[0,1] \in [0,1] \end{cases}$

$\begin{cases} \text{3. While } G < G_{max} \\ \forall i \leq NP \begin{cases} \text{4. Mutate and recombine:} \\ \quad \text{4.1 } r_1, r_2, r_3 \in \{1, 2, \ldots, NP\}, \text{randomly selected, except : } r_1 \neq r_2 \neq r_3 \neq i \\ \quad \text{4.2. } j_{rand} \in \{1, 2, \ldots, D\}, \text{randomly selected once each } i \\ \quad \text{4.3 } \forall j \leq D, u_{j,i,G+1} = \begin{cases} x_{j,r3,G} + F \cdot (x_{j,r1,G} - x_{j,r2,G}) \\ \text{if } (rand_j[0,1] < CR \vee j = j_{rand}) \\ x_{j,i,G} \text{ otherwise} \end{cases} \\ \text{5. Select} \\ \quad \vec{x}_{i,G+1} = \begin{cases} \vec{u}_{i,G+1} \text{ if } f(\vec{u}_{i,G+1}) \leq f(\vec{x}_{i,G}) \\ \vec{x}_{i,G} \text{ otherwise} \end{cases} \end{cases} \\ G = G + 1 \end{cases}$

Pseudocode for DE, especially for DERand1Bin, is shown in Table 2 An example of DE is demonstrated on Fig.5.

SOMA () is a stochastic optimization algorithm, modelled based on the social behavior of competetive-cooperating individuals [34]. It was chosen

**Fig. 5** Diferential
evolution
(http://www.icsi.berkeley.
edu/~storn/code.html)

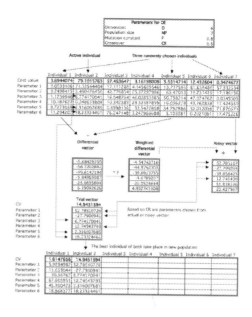

because it has been proved that this algorithm has the ability to converge
towards the global optimum [34]. SOMA works on a population of candidate solutions in loops, called migration loops. The population is initialized
by being randomly and uniformly distributed over the search space at the
beginning of the search. In each loop, the population is evaluated and the
solution with the lowest cost value becomes the leader. Apart from the leader,
in one migration loop, all individuals will traverse the searched space in the
direction of the leader. Mutation, the random perturbation of individuals, is
an important operation for evolutionary strategies. It ensures the diversity
among all the individuals and it also provides a means to restore lost information in a population. Mutation is different in SOMA as compared with
other evolutionary strategies. SOMA uses a parameter called PRT to achieve
perturbations. This parameter has the same effect for SOMA as mutation
for GA. The novelty of this approach lies in that the PRT vector is created
before an individual starts its journey over the search space. The PRT vector
defines the final movement of an active individual in the search space. The
randomly generated binary perturbation vector controls the permissible dimensions for an individual. If an element of the perturbation vector is set to
zero, then the individual is not allowed to change its position in the corresponding dimension. An individual will travel over a certain distance (called
the PathLength) towards the leader in a finite number of steps in the defined
length. If the PathLength is chosen to be greater than one, then the individual will overshoot the leader. This path is perturbed randomly. Pseudocode
for SOMA is shown in Table 3.

An example of SOMA is demonstrated on Fig.6.

For exact descriptions of the algorithms, see [44] for DE, [34] for SOMA, and [39]; [48] for GA which is not described here due to its popular awareness.

**Table 3** SOMA pseudocode

Input:$N, Migrations, PopSize \geq 2, PRT \in [0,1], Step \in (0,1], MinDiv \in (0,1],$
PathLength $\in (0,5], Specimen$ with upper and lower bound $x_j^{(hi)}, x_j^{(lo)}$

Initialization:
$$\begin{cases} \forall i \in PopSize \wedge \forall j \in N : x_{i,j,Migrations=0} = x_j^{(lo)} + rand_j[0,1] \bullet (x_j^{(hi)} - x_j^{(lo)}) \\ i = \{1,2,\ldots,Migrations\}, j = \{1,2,\ldots,N\}, Migrations = 0, rand_j[0,1] \in [0,1] \\ \begin{cases} While Migrations < Migrations_{max} \\ \forall i \leq PopSize \begin{cases} While t \leq PathLength \\ if rnd_j \leq PRT then PRTVector_j = 1 \text{ else } 0, j = 1,\ldots,N \\ x_{i,j}^{ML+1} = x_{i,j,start}^{ML} + (x_{L,j}^{ML} - x_{i,j,start}^{ML})t PRTVector_j \\ f(x_{i,j}^{ML+1}) = \text{if } f(x_j^{ML}) \leq f(x_{i,j,start}^{ML}) \text{ else } f(x_{i,j,start}^{ML}) \\ t = t + Step \end{cases} \\ Migrations = Migrations + 1 \end{cases} \end{cases}$$

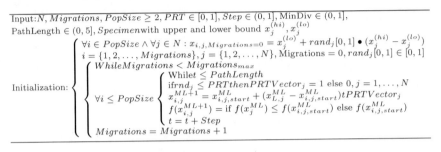

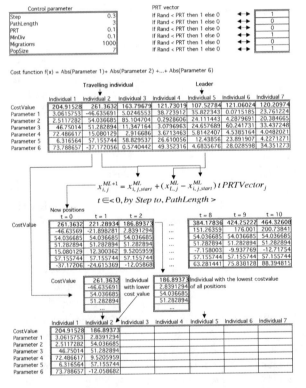

**Fig. 6** SOMA algorithm (http://www.fai.utb.cz/people/zelinka/soma)

## 2.3  Cost Function in Bioinformatics

An appropriate cost function is one of the most important points in evolutionary computation. When the cost function is not properly defined, the obtained results are usually wrong. Nature of EAs is to minimize or maximize. The cost function thus has to be defined as a minimization or maximization problem. We will show an example of lattice models of protein structure, where the cost function evaluates the number of hydrophobic - hydrophilic amino acid interactions.

## 2.4  Evolutionary Algorithms and Lattice Models of Protein Folding

The problem of predicting protein structure from sequence can be approached from different angles. In addition to an atomic-scale physical-chemistry energy-minimalization approach, one can take a more coarse-grained view of proteins. Amino acids can be considered rigid, while only angles between neighboring amino-acids are allowed to be discretized, so that predicting protein structure becomes a combination and optimalization problem from the domain of computer science rather than a chemical one. Stating the problem this way has lead to a relatively well-defined protein structure prediction (PSP) problem defined and solved using lattice models of protein structure [58],[16]. For example in a cubic lattice model, connections between amino-acids can take one of the six possible principal directions on a cubic equally-spaced grid (Fig.7B). Simplifying even the amino-acid alphabet from 20 to 2 possible symbols (H - hydrophobic (non-polar); P - hydrophilic (polar)) leads to the problem of placing a sequence of Hs and Ps on the grid, so that the number of H-H and P -P neighbors is maximized at the expense of H-P neighbors. The problem has a simpler version on a 2-D lattice (Fig.7A) that relates to the 3-D version of the problem. It has been shown that this problem of global minimum search is NP-complete [14]. As such, it is ideally suited for the application of evolutionary algorithms and this has been done by many authors, to find feasible sub-optimal solutions. The complexity of the problem limits the size of the protein that can be predicted in this manner. So far the best prediction seems to be that of [11], for a 60-amino acid ribosomal protein L20, although these authors considered individual atoms and used a richer energy function.

In the past, only small or artificial proteins were analyzed, however, recent works employing lattice models approximate larger and more complex proteins [11].

Search for the optimal arrangement of a sequence in a lattice model happens to be solvable efficiently with specially tailored evolutionary algorithms.

Because proteins are molecules coded by linear sequences of their corresponding genes, it is quite possible that the approach using EAs is so

A)                                                B)

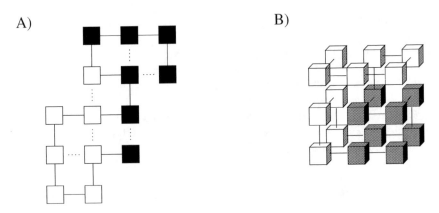

**Fig. 7** Illustration of a A) 2-D and B) 3-D HP lattice model of protein structure.
The objective of optimization is to maximaize contacts between nodes of the same
type (dotted lines) and minimize other interactions. Full lines model the peptide
backbone

successful for a reason. Perhaps nature has been solving the same kind of optimization problems on real protein molecules using the very same operations
on genomic sequences that we use in nature-inspired EAs.

## 3   Similarity Measures for Proteins

### 3.1   Sequence Similarity

As mentioned before, in live cells, proteins and their coding sequences are
passed from generation to generation from one cell to another. Life as we know
it depends on their interdependent functions. Ironically, life as we know it
also depends on the lack of constancy in their precise structure and function,
allowing evolution to introduce slight changes in individuals that lead to
speciation and more complexity in nature. As a result, many proteins share
the same structural and functional theme, because they evolved from the
same ancestor gene/protein. Such proteins are called homologs (orthologs, if
they exist in two different and paralogs if they are a result of gene duplication
that remains within the same species).

Because molecular biologists must mostly rely on living species to study
DNA and proteins, identification of homologs can best be done by comparison of biological sequences and structures, assuming that the amounts of
differences seen today are proportional to the time passed from the dupli-
cation/speciation event. A whole discipline of phylogenetics or evolutionary
molecular biology studies this problem.

Biological sequence comparison can therefore be best viewed as a method
to detect homology. As such, it provides us with a powerful, but often unreli-
able tool - transitive structure and function assignment. If sequence similarity

between two proteins is high enough to suggest homology, it is quite likely that the two proteins still share the same fold and general function. Mutations present in the sequences and structures of the two proteins would be best thought of fine-tunings of the given protein to its new environment. In extreme, but frequent cases, one of the homologs may become redundant and free to take on a new function, or to dissappear from the organism altogether.

To compare two sequences, molecular and computational biologists most often rely on an edit-distance-like metric, where each edit operation represents a mutation. The advantage of this method is that similarity can be efficiently calculated by a dynamic-programming algorithm, using an amino-acid similarity matrix to give appropriate weights to likely and less-likely mutations. Today BLOSUM and PAM family of matrices are the most commonly used [66],[57].

This method of comparing sequences has gained extreme popularity with biologists. Its popularity actually reached a point where it is used even outside its proper scope. Many proteins are homologs that drifted further apart in their structure and function, accumulating more dramatic changes than just single-aminoacid substitutions or deletions/insertions. Such proteins may share only a limited portion of their sequence/structure/function. Still, big changes may have happened relatively quickly in evolution, for example by re-translocation of a complete gene fragment into a suitable insertion site. What may seem as a lengthy change of many aminoacids to the previous model, may have been a rapid high-level switch of already available (to evolution) functions.

We will list a few alternatives to the highly popular Needleman-Wunsch and Smith-Waterman algorithms and leave it mostly to the reader to speculate about how well they may or may not model the evolutionary process.

*N-gram composition.* One of the main problems of using a form of simple edit distance to compare protein sequences is its lack of respect for medium-scale rearrangements, such as multiple reversions or translocations. The metrics of n-gram composition, so popular for comparing documents and other human-language strings is an attractive alternative to Needleman-Wunsh global similarity scores used by molecular biologists. Let us have four sequences

A = MASAFNMDHDIL original sequence
B = MASAYNMDHEIL two point mutations
C = FNMDMASAHDIL single segmental translocation
D = FAHDINMDMASL double segmental translocation

Sequence B contains two "point mutations" compared to A. Sequence C contains a single translocation (the first four aminoacids are moved to the middle of the string). Sequence D contains an additional translocation.

| | distance N-W | n-gram | mutation type |
|---|---|---|---|
| AB | 2 | 4 | point |
| AC | 8 | 2 | single translocation |
| AD | 11 | 5 | double translocation |

The above table clearly shows that the edit distance-based Needleman-Wunsch score penalizes translocations heavily, while n-gram composition is a much better measure of the number of elementary changes applied to the original sequence.

Analysis of protein sequences based on their n-gram composition has been successfully used by [30] [52] [7]

*Markovian similarity.* The above example modeled protein sequences as sets of independent n-grams. In reality, the consecutive n-grams depend on preceding n-grams. This dependency can be modelled as a Markov chain, assigning probabilities to amino acids based on previously read sequence. This is just another way of saying that protein sequences are localy constrained. Near-neighbors in the sequence influence each other, often forming one of several typical types of *secondary structure* ($\alpha$-helix, $\beta$-strand, turns, etc.). Coiled regions have no recognizable secondary structure. Probabilities associated with a Markov model of helical stretches in proteins differ from probabilities associated with other types of secondary structure [36].

*Compressibility or algorithmic distance.* The idea that sequences that can be generated by similar Markov models ought be similar can be taken one step further. Any kind of sequence model can be used to compress sequences, simply by reducing the uncertainty about the next character in the sequence. We can use this knowledge to recode the sequence (for example by arithmetic coding) so as to compress it as much as possible. Intuitively, if our sequence model comes from a similar sequence, compression will be more efficient. This leads to a simple scheme for calculating sequence similarity above a specific sequence model: take two sequences, concatenate them and compress the resulting sequence. Similarity should be a function of how well the concatenated sequence can be compressed, compared to each of the sequences alone [55],[54].

## 3.2  *Structural Similarity*

Two main approaches to comparing structural similarity can be traced in literature. One considers the coordinates of individual atoms or some representative coordinates of amino-acids (Carbon-$\alpha$, center). The similarity of two proteins defined by such coordinates can then be obtained from the distance between two structures A and B given by

$$D = \sqrt{\sum_{i=1}^{N} \left( d(x_i^A, x_i^B) \right)^2}$$

where $d(x_A, x_B)$ is the distance between i-th aminoacids or atoms, N is the total number of amino acids or atoms and A and B are the tw proteins.

If one allows for certain amount of discretization in the structural data, it is possible to convert three consecutive aminoacids into a vector or sterical angle of the first two and the last two points in space, bin the values of the calculated angles into discrete intervals and label each interval by a symbol. The structure then becomes a sequence of symbols and structural similarity can be assessed using algorithms for sequence similarity. This appraoach has the advantage of faster calculations on sequences than is possible with large sets of floating-point coordinates.

## 4   Clustering Protein Segments

In this section, we look at methods that group protein fragments into related subgroups based on their sequence, structure, or both. Many different approaches to clustering of proteins have been undertaken so far. Depending on the entity clustered, whole molecules may be grouped together, or only segments of proteins may be used (Fig.8). Based on the similarity metrics used, clustering can follow sequence similarity, structural similarity, or even ability to form some kind of pairs in the sequences or in the structures. Each kind of similarity can be calculated in one of many different ways (global sequence similarity according to Needleman-Wunsh algorithm, n-gram content, composition, etc.).

**Fig. 8** Three distinct modes of clustering protein structures (or sequences). A) Entire proteins are clustered based on similarity B) Protein segments are clustered based on similarity and C) protein segments are clustered based on compatibility

## *4.1   Clustering by Sequence Similarity*

Entire protein sequences can be compared by sequence similarity and clustered into groups sharing important sequence motifs. This has been done previously in many different ways, helping us to identify important protein families. This approach leads to a network of proteins connected by similarity. The topology of such network would be undisputable, if distantly related proteins all shared some common "ancient" motif at all levels of detail. However, the same problem that complicates sequence similarity calculation for distant homologs, complicates clustering of protein sequences. If a protein contains sequential motifs from several protein families, how should it be properly grouped and classified?

As in previous sections, we focus on methods that work on protein segments rather than entire sequences. What metrics are appropriate for such methods? If we analyze many different proteins and use Needleman-Wunsch global sequence similarity to compare their arbitrary fragments, we position ourselves to evade the "distant homolog curse". Rather than finding mixed sequences that are difficult to cluster, we isolate individual protein fragments to cluster just lower-level functional motifs. Exercises like that have been carried out by Trifonov [23]. Clustering lead to large clusters of protein fragments, which often followed the same pattern of hydrophobic and hydrophillic amino-acids. This lead Trifonov to the conclusion, that ancient life may have been made of proteins that were made of only two amino-acids, one hydrophobic and one hydrophillic [24].

## 4.2   Combined Sequence-Structure Clustering

The previous methods of working with protein fragments have some serious drawbacks. Sequence similarity is only useful to the point where some features of the ancient sequence are still recognizable after numerous mutations. Sometimes a seemingly unrelated sequence of aminoacids remains in identical structure and still, no sequence similarity method is able to recognize the homology.

A way around this problem, is to use similarity metrics in the structure space, rather than sequence space. That way we can identify larger groups of sequences that could never be arrived at by sequence clustering. But what if fragments clustered by structure had no common sequenctial motif?

It turns out, that this problem has a very elegant solution, that led to some of the best structure prediction methods available today. Baker, Bystroff and Han realized [15] they can iteratively cluster fragments by sequence and by structure until the only clusters remaining are those, where there is a 1:1 relationship between sequence and structure. They call the structural representation of the center of their cluster a "paradigm". This paradigm is basically an average structure for the given group of fragments, while its sequence properties are represented by a sequence profile. The profile simply defines the likelihood of encountering individual aminoacids at each of the positions in the fragment.

# 5 Useful Analogies

Every scientific discipline proceeds partly on its own, developing its own tools and approaches to its particular problems. Although a necessity, it is also critical for any such discipline to occasionally adopt approaches useful in other areas of science. We determined several contact-planes between biological sequence analysis and computational linguistics, game theory and image analysis.

## 5.1 *Biological Sequence as a Language*

One way to look at biological sequences is to think of them as a message for the next generation. In cells, this message is processed, it materializes in proteins and carries out its meaning. Something similar happens in human language. A message in the form of sound waves (or written text) finds its way into the brain where it gets processed and materializes in a set of activated neurons. Both messages are written in a language defined by proper and inproper combinations of lower-level symbols arranged in sequence. Both molecular structures and neural systems are multidimensional structures created by the sequence.

Analogical terms and structures between biological sequences and natural language are shown in Table 4.

**Table 4** Biological sequence - human language analogy examples

| Linguistic domain | Biological domain |
|---|---|
| sentence | protein |
| phrase | domain |
| word | secondary structure |
| letter | amino acid |
| meaning (sentence) | function (protein) |
| meaning (of a word) | binding (of a structure) |
| set of neurons | molecule |
| learning | metabolism |

# 6 Statistical Sequence Models

Whenever we deal with a set of sequences, such as a cluster of protein fragments with a non-random aminoacid composition, we need to be able to characterize this "non-randomness" of the sequences. This section reviews some of the methods applicable.

## 6.1  Profiles

Perhaps the most intuitive way of defining a group of related sequences is by their sequence profile. A profile is a matrix with 20 rows (one for each aminoacid) and one column for every possible position in the sequences. The matrix elements are probabilities of a given aminoacid occuring at the given position. Such model is quite useful when positions within a motif or a fragment can be fixed. Fortunately, this is often the case.

For example, many proteins are capable of binding nucleotides in cells. Nucleotides are building blocks of DNA, but they also serve as regulatory and energy-providing molecules in cells. As such, they must be recognized by many different proteins. The structure of a nucleotide is always the same, regardless of its source, it is not surprising then to find motifs in proteins that are part of this recognition. For example, the well-known Rossman motif satisfies the following aminoacid profile: GX[GS]XX[GSTA] [68].

## 6.2  Markov Chains

While profiles treat each position in the aminoacid sequence as independent of all the other positions, often it is impossible to predict aminoacids at individual positions without reference to previous positions. Situations like these arise in alfa- a nd beta- secondary structure, where short-distance relationships between aminoacids contribute to the formation of a particular secondary structure.

A common way to model the sequence in such cases is the use of Markov chains. The set of probabilities of encountering one aminoacid after another forms the basis of a Markov model. Different sequences can be represented by different Markov models.

First- and second-order Markov models are usually most useful in general cases. Specific types of protein fragments may be modelled as higher-order Markov chains.

## 7  Segmentation

It should be noted, that many of the previous topics considered proteins to be made of shorter fragments that are combined to provide a higher-level function. It is just a short step from there to ask ourselves, how we could possibly identify such fragments by analyzing the sequences or structures of known proteins. Several methods of protein segmentation exist already, however we see a number of possible directions that have not been explored in great detail yet.

In this section we review some of these method and propose a few as well.

## 7.1   Segmentation by Competing Sequence Models

We have already seen that different types of protein secondary structure can be successfully predicted from sequence using Markov models [36]. A model trained to recognize sequences charateristic of an α-helix will perform poorly on β-sheet sequences and vice versa. Generalizing to any kind of model, from simple statistical distributions to complex neural-network models, we can always detect sequence segments compatible with specific models or their particular versions. Among early applications of this type to genomic data are [28], [41].

## 7.2   Information-Theorethical Segmentation

Protein sequences contain a variety of short-, medium- and long-range cor-rellations that most likely reflect different chemical, physical, structural and functional interactions in the protein molecule. Given a large set of sequences such as those present in the Uniprot database or in the proteome of a specific organism, we can analyse any sequence by comparing its composition with the selected set. Typical combinations of amino acids over different distances can be detected by calculating mutual information between n-grams or other short patterns of aminoacids in the analysed sequence. Mutual information

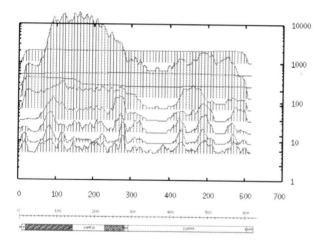

**Fig. 9** Information-theory based analysis of the ARR21 protein sequence shows that areas in the sequence spanned by a high number of segment pairs exceed-ing specific mutual-information threshold (red graphs) coincide with described do-main composition of the protein (below). Different red graphs show information for segment pairs spanning the sequence at a different distance (from bottom up: 4,8,16,32,64,128,256,512,1024 amino-acids)

has been successfully used as a segmentation measure in the area of syllable-
and word- segmentation of Japanese and Chinese texts [40],[65].

We applied such MI-based measure to protein data in a preliminary study
and found it to correlate with domain boundaries [53] (Fig.9).

Similar approach has been successfully used to segment membrane proteins. These proteins contain segments that span the hydrophobic membrane
interior but also stretches of amino acids exposed to the hydrophilic regions
of cells on both sides of the membrane. It turns out that n-grams in these
sequences can reveal boundaries between these segments without relying on
any structural information whatsoever [18] [30] [52] [7] .

## 8 Conclusions

We presented the problem of protein sequence analysis and that of mining
protein sequences for structural information. This area of research is currently
very active, with researchers abandoning traditional sequence similarity approaches used by molecular biologists for decades, turning to applications
of information theory, evolutionary computing, combined sequence-structure
clustering and language-inspired segmentation. While it is unclear whether a
universal structure prediction method can be found, these avenues of research
are already bringing their fruit by bridging the gap between bioinformatics
and other disciplines, such as computational linguistics, image analysis, game
theory and data-mining.

**Acknowledgements.** This work was supported by Grant no. MSM 7088352101 of
the Ministry of Education of the Czech Republic and by grant of the Grant Agency
of the Czech Republic GACR 102/06/1132.

## References

1. http://www.ornl.gov/sci/techresources/Human_Genome/home.shtml
2. http://www.pdb.org
3. http://www.uniprot.org
4. http://scop.mrc-lmb.cam.ac.uk/scop/count.html
5. http://www.geneontology.org
6. Scop a structural classification of proteins database for the investigation of
   sequences and structures. J. Mol. Biol. 247, 536–540 (1995)
7. BoganMarta, A., Hategan, A., Pitas, I.: Language engineering and information
   theoretic methods in protein sequence similarity studies. In: Computational
   Intelligence in Medical Informatics, pp. 151–183 (2008)
8. Bruce, A., Johnson, A., Lewis, J., Raff, M., Roberts, K., Walter, P.: Molecular
   Biology of the Cell, 4th edn. Garland (2002)
9. Harrison, A., Pearl, F., Sillitoe, I., Slidel, T., Mott, R., Thornton, J., Orengo,
   C.: Recognizing the fold of a protein structure. Bioinformatics 19(14), 1748–
   1759 (2003)

10. McPherson, A.: Introduction to Macromolecular Crystallography. John Wiley Sons, Chichester (2003)
11. Schug, A., Wenzel, W.: An evolutionary strategy for all-atom folding of the sixty amino-acid bacterial ribosomal protein l20. Biophys J (2006)
12. Dunker, A.K., Lawson, J.D., Brown, C.J., Williams, R.M., Romero, P., Oh, J.S., Oldfield, C.J., Campen, A.M., Ratliff, C.M., Hipps, K.W., Ausio, J., Nissen, M.S., Reeves, R., Kang, C.H., Kissinger, C.R., Bailey, R.W., Griswold, M.D., Chiu, W., Garner, E.C., Obradovic, Z.: Intrinsically disordered protein. Journal of Molecular Graphics and Modelling 19(1), 26–59 (2001)
13. Yang, A.S., Honig, B.: An integrated approach to the analysis and modeling of protein sequences and structures. i. protein structure alignment and quantitative measure for protein structural distance. J. Mol. Biol. 301(3), 665–678 (2000)
14. Berger, B., Leight, T.: Protein folding in the hydrophobic-hydrophilic (hp) model is np-complete. Journal of Computational Biology 5(2), 27–40 (1998)
15. Bystroff, C., Baker, D.: Prediction of local structure in proteins using a library of sequence-structure motifs. Journal of Molecular Biology 281, 565–577 (1998)
16. Cotta, C.: Protein structure prediction using evolutionary algorithms hybridized with backtracking (2003)
17. Anfinsen, C.B.: Principles that govern the folding of protein chains. Science 181(96), 223–230 (1973)
18. Beeferman, D., Berger, A., Lafferty, J.: Statistical Models for Text Segmentation. Machine Learning, special issue on Natural Language Learning 34, 177–210 (1999)
19. Howard, D., Benson, K.: Evolutionary computation method for pattern recognition of cisacting sites. Biosystems 72(12), 19–27 (2003)
20. Voet, D.: Biochemistry. Wiley, Chichester (2004)
21. Dubey, V.K., Monu, P., Jagannadham, M.V.: Snapshots of protein folding problem implications of folding and misfolding studies. Protein and Peptide Letters 13(9), 883–888 (2006)
22. Keedwell, E., Narayanan, A.: Intelligent Bioinformatics The Application of Artificial Intelligence Techniques to Bioinformatics Problems. Wiley, Chichester (2005)
23. Trifonov, E.N.: Segmented structure of protein sequences and early evolution of genome by combinatorial fusion of dna elements. J. Mol. Evol. 40, 337–342 (1995)
24. Trifonov, E.N., Berezovsky, I.N.: Evolutionary aspects of protein structure and folding. Current Opinion in Structural Biology 13(1), 110–114 (2003)
25. Van Dongen, S., Enright, A.J., Ouzounis, C.A.: An efficient algorithm for largescale detection of protein families. Nucleic Acids Res. 30, 1575–1584 (2002)
26. Bucher, P., Hulo, N., Sigrist, C.J., Hofmann, K., Falquet, L., Pagni, M., Bairoch, A.: The prosite database, its status in 2002. Nucleic Acids Res. 30, 235–238 (2002)
27. Pearl, F.M.G., Bennett, C.F., Bray, J.E., Harrison, A.P., Martin, N., Shepherd, A., Sillitoe, I., Thornton, J., Orengo, C.A.: The cath database an extended protein family resource for structural and functional genomics. Nucleic Acids Research 31(1), 452–455 (2003)
28. Churchill, G.A.: Hidden markov chains and the analysis of genome structure. Computers in Chemistry (16), 107–115

29. Fogel, G.B., Corne, D.W.: Evolutionary Computation in Bioinformatics. Morgan Kaufmann, San Francisco (2002)
30. Singh, G.B., Singh, H.: Functional proteomics with biolinguistic methods. Engineering in Medicine and Biology Magazine, IEEE 24(3), 73–80 (2005)
31. Rule, G.S., Hitchens, T.K.: Fundamentals of Protein NMR Spectroscopy. Springer, Heidelberg (2006)
32. Liu, H., Liu, J.: Clustering Protein Interaction Data Through Chaotic Genetic Algorithm. LNCS. Springer, Heidelberg (2006)
33. Beyer, H.G.: Theory of evolution strategies (2001)
34. Zelinka, I.: SOMA Self Organizing Migrating Algorithm. In: New Optimization Techniques in Engineering, pp. 167–218. Springer, New York (2004)
35. Shindyalov, I.N., Bourne, P.E.: Protein structure alignment by incremental combinatorial extension ce of the optimal path. Protein Engineering 11(9), 739–747 (1998)
36. Martin, J., Gibrat, J.F., Rodolphe, F.: Choosing the optimal hidden markov model for secondary structure prediction. IEEE Intelligent Systems 20(6), 19–25 (2005)
37. Watson, J.D., Baker, T., Bell, S.P., Gann, A.: Molecular Biology of the Gene, 6th edn. The Benjamin Cummings Publishing Co., Inc., Menlo Park (2007)
38. Gibrat, J.F., Madej, T., Bryant, S.H.: Surprising similarities in structure comparison. Current Opinion in Structural Biology 6, 377–385 (1996)
39. Holland, J.H.: Adaptation in Natural and Artificial Systems. Univ. Michigan Press, Ann Arbor (1975)
40. Chang, J.S., Lin, Y.C., Su, K.Y.: Automatic construction of a chinese electronic dictionary. Technical report, National Tsing-Hua University, Behavior Design Corporation (1995)
41. Braun, J.V., Muller, H.G.: Statistical methods for dna sequence segmentation. Statist. Sci. 13(2), 142–162 (1998)
42. Price, K.: Genetic annealing. Dr. Dobbs Journal, 127–132 (October 1994)
43. Price, K.: Differential evolution a fast and simple numerical optimizer, pp. 524–527. IEEE Press, New York (1996)
44. Price, K.: An Introduction to Differential Evolution, New Ideas in Optimization, pp. 79–108. McGrawHill, London (1999)
45. Wuthrich, K.: Protein structure determination in solution by NMR spectroscopy. J. Biol. Chem. 265(36), 22059–22062 (1990)
46. Coward, E., Vingron, M., Krause, A., Haas, S.A.: Systers, genenest, splicenest exploring sequence space from genome to protein. Nucleic Acids Res. 30, 299–300 (2002)
47. Zdobnov, E.M., Apweiler, R., Kriventseva, E.V., Fleischmann, W.: Clustr a database of clusters of swissprottrembl proteins. Nucleic. Acids Res. 29, 33–36 (2001)
48. Davis, L.: Handbook of Genetic Algorithms. Van Nostrand Reinhold, Berlin (1996)
49. Holm, L., Sander, C.: Touring protein fold space with dalifssp. Nucleic. Acids Res. 26, 316–319 (1998)
50. Slabinski, L., Jaroszewski, L., Rodrigues, A.P.C., Rychlewski, L., Wilson, I.A., Lesley, S.A., Godzik, A.: The challenge of protein structure determination lessons from structural genomics. Protein Sci. 16(11), 2472–2482 (2007)
51. Dorigo, M., Stutzle, T.: Ant Colony Optimization. MIT Press, Cambridge (2004)

52. Ganpathiraju, M., Weisser, D., Rosenfeld, R., Carbonell, J., Reddy, R., Klein-Seetharaman, J.: Comparative ngram anaylsis of wholegenome protein sequences. In: Proceedings of the human language technologies conference (2002)

53. Lexa, M., Valle, G.: Combining rapid word searches with segment-to-segment alignment for sensitive similarity detection, domain identification and structural modelling. In: BITS 2004 conference, Padova, Italy, March 23-26 (2004)

54. Li, M., Vitanyi, P.M.B.: Applications of algorithmic information theory. Scholarpedia 2(5), 26–58 (2007)

55. Li, M., Chen, X., Li, X., Ma, B., Vitanyi, P.: The similarity metric. In: Proceedings of the 14th annual ACM-SIAM symposium on discrete algorithms (SODA), pp. 863–872 (2003)

56. Lones, M., Tyrrell, A.: Regulatory motif discovery using a population clustering evolutionary algorithm. In: IEEE ACM Transaction on Computational Biology and Bioinformatics, pp. 403–414

57. Dayhoff, M.O., Schwartz, R.M., Orcutt, B.C.: A model for evolutionary change in proteins

58. Krasnogor, N., Hart, W.E., Smith, J., Pelta, D.A.: Protein structure prediction with evolutionary algorithms (1999)

59. Bray, J.E., Todd, A.E., Martin, A.C., Lo Conte, L., Orengo, C.A., Pearl, F.M., Thornton, J.M.: The cath database provides insights into protein structure-function relationships. Nucleic Acids Res. 27, 275–279 (1999)

60. Schneckener, S., Schonhuth, A., Schomburg, D., Pipenbacher, P., Schliep, A., Schrader, R.: Proclust improved clustering of protein sequences with an extended graphbased approach. Bioinformatics 18(suppl. 2), S182–S191 (2002)

61. Konig, R., Dandekar, T.: Improving genetic algorithms for protein folding simulations by systematic crossover. European Molecular Biology Laboratory, Heidelberg

62. Eberhart, R.C., Kennedy, J.: A new optimizer using particle swarm theory, Nagoya, Japan, pp. 39–43 (1995)

63. Weaver, R.F.: Molecular Biology, 4th edn. MacGrawHill (2007)

64. Ellis, R.J., van der Vies, S.M.: Molecular chaperones. Annual Review of Biochemistry 60(1), 321–347 (1991)

65. Ando, R.K., Lee, L.: Mostly-unsupervised statistical segmentation of japanese: Applications to kanji. Technical report, Cornell University (1999)

66. Henikoff, S., Henikoff, G.J.: Amino acid substitution matrices from protein blocks. Proc. Natl. Acad. Sci. USA 89(22), 10915–10919 (1992)

67. Kirkpatrick, S., Gelatt, C.D., Vecchi, M.P.: Optimization by simulated annealing. Science 220(4598), 671–680 (1983)

68. Rao, S., Rossmann, M.: Comparison of super-secondary structures in proteins. J. Mol. Biol. 76(2), 241–256 (1973)

69. Fleischer, H., Portugaly, E., Bilu, Y., Linial, N., Linial, M., Sasson, O., Vaaknin, A.: Protonet hierarchical classification of the protein space. Nucleic Acids Res. 31, 348–352 (2003)

70. Altschul, S.F., Madden, T.L., Schaffer, A.A., Zhang, J., Zhang, Z., Miller, W., Lipman, D.J.: Gapped blast and psiblast a new generation of protein database search programs. Nucleic Acids Res. 25, 3389–3402 (1997)

71. Back, T., Fogel, D.B., Michalewicz, Z.: Handbook of Evolutionary Computation. Institute of Physics, London (1997)

72. Smith, T.F., Waterman, M.S.: Identification of common molecular subsequences. J. Mol. Biol. 147, 195–197 (1981)

73. Cerny, V.: Thermodynamical approach to the traveling salesman problem an efficient simulation algorithm. J. Opt. Theory Appl. 45(1), 41–51 (1985)
74. Pearson, W.R.: Rapid and Sentive Sequence Comparison with PASTP and FASTA, vol. 183, pp. 63–98 (1990)
75. Hu, Y.J.: Biopattern discovery by genetic programming, pp. 152–157 (1998)
76. Linial, M., Yona, G., Linial, N.: Protomap automatic classification of protein sequences and hierarchy of protein families. Nucleic Acids Res. 28, 49–55 (2000)
77. Chu, Y.W., Sun, C.T.: A hybrid genetic algorithm approach for protein secondary structures. In: Intelligent Control and Automation, WCICA, The Sixth World Congress, vol. 1, pp. 3320–3324 (2006)

# A Clustering Genetic Algorithm for Genomic Data Mining

José Juan Tapia, Enrique Morett, and Edgar E. Vallejo

**Abstract.** In this chapter we summarize our work toward developing clustering algorithms based on evolutionary computing and its application to genomic data mining. We have focused on the reconstruction of protein-protein functional interactions from genomic data. The discovery of functional modules of proteins is formulated as an optimization problem in which proteins with similar genomic attributes are grouped together. By considering gene co-occurrence, gene directionality and gene proximity, clustering genetic algorithms can predict functional associations accurately. Moreover, clustering genetic algorithms eliminate the need for the *a priori* specification of clustering parameters (*e. g.* number of clusters, initial position of centroids, etc.). Several methods for the reconstruction of protein interactions are described, including single-objective and multi-objective clustering genetic algorithms. We present our preliminary results on the reconstruction of bacterial operons and protein associations as specified by the DIP and ECOCYC databases.

## 1 Introduction

Modern experimental techniques in molecular biology are producing an enormous amount of data. The efficient organization and interpretation of this data is rapidly becoming crucial for the success of genome sequencing projects and similar genomic data generation efforts. In effect, our potential to answer fundamental biological questions will rely heavily on the development

José Juan Tapia and Edgar E. Vallejo
Instituto Tecnológico y de Estudios Superiores de Monterrey
e-mail: {A00466859,vallejo}@itesm.mx

Enrique Morett
Instituto de Biotecnología, UNAM
e-mail: emorett@ibt.unam.mx

A. Abraham et al. (Eds.): Foundations of Comput. Intel. Vol. 4, SCI 204, pp. 249–275.
springerlink.com                                    © Springer-Verlag Berlin Heidelberg 2009

of robust data mining methods capable of producing meaningful transitions from data to knowledge.

Particularly, the development of computational methods for inferring protein function and protein-protein functional interactions from sequence and genomic data is now considered a fundamental problem in bioinformatics and computational biology. For instance, the determination of unknown protein interactions in functional pathways and perhaps, their association with diseases, are at the core of what is required to enable the understanding of biology at unprecedented detail.

Traditional homology-based computational methods are useful to assign function to proteins by establishing similarity of the underlying sequences and structures with others with previously assigned function. However, when similarity of is not sufficiently significant, alternative approaches have been considered.

A collection of non homology-based computational methods have been developed to predict protein function and functional interactions between proteins, using data attributes obtained from whole genome sequences of different species [6]. These methods are called based on genomic-context, and they include gene co-occurrence (phylogenetic profiles), gene fusions (Rosetta Stone), gene co-expression, and neighborhood conservation, among others. It has been demonstrated that these methods are capable of producing meaningful predictions [1][12]. Moreover, due to the increasing availability of complete genomes from diverse organisms these methods hold the promise of better performance.

Although the predictions provided by most genomic-context approaches have proven to be useful, the integration of different genomic attributes within a general purpose, comprehensive algorithm have not yet been proposed, to our knowledge. Here, we conceive the reconstruction of functional modules of proteins as a clustering problem –the grouping of proteins with similar collections of genomic attributes.

We have previously developed and tested different clustering algorithms to be used with phylogenetic profiles. We have taken several approaches that have given us promising results by identifying transitive functional relationships, then look at these features to detect functional modules of proteins using the Bond Energy Algorithm [37]. We have also used Fuzzy $C$-means to extend the coverage of predictions to include pairs of interacting proteins with relatively dissimilar phylogenetic profiles [8].

However, we have found it most helpful is to develop clustering methods based on evolutionary algorithms. In our previous experiments, clustering genetic algorithms have yielded competitive results on clustering phylogenetic profiles [33]. Moreover, they have proved capable of evolving appropriate clustering parameters, such as the number of clusters and the initial position of centers, among others. The empirical determination of appropriate clustering parameters can be extremely difficult in practice. Therefore, this capability of clustering genetic algorithms have simplified our experiments greatly.

In addition, we believe the consideration of different sources of genomic data simultaneously holds much promise to improve the significance of data mining in bioinformatics generally. For instance, gene directionality, namely the direction in which genes are transcribed, could be very useful for discarding false positives. For example, the transcriptional machinery of bacteria usually expresses groups of adjacent genes –operons, in response to a specific environmental stimulus. Even though a pair of genes are present in the same organisms simultaneously, thus satisfying the gene co-occurrence condition, and appear in the same gene neighborhood (gene-adjacency condition), they would not be transcribed together unless they have the same directionality. Therefore, such a pair of proteins would not typically hold a functional association. Similar examples exist for not discarding false negatives. As shown, several genomic attributes can be simultaneously considered to improve the accuracy of the discovery of protein-protein functional interactions.

In this chapter, we propose a new framework for the reconstruction of functional modules of proteins. In our model, the discovery of protein-protein interactions is formulated as a multi-objective optimization problem [2]. In this view, clustering is conducted by considering different objectives in correspondence to different genomic attributes. In addition, we extended the single-objective clustering genetic algorithms to consider multiple objectives. We present a series of experiments using three different attributes: phylogenetic profiles, gene directionality and gene positional distance. We used gene co-occurrence patterns provided by the COG database [36], as well as gene directionality and intergenic distance data, as specified by GeCont [21].

Experimental results demonstrated that the proposed method is capable of producing competitive results as validated by a comprehensive set of *E. coli* operons. The proposed algorithm outperforms traditionally used clustering methods, such as $k$-means and hierarchical clustering by predicting functional relationships between proteins with higher accuracy.

## 2   Previous Work

### 2.1   *Using Genetic Algorithms as a Clustering Method in Bioinformatics*

Genetic Algorithms have been previously used as an aid in different clustering processes to automatically determine many of the parameters these algorithms need. One of the most common uses is to calculate the number of centroids an algorithm will work with, or helping the algorithm to avoid local minima [40]. However the use of Genetic Algorithms as a clustering process in and of itself has only raised interest in recent years.

Grouping Genetic Algorithms were first developed by Falkenauer in 1998 when applicated to the optimization of production lines, and from then it has been applied to multiple areas, because of its ability to seemingly find the optimal clustering of landscapes where classic clustering algorithms have serious problems to work with great success, such as Intrusion Detection Systems[15] and Control systems [20].

On the other hand, Multiple Objective Genetic Algorithms have seen a wider use in Bioinformatics, with many successful applications in areas such as System Optimization [3, 17], and inverse problems [32]. However its combination with Grouping Genetic Algorithms hasn't seen much development. One of the first approaches was proposed by Deb [4], when he applied applied a multi-objective optimization algorithm in order to classify cancer data. Deb reported that he had obtained satisfactory results in obtaining optimal clusters with minimal false positives and sizes. Similar approaches have been tried more recently by Facelli and others [7].

However, in those works the multiple objective functions have only been used as a measure to introduce multiple computational clustering characteristics into the classification process. Deb used fitness functions that minimized both the cluster size and the false positives. Although the introduction of such data is of great help for the classification process of the proposed data, we think that it would also be interesting to introduce various Biological data in the form of different fitness functions in order to not only increasing the computational congruence of our clusters, but also of their Biological relevance.

## 3   Biological Context

Through this section we will proceed to explain the different biological concepts we introduced into our algorithm, that is, what phylogenetic profiles and genomic directionality and gene positional distance stand for. It is necessary to understand how they are related to the functional coupling of proteins in order to understand the full biological significance of the results one can obtain.

### 3.1   Phylogenetic Profiles

Phylogenetic profiles are a technique used to describe the presence or absence of a specific protein across a group of different species. Normally we work with phylogenetic profiles in large groups, their representation being similar to the example in Table 1, where each column represents an specific species and each row one of the proteins we want to analyze. Formally, the construction of phylogenetic profiles begins with a collection of $k$ completely sequenced genomes $G$ from different organisms and a collection of $l$ proteins $P$ of interest. For each protein $p_i$, a phylogenetic profile is represented as a $k$-length binary string $s = s_1 s_2 \cdots s_k$ where $s_j = 1$ if protein $p_i$ is present in genome $g_j$ and $s_j = 0$ if protein $p_i$ is absent in genome $g_j$.

In Table I we can see a reduced version of how Phylogenetic profiles are normally represented. In this example, we are comparing 8 proteins $(p_1, p_2...p_8)$ indicating their presence (or absence) in 5 genomes $(g_1, g_2...g_5)$ from different species. It is from this way of arranging the data that useful information can be mined. For example, we can note that all proteins $p_1, \ldots, p_8$ are present in genome $g_1$, and that $p_3$ and $p_6$ tend to appear in the same species.

**Table 1** Example of phylogenetic profiles

| Protein | $g_1$ | $g_2$ | $g_3$ | $g_4$ | $g_5$ |
|---------|-------|-------|-------|-------|-------|
| $p_1$ | 1 | 1 | 0 | 1 | 1 |
| $p_2$ | 1 | 1 | 1 | 0 | 1 |
| $p_3$ | 1 | 0 | 1 | 1 | 1 |
| $p_4$ | 1 | 1 | 0 | 0 | 0 |
| $p_5$ | 1 | 1 | 1 | 1 | 1 |
| $p_6$ | 1 | 0 | 1 | 1 | 1 |
| $p_7$ | 1 | 1 | 1 | 0 | 1 |
| $p_8$ | 1 | 0 | 0 | 1 | 1 |

Functional coupling of proteins is then inferred by clustering proteins according to the intrinsic similarities of the underlying phylogenetic profile patterns. It is often concluded that proteins associated to the same cluster are functionally related. For the example shown in table I, a functional association between proteins $p_2$ and $p_7$ would be inferred by this method as they posses identical phylogenetic profiles.

The logic underlying this reasoning is that proteins with similar phylogenetic profiles are likely to interact in performing some biological process. In effect, there should be an evolutionary pressure acting on a group of proteins in order to preserve a function that confers an advantage to the organisms.

## Data set

One of the most well known Phylogenetic Profiles databases is provided by the Cluster of Orthologous Groups of proteins[34], which we have previously used coupled with other clustering algorithms with great success[8, 37]. This database consists of a collection of conserved protein families (COGs) that are presumed to be orthologous.

Particularly, we relied on the 43 completely sequenced genomes COG initial version, in which phylogenetic profiles are represented as 26-length binary strings. These genomes come from three different domains: 6 *archaea*, 19 *bacteria* and 1 *eukaryota*. In constructing phylogenetic profiles, 43 genomes were collapsed into 26 representative genomes. The reason was that some of these genomes are very similar to each other or they belong to a subspecies of other organism represented in the database. The organisms represented in the COG database are listed in Table 2.

**Table 2** Genomes represented in COG database

| ID | Description | Domain |
|---|---|---|
| A | *Archaeoglobus fulgidus* | archaea |
| O | *Halobacterium sp. NRC-1* | archaea |
| M | *Methanococcus jannaschii* | archaea |
|   | *Methanobacterium thermoautotrophicum* | archaea |
| P | *Thermoplasma acidophilum* | archaea |
|   | *Thermoplasma volcanium* | archaea |
| K | *Pyrococcus horikoshii* | archaea |
|   | *Pyrococcus abyssi* | archaea |
| Z | *Aeropyrum pernix* | archaea |
| Y | *Saccharomyces cerevisiae* | eukaryote |
| Q | *Aquifex aeolicus* | bacteria |
| V | **Thermotoga maritima** | bacteria |
| D | *Deinococcus radiodurans* | bacteria |
| R | *Mycobacterium tuberculosis* | bacteria |
|   | *Mycobacterium leprae* | bacteria |
| L | *Lactococcus lactis* | bacteria |
|   | *Streptococcus pyogenes* | bacteria |
| B | *Bacillus subtilis* | bacteria |
|   | *Bacillus halodurans* | bacteria |
| C | *Synechocystis* | bacteria |
| E | *Escherichia coli K12* | bacteria |
|   | *Escherichia coli O157* | bacteria |
|   | *Buchnera sp. APS* | bacteria |
| F | *Pseudomonas aeruginosa* | bacteria |
| G | *Vibrio cholerae* | bacteria |
| H | *Haemophilus influenzae* | bacteria |
|   | *Pasteurella multocida* | bacteria |
| S | *Xylella fastidiosa* | bacteria |
| N | *Neisseria meningitidis MC58* | bacteria |
|   | *Neisseria meningitidis Z2491* | bacteria |
| U | *Helicobacter pylori 26695* | bacteria |
|   | *Helicobacter pylori J99* | bacteria |
|   | *Campylobacter jejuni* | bacteria |
| J | *Mesorhizobium loti* | bacteria |
|   | *Caulobacter crescentus* | bacteria |
| X | *Rickettsia prowazekii* | bacteria |
| I | *Chlamydia trachomatis* | bacteria |
|   | *Chlamydia pneumoniae* | bacteria |
| T | *Treponema pallidum* | bacteria |
|   | *Borrelia burgdorferi* | bacteria |
| W | *Ureaplasma urealyticum* | bacteria |
|   | *Mycoplasma pneumoniae* | bacteria |
|   | *Mycoplasma genitalium* | bacteria |

The number of that can be represented by a 26-length binary vector is $2^{26} = 67,108,864$. This number is far greater than the number of known proteins. The COG database provides a collection of 3,307 phylogenetic profiles, which are fed into our clustering algorithm.

## 3.2 Genetic Context

### Data set

For our experiments we used a gene database provided by the Biotechnology Institute from the National Autonomous University of Mexico (UNAM). This database was originally used as a backend for GeCont[21], a tool that graphically displays the gene context of a certain protein the user is interested in. As such, the database is quite rich in terms of the information we need, such as including the proteome of different species, starting and ending nucleotide, scientific name, corresponding number ID from the COG database etc. We used this database as our source of information for the Genomic Directionality and the Intergenic Distance fitness functions.

## 4 Computational Context

## 4.1 Grouping Genetic Algorithms (GGA)

Genetic algorithms can be defined as a search technique whose algorithm is based on the mechanics of natural selection and genetics. It has been successfully used in realms as diverse as search, optimization, and machine learning problems, since they are not restricted by problem specific assumptions like continuity, existence of derivatives, unimodality and similar.[9]. In rough terms, a genetic algorithm searches a landscape of possible solutions to a specific problem. Initially those solutions are normally randomly generated solutions, so they won't perform well. However, no matter how bad, there will be small segments of our solution collection that will be near our desired solution, that is partially correct answers. Genetic Algorithms exploit this characteristic by recombinating and progressively creating better solutions, so that by the end of the algorithm we have achieved one solution that is near or is what we seek for, the optimal solution

However, when they were first applied to clustering, Genetic Algorithms did not perform as well as they had done for other type of optimization problems. In the first proposed approach, the design was similar to that shown in Table 3. For a group of elements $e_k$ — $k = 1..8$ and clusters $g_i$ — $i = 1..4$, each gene was represented by a element-group pair, expressing which groups each elements belong to.

However, this type of representation demonstrated having a very poor capability of converging into useful information. Structures and clusters where often eliminated by the way the space search is explored by Genetic

**Table 3** Example of early clustering genetic algorithms

| Element | Cluster |
|---|---|
| $e_1$ | $g_1$ |
| $e_2$ | $g_3$ |
| $e_3$ | $g_4$ |
| $e_4$ | $g_2$ |
| $e_5$ | $g_1$ |
| $e_6$ | $g_3$ |
| $e_7$ | $g_2$ |
| $e_8$ | $g_1$ |

Algorithms. As a response to this, Emmanuel Falkenauer developed the Grouping Genetic Algorithms (GGA)[30]. Falkenauer proposed a design where the chromosomes were not a set of elements which conformed a solution, but that the genes themselves represented the groups, and as such the chromosomes were a collection of groups, instead of directly containing the elements.

**Basic Structures**

Figure 1 depicts the hierarchical structure of the genetic algorithm we used. Although it is remarkably similar to that of Classic Genetic Algorithms (GA), there are some differences that are worth noticing.

*Population*

The population structure that was used, as with GGA, is a collection of solutions our current generations is comprised of. It takes a series of Cog's to be sorted and the fitness functions data to be used as an input. It is important to note that the algorithm does not need to be fed the number of groups that the Cog's will be sorted into as an input, the algorithm will determine that by itself, this fact gives it an important edge over other classic grouping algorithms. The initial population is constructed from a random distribution, and further populations are the product of applying evolution operators to the previous ones.

*Chromosome*

As we mentioned earlier, a chromosome within the GGA context refers to a proposed arrangement of the elements into different clusters. Other than sharing the elements they arrange, each clusters has its own number and size of clusters.

*Group Genes*

A group gene in our algorithm is defined as a limited collection of Cog's (a cluster), and a set of special structures called the Consensus Structures

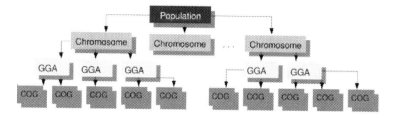

**Fig. 1** Structure of a population

(CoSt), which are the centroids of the members of a group according to the different fitness structures that we will be evaluating as described earlier.

## Genetic operators

Traditionally Genetic Algorithms exploration and exploitation operators can be summed up into two classes: *Crossover* and *Mutation*. Crossover in clustering genetic algorithms is normally done by exchanging clusters between different chromosomes, and rearranging the possible inconsistencies afterwards. However, preliminary computational experiments showed us that in our particular problem crossover does not effort exert enough exploratory capabilities, on the contrary, more than often it is disruptive, not to mention that the computational complexity of a crossover was higher than any given mutator, and as such only slows down the convergence process. Because of these reasons we only included a collection of mutation operators. For the mutation operators, we used two operators that are normally used in the clustering genetic algorithms, and we designed other three for our specific application that are described below.

- *RandomNewGroup*: Grabs random members from different groups and creates a new group with them.
- *RandomDeleteGroup*: Selects a random group and disperses its members through different groups in order to delete it.
- *MergeStrong*: Searches two groups whose centroids (according to a random fitness function) are very close in terms of the distance between them, and them proceeds to merge them into a single group.
- *RehashWeak*: Searches for one group whose variability between its members is considerably greater (surpasses some threshold) than the average of the whole chromosome, and scatters its members across the rest of the clusters within one chromosome.
- *Weeding*: Searches for outliers (according to a random fitness function) within a group and reallocates them to a group whose centroid is reasonably near from the outliers.

Normally the mutators are preferred to be entirely random in its application since theoretically this maximizes the search space exploratory

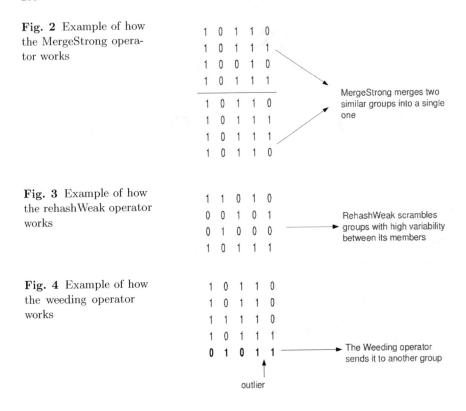

**Fig. 2** Example of how the MergeStrong operator works

MergeStrong merges two similar groups into a single one

**Fig. 3** Example of how the rehashWeak operator works

RehashWeak scrambles groups with high variability between its members

**Fig. 4** Example of how the weeding operator works

The Weeding operator sends it to another group

outlier

capabilities of Evolutionary Algorithms. However, here we decided to introduce directed mutators that worked for specific cases, since there are specific regions from the search space that we are interested in mining. That is, we are further increasing the searching capabilities of the algorithm into making it specifically explore the regions we are interested in.

## Using a single evaluation function

As it was explained before, the initial population will be constructed from an uniform distribution. Once the first generation was generated , a series of evaluation functions were applied. Which and how many evaluation functions should be used was determined after a number of empirical tests in which various techniques were tried, included but not limited to, doing a $n$ to $n$ correlationship comparison of the different species all the Cog's within a group gene appeared in, or creating an artificial type of COG comprised of the average of all the Cog's included in a group, and compare the Cog's against that average. the best metric that was found was creating a CoSt akin to that used in motif analysis. The fitness operation would then only be applied $n$ times (one for each member of the group against the CoSt) reducing the evaluation time by one polynomial order. This also proved having a high performance and efficiency in doing an exploratory work across the search space, not showing any significant inferiority to doing a COG per COG comparison. An increasingly bigger bonus and penalization was given for every COG that was sufficiently near (or far) from the CoSt. This was done in order to award those groups whose members were generally close to each other, against those groups which merely had a large number of members. The process that computes the fitness for each Genetic Group is described in Algorithm 1. In the algorithm, the values 8 and 12 against which the COG evaluation are compared against were empirically obtained values that determine whether the variability of a specific COG is good or bad, and the *pFactor* and *nFactor* factors act in a pendulum effect. The more 'good' Cog's a group has, the less the differences will matter. This is done in order to further reward the groups with little differences. The same effect works in an opposite manner if a group has too many 'bad' members in its cluster.

---

**Algorithm 1.** Evaluation function for the single objective genetic algorithm

---

$pFactor \leftarrow 1$
$nFactor \leftarrow 1$
**for all** Cog's within the group **do**
   COGEvaluation = differences with the centroid
   **if** $COGevaluation < 5$ **then**
      COGevaluation *= pFactor
      pFactor-=0.05
   **else if** $COGevaluation > 10$ **then**
      COGevaluation *= nFactor
      nfactor+=0.02
   **end if**
   groupEvaluation += COGevaluation
**end for**
**return** groupEvaluation

---

## Selection operators

There are several operators in the Genetic Algorithms literature that we could have used, including but not limited to, Roulette, Boltzmann probability

**Table 4** Example of a phylogenetic profile centroid

| Protein | $g_1$ | $g_2$ | $g_3$ | $g_4$ | $g_5$ |
|---|---|---|---|---|---|
| $p_1$ | 1 | 1 | 0 | 1 | 1 |
| $p_2$ | 1 | 1 | 1 | 0 | 1 |
| $p_3$ | 1 | 0 | 1 | 1 | 1 |
| $p_4$ | 1 | 1 | 0 | 0 | 0 |
| $p_5$ | 1 | 1 | 1 | 0 | 1 |
| *centroid* | 1 | 0.8 | 0.6 | 0.4 | 0.8 |

distribution, Tournament, Elitism. In the end,preliminary experiments demonstrated that a combination of both Tournament and Elitism had the best results for our particular application.

## Integrating the algorithm

Now that we have described each of the parts of the algorithm, we will proceed to explain how it all connects. It is interesting to note that we optimized the algorithm to be able to fully occupy the capacity of the 4 Xeon 3.2GHz server we run our program into. We parallelized the algorithm as much as possible, with the added advantage over other types of parallel genetic algorithms that the different population sections have a common memory. As such, when running the generateCentroids, evaluate and evolve operations we divided the population between each core, to later merge it again.

---

**Algorithm 2.** Overview of the complete algorithm

---

population ← randomGenerator
evaluator ← profileDistanceFitness
operator ← {RandomNewGroup,RandomDeleteGroup,MergeStrong}
operator ← {RehashWeak,SplitStrong,Weeding}
evolution ← {Elitism, Tournament}
population.generateCentroids
population.evaluate(evaluator)
**while** population.averageEvaluation > threshold **do**
    population.evolve(evolution,operator)
    population.generateCentroids
    population.evaluate(evaluator)
**end while**
**return** population

---

The so called threshold that ends the cycle is normally considered a cumulative change over a number of generations. When over the generations the evaluation average does not show a significant change, the evaluation will finish.

## 5  Results Using a Single Objective Evaluation Function

We conducted a large series of computational experiments on obtaining the clusters on using GGA. Through this experiments, we obtained two sets of results. The Grouping Genetic algorithm gave us both the number of groups, and created the respective clusters using only lists indicating the co-occurrence of the phylogenetic data input. The data was validated using the ECOCyC and DIP databases, against the clusters proposed by the $k$-Means method. It is important to note that for the algorithm determining the number of groups and determining the composition of the clusters is something it does at the same time, however for the sake of reporting we will be separating them.

## 5.1 Number of Groups

Throughtout the numerous number of computational experiments we made, the final number of clusters was pretty consistent. In figure 5 we can see a graph that represents how the groups evolved throughout a series of experiments, with the brackets representing the highest variations from the average. As we can see, towards the end the algorithm always stabilized around the 120 groups area. The ideal number groups determined by the algorithm is a number between 110 and 120 groups, however, as we can see below, the speed of the algorithm was highly dependent on how the data was initialized in the first random construction. Despite this, the algorithm always arrived at the same range.

**Fig. 5** Average run of the GGA engine

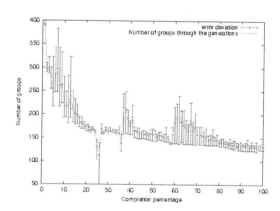

## 5.2 Running Time

In our initial version each generation took approximately 10 seconds to be completed, from the very start of the evaluation to the generation of the next one, our algorithm runtime would ranged between 15 and 20 minutes. The algorithm was later optimized for the fitness, mutation, crossover, and special functions to be run in parallel so that we could use the server's multi processors capabilities as much as possible, and we could see an increase in speed in an order of 5 reducing the clustering time to as little as 4 minutes.

Still this can be still be seen as a time much bigger than that shown by classic grouping algorithms (specially when compared to algorithms like $k$-means, however we have to take into account that our GGA algorithm has a very flexible fitness functions that permitted us not only to include co-occurrences, but several other factors that helps us in refining our groups,

**Table 5** Percentage of correct EcoCYC/DIP group assignments

| Run | DB matches |
|---|---|
| Total DB entries | 346 |
| GGA Average Matches | 256.7 |
| $k$-means Average Matches | 203 |

something that would be extremely complicated to implement in an algorithm such as $k$-means.

## 5.3   Validation of Results

The significance of the data mined by automated methods, specially when those are intended to discover functionality which is unknown as of yet, is often difficult to assess. In the validation phase we compared our results against known Biological databases, such as ECOCyC and DIP. Similarly, the

**Table 6** Excerpt from the GGA group where the *purEK* pair is located

| COG | Phylogenetic profile |
|---|---|
| COG0581 | 11101101111111111110110011 |
| COG0413 | 01001111111011111011110000 |
| COG0458 | 11110011111111111111110000 |
| COG0065 | 11101011111111111111110000 |
| COG0685 | 00011111110111111111110000 |
| COG0035 | 01101111111111111101110011 |
| COG0104 | 11111011111111111111110000 |
| COG0287 | 11110111111111111111110000 |
| COG0352 | 10011011011111111111110000 |
| COG0157 | 11101011011111111111110000 |
| COG0079 | 11111111111111111111110000 |
| COG0045 | 11110111011011111111111100 |
| COG0382 | 11111111011011111111111100 |
| COG0714 | 11111111111011111010010010 |
| COG0807 | 10010011111111111111110100 |
| **COG0026** | 01010011111111111111010000 |
| COG0251 | 01001111111111111111110000 |
| **COG0041** | 11111011111111111111110000 |
| COG1011 | 01111011111110111111010010 |
| COG0226 | 11101101111111111110110011 |
| COG0778 | 11111101111111111111110000 |
| COG0127 | 11111111111111111111110110 |
| COG0512 | 11111111111111111111110000 |
| COG0059 | 10101011111111111111110000 |
| COG1985 | 11100111111111111111110100 |

**Table 7** Excerpts from the *k*-means group were the *purEK* disjointed pairs were located

| COG | COG |
|-----|-----|
| COG0364 | COG0034 |
| COG0176 | COG0461 |
| COG0337 | COG0284 |
| COG0140 | COG0151 |
| COG0703 | COG0150 |
| COG0590 | COG0152 |
| COG0299 | **COG0041** |
| **COG0026** | COG0047 |
| COG0801 | COG0046 |
| COG0161 | COG0108 |
| COG0502 | COG0294 |
| COG0132 | COG0476 |
| COG1539 | COG0001 |

algorithm was found to identify several biological patterns whose correlation has already been experimentally proven previously.

### DIP and EcoCyC comparison

As we mentioned, we used the EcoCyC and DIP database as a means to assert the expressiveness of the results thrown by our algorithm and *k*-means. The entries found in DIP and EcoCYC entries are a good way to measure the usefulness of our non-co-occurrence related metrics, since many pairs in that list does not show that type of relationship, as proven by the direct approach *k*-Means takes. As it is shown by the validation process, GGA demonstrated to have a higher proficiency at finding these kind of relations.

### Operon comparison

To further prove the biological expressiveness of the results obtained, they were compared against a number of operons. Operons can be defined as a functioning unit of key nucleotide sequences including an operator, a common promoter, and one or more structural genes, which is controlled as a unit to produce messenger RNA (mRNA), in the process of transcription. They are normally found on prokaryotes, which is quite convenient when we consider that 80% of our reference database is formed up by such type of beings. [5]. In consequence, an apt grouping mechanism would have to group together this patterns, and hence it becomes a good way to our validate our data. Our first comparison was in respect to the *purEK* operon (which is a combination of Cog's COG0026 and COG0041); the results obtained for our GGA algorithm are shown in Table 6.

**Table 8** Excerpt from the GGA group where the *Leu* and *SpeED* operon pair are located

| COG | Phylogenetic profile |
|---|---|
| **COG0473** | 1010101111111111111111111000 |
| **COG0119** | 1011101111101111111111110000 |
| **COG0065** | 1110101111111111111111110000 |
| **COG0066** | 1010101111111111111111110000 |
| COG0104 | 1111110111111111111111110000 |
| COG0157 | 1110101110111111111111110000 |
| COG0152 | 1111101111111111111111111000 |
| COG1989 | 1110110111011111111111110000 |
| COG0581 | 1110110111111111111110110011 |
| COG0778 | 1111110111111111111111110000 |
| COG0147 | 1111111111111111111111110000 |
| COG0512 | 1111111111111111111111110000 |
| COG0714 | 1111111111101111010010010010 |
| COG0547 | 1111111111111111111111110000 |
| COG0159 | 1111111111111111111111110100 |
| COG0133 | 1110101111111111111111110100 |
| COG0421 | 1011111111010110011110000 |
| COG1586 | 1011110110001011001000000000 |

An excerpt of the $k$-means output is shown in Table 7. This method failed to find *purEK* (as in , this method lacks the means to correlate beyond the co-appearance factor.

Similarly, we compared the results of the *Leu E. coli* operon, which includes entries COG0066, COG0473, COG0119 and COG0065. The results of this comparison are shown in Table 8

**Table 9** Exerpt from the GGA group where the *hflA* operon pair is located

| COG | Phylogenetic profile |
|---|---|
| COG0519 | 1111111111111111111111110110 |
| **COG0330** | 1111111111111111111111111010 |
| COG0611 | 1111110100101111111110000 |
| COG1841 | 1111111111110111111011010 |
| COG0475 | 1011111101111111111111010 |
| COG0825 | 1100101111111111111111100 |
| COG1120 | 1111100011111111101110001 |
| **COG2262** | 0110110111111111111010100 |
| COG0493 | 0001111111111111010111010 |
| COG0473 | 1010101111111111111111000 |
| COG0026 | 0101001111111111111010000 |

A comparison against the *hflA* operon was made as well. This operon consists of a putative *GTPase* (COG2262) and a putative integral membrane protease (COG0330). Our algorithm correctly assigned these two proteins to the same group, as shown in Table 9. On the other hand, $k$-means was not able to recognize the relationship between these two proteins, and separated them into two groups.

The *SpeED* operon also showed to have better results in our engine than $k$-means (COG1586 and COG421), as shown in Table 8.

# 6   Preliminary Discussion

The results mined by our algorithm were of more expressiveness and biological significance than those obtained by more classic grouping algorithms, as shown by the comparison data. This was a byproduct of the exploratory capabilities of every Genetic Algorithm, that allow us to explore irregular landscapes with great success, and where other traditional algorithms fail. We consider that many biological problems fall under this type, and as such Genetic Algorithms present us with a great framework to tackle this type of search spaces.

Moreover, the fact that the algorithm requires minimal input is another tremendous advantage. Not even the number of groups is of great importance, since this is evolved by the algorithm itself. When running this type of gene function discovery algorithms, we do not possess much more information than the bare minimum, and as such, the number of groups, the distance between the centroids of each group and other factors classical grouping algorithms tend to ask for, is something we traditionally determine either by using some other heuristic that will try to approximate this values, or by mere trial and error.

As such, an algorithm that makes the determination of this initial parameters as part of the search heuristic is of tremendous help. GGA's have the advantage that they will not only determine the number of groups and their composition as it has been discussed several times before, but that it has the flexibility as well to determine a myriad of other values.

However, there are still some disadvantages that need to be addressed in this algorithm. Although the results are computationally consistent in terms of the correct clustering of phylogenetic profiles, they are still biologically insufficient in the information they possess. There are severall ways to solve this problem. One of the approaches that has been tried is that of including more biological information into the searching process, like STRING[14] does. STRING uses various information sources independently in order to determine relationship levels between a network of proteins, and then present the results to the user in a table. That is, an *a posteriori* analysis.

What we propose with the second part of our work is to integrate all this analysis in an *a priori* approach through Multi Objective Genetic

Algorithms. Our hypothesis is that integrating multiple biological information into the genetic algorithm would direct us to a strong clustering technique, and ultimately give us better results.

# 7 Computational Context II

We will proceed to introduce the concept of Multi-Objective Algorithms and how do they work. After that, we will explain our implementation and the fitness functions we used for the clustering process.

## 7.1 Introducing Multi-Objective Genetic Algorithms

Classic Genetic Algorithms, as defined by Holland[10] consider a single parameter to optimize, and thence a single fitness function to use. While this is usefull for a limited set of problems, most of the problems of nature are restricted by multiple constraints, or must meet a diverse number of requirements at the same time. At the same time, multi-objective problems have always presented one of the greatest challenges for algorithm designers. Because of this alternative approaches, like that presented by Multi Objective Genetic Algorithms (MOGA) have gained much interest in recent years. [2].

The MOGA literature tell us multiple ways into which the many fitness functions can be integrated into a single scheme. However they can be grouped into two main categories: *a priori* techniques and *a posteriori* techniques. The former ones need to have the importance of our objectives defined the search is done. These include linear aggregations techniques, lexicographic techniques (which give a priority to different objectives) and so on. These techniques have been shown to be weak on multiple comparison works, since they require a predetermination of the objective priority, which means we are more or less arbitrarily limiting the search space, and as such we cannot find all the optimal solutions for our problem. On the other hand a posteriori techniques implicitly searches for the most optimal solutions that optimize all of our objective functions. There are multiple examples of these kind of algorithms, from which we decided to use the Niched Pareto Genetic Algorithms.

### Pareto optimum

When working with multi-objective systems, the concept of optimum changes because we have to find good compromises or trade offs between our objectives. For example, not all of our objectives may be directly linearly related, or they may actually be contradictory, yet we still have to maximize (or minimize, depending on the problem) both of them as much as possible.

The most accepted definition of optimality is the one proposed by Vilfredo Pareto in 1896, that tell us that the vector or chromosome $\mathbf{x}^*$ is Pareto optimal if there exist no feasible vector $\mathbf{x}$ within our population that would decrease some criterion without causing a simultaneous increase in at least other criterion.

We also need to introduce the definition of Pareto Dominance. A vector $\mathbf{u} = \{u_1, ..., u_k\}$ (where each member is a particular fitness value) is said to dominate another vector $\mathbf{v}$ when $\mathbf{u}$ is partially less than $\mathbf{v}$, that is, at least one the members in $\mathbf{u}$ is strictly less than its equivalent member in $\mathbf{v}$ (for a minimization problem) and the rest of the members in $\mathbf{u}$ are equal or less than their counterparts.

Conversely, Pareto non-dominance of a vector $\mathbf{u}$ to a vector $\mathbf{v}$ means that $\mathbf{v}$ does not dominate $\mathbf{u}$, or in plain words, we cannot find a value in $\mathbf{v}$ that is less than a value in $\mathbf{u}$ without also finding another fitness in $\mathbf{v}$ that is bigger than its $\mathbf{u}$ counterpart. (For more information we refer the readers to [2].)

Over these definitions we create the Pareto Set ($P^*$) which is the collection of all the vectors in our population that are non-dominated by the rest of the population.

The evaluated objective vectors of $P^*$ build up what is known as the Pareto Front ($PF^*$). Geometrically the $PF^*$ can be defined as the bounder of the design region, or in the locus of the tangent points of the objective functions.

So in the end what a Multi Objective Genetic Algorithm give us in the end is not a single solution, but a collection of the best solutions that can be obtained given different weights for each parameter. From this point, it is important to consider the context, or what we want in order to decide which of this solutions we are going to consider the "definite" one for our application.

## Niche techniques

Niching methods are a series of techniques developed to maintain diversity within a population. It is often the case that when the Pareto Front is being built, the population has an early convergence towards some zone of the Pareto Front, leaving whole regions untouched. Obviously this goes against the spirit of creating the whole Pareto Set, so the niching methods were introduced as a set of techniques to preserve variability. [24]

## Fitness sharing

Fitness sharing refers to a technique in which we analyze a series of candidates to determine which of them is localized in a less crowded area. By doing this we avoid having all the chromosomes in one area and ensure variability. We do this by applying the equation shown below, where $d_{ij}$ refers to the distance between chromosomes $i$ and $j$, and $\sigma_{share}$ refers to the distance threshold, a parameter we introduce to the algorithm that defines what is the maximum distance for two chromosomes to be considered near each other. We apply this formula to our candidates, and the one which has the smallest value is located in the least crowded area, and as such the one that holds the higher exploratory value.

$$\sum_{i=0}^{N} sh(d_{ij})$$

$$sh(d_{ij}) = \begin{cases} 1 - (d_{ij}/\sigma_{share}) & d_{ij} \leq \sigma_{share} \\ 0 & otherwise \end{cases}$$

## 7.2 Multi Objective Fitness Evaluation Functions

For our multi objective computational experiments we used three distinct fitness functions. A fitness function that clusterized data using phylogenetic profiles, another that clusterized data using gene transcription directionality, a third one that considered intergenic distance between the members of a cluster and a final one that rewarded cluster connectivity.

### Phylogenetic profiles

Because of the modular design of our implementation, we directly used the Phylogenetic Profiles Distance fitness function we had used for the previous version of the algorithm and just added the other two to the engine.

### Transcriptional directionality

In a very similar approach as the one used for phylogenetic profiles, we constructed a centroid of the group considering the number of groups that had a left or right directionality across the set of species indicated in the phylogenetic profiles as containing the proteins in the cluster, and added it as an additional field to each Group Gene. From here it becomes a task of minimizing the distance to the centroid, or minimizing the differences in directionality of the proteins found in a particular specie, among all the species where our set of proteins are present. As we can see, both the phylogenetic profiles and transcription directionality fitness functions tend to create smaller groups.

### Intergenic distance

In order to implement a Intergenic distance fitness function we first added a $n$x2 array (where n is the number of phylogenetic profiles) that would contain the information of where the transcribed protein original information started and finished, for each of the $n$ species the array contains information from. (Table 10 contains a reduced example of this field). This information would be later used by the fitness function. Which would create a list of this fields

**Table 10** Example of a context array field from an hypothetical COG

|                 | $s_1$ | $s_2$ | $s_3$ | $s_4$ | $s_5$ |
|-----------------|-------|-------|-------|-------|-------|
| Starting Point  | 32    | 16    | 17    | 34    | 12    |
| Finishing Point | 45    | 28    | 21    | 45    | 34    |

**Table 11** Example field of the context 3 dimensional array

|                 | $p_1$ | $p_2$ | $p_3$ | $p_4$ | $p_5$ |
|-----------------|-------|-------|-------|-------|-------|
| Starting Point  | 32    | 49    | 60    | 90    | 160   |
| Finishing Point | 45    | 54    | 73    | 99    | 189   |

for each of the members of the cluster it is analyzing, and arrange them on a species-starting/ending point- protein hierarchy in a $nx2xk$ 3-dimensional array, where $k$ is the number of members in the cluster. We can see an example for a single species $s_n$ on Table 11. The third dimension is formed by forming a list of these for a set of species. From here, it becomes a matter of adding up all the distances between transcription points.

### Cluster size

This fitness function was introduced in order to balance the divisive nature of the other fitness functions. Since they tend to minimize distance, one of the approaches the genetic engine could take would be that of making just unitary groups. In order to counter that tendency the cluster size fitness function was introduced, a fitness which rewards those groups that despite having a low variability have a greater size than the average of the population.

## 7.3 Implementation of the Evaluation Function

We used a niched Pareto algorithm as described by Horn[11], which combines both a Pareto optimum next generation construction with a fitness sharing scheme to maintain variability. In a broad sense we use a tournament scheme in order to select those members of the current population which will pass to the next population. The winners are selected by pitting to members of the population against a sample subset of the same population, and selecting the one which is non dominant in respect to their competitors . In case we find that the competitors are tied, we use a fitness sharing scheme to select those that will pass to the next generation. Through this method we can progressively construct better Pareto Fronts.

Our implementation is done as shown in Algorithm 3. For a parent population *parent*, and a target population *offspring*.

**Algorithm 3.** Algorithm for the creation of a new population

```
 1: while offspring.size < threshold do
 2:     c1 ← parent.randomMember
 3:     c2 ← parent.randomMember
 4:     list[] ← parent.randomMembers
 5:     tournament(c1,c2,list)
 6:     if c1 is dominated and c2 is nonDominated then
 7:         offspring.add(c2)
 8:     else if c2 is dominated and c1 is nonDominated then
 9:         offspring.add(c1)
10:     else
11:         c3 ← fitSh(c1,c2)
12:         offspring.add(c3)
13:     end if
14: end while
15: return offspring
```

# 8  Results

## 8.1  *Formation of the Pareto Front*

The algorithm succeeded in conforming a viable Pareto Front that combines in different degrees the different fitness functions we used as shown in Fig. 7. (the intergenic distance fitness function of each chromosome is not included in the plot, however we have to remember that the result vector is a 4-dimensional value). Although further selection from this point is a task where a proper Biologist should take the decision of which criteria combination is more viable, we selected the element from the Pareto that maximizes the profile distance, genome directionality and intergenic distance fitness functions for comparison purposes.

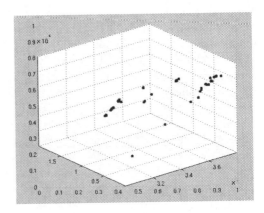

**Fig. 6** Initial random distribution of the population

**Fig. 7** Pareto Front
formation. The curve is
bent towards the viewer

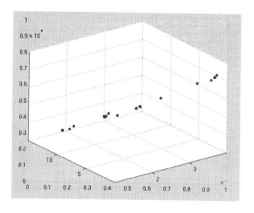

## 8.2  Execution Time

Despite we having to deal with four evaluation functions instead of one as in our previous version, the computational complexity of the Genomic directionality and Connectivity fitness functions is equally linear - $O(n)$ -, while the Intergenic Distance function is $O(n^2)$, being the overall complexity of the algorithm of a low polynomic order. As such, the only real bottleneck of the overall algorithm is the number of generations it takes to reach convergence.

## 8.3  Validation of Results

As shown earlier in this chapter, our original algorithm had already been demonstrated to work better than classic clustering algorithms like $k$-means when identifying known functional interactions between proteins, so we will use it as a reference algorithm to demonstrate that introducing more biological information to the clustering engine has a clearly positive effect on the quality of the clusters.

We ran our algorithm under a different number of modes modes, combining phylogenetic profile distance(PP), genomic directionality(GD) and intergenic distance(ID) (always combined with the cluster size objective function). For our tests we compared against the same operon set as the one shown earlier.

In our first run, the algorithm running only with phylogenetic profile distances, which is the same algorithm that was shown earlier with some minor optimizations. Although the clustering itself was not bad, when compared to real biological data it was not able to detect many operons (In specific it was able to consistently detect 6 operons. *clpPX* (clp Protease), *glmUS* (used for amino sugar biosynthesis), *SpeED*, *hFLA*, *leu E. coli*, and *purEK*. In our second run we only considered genomic directionality (GD). It is interesting to note that although the performance of this fitness function was slightly better than that shown by phylogenetic profiles, the overall convergence time for the algorithm in this run was much larger, being one of the slowest modes.

**Table 12** Results chart

| Mode | Operons Found | DIP-EcoCyc |
|---|---|---|
| PP | 22 | |
| 61 ID | 14 | 37 |
| PP & GD | 90 | 88 |
| PP & ID | 110 | 87 |
| ID & GD | 20 | 58 |
| **PP & ID & GD** | **125** | **100** |
| All with cross-over | 30 | 50 |
| $k$-means with PP | 45 | 103 |
| $k$-means with ID | 47 | 78 |
| $k$-means with both | 46 | 85 |
| Total | 442 | 346 |

**Table 13** Excerpt from the group where the DNA biosyntesis operon is located

| | | | | |
|---|---|---|---|---|
| $p_1$ | COG0576 | 0111001111111111111111111111 | -010–10000010101101111001 | |
| $p_2$ | COG0210 | 0111001111111111111111111111 | -110–00011011010001101111 | |
| $p_3$ | COG0484 | 0111001111111111111111111111 | -010–01000010001100010001 | dnaJ |
| $p_4$ | COG0443 | 0111001111111111111111111111 | -010–01100011111101100000 | dnaK |
| $p_5$ | COG0190 | 0101001111111111111111111111 | -1-0–10101000101100111000 | |
| $p_6$ | COG0166 | 0111011111111111111111110111 | -101-10011010101111001-001 | |

Moreover, the engine showed severe difficulties in creating consistent groups. (In general this seems to be a characteristics of all the mods that lacked the phylogenetic profiles objective function). This mode was able to consistently detect 8 operons. In our third run we used both objective functions. This time around the algorithm was able to correctly identify over 15 operons, including the *dnaK* operon (made up from dnaK (COG0443) and dnaJ (COG0484), the group is shown in Table 13. It was also able to identify the *carAB* operon and the *fixABC* operon, in addition to the ones discovered by the previous algorithms and 10 others, which shows the robustness of the algorithm since some of these operons have considerable distance in terms of their phylogenetic profile but they are near each other in terms of their directionality and viceversa, completing each other in its space search. The overall table of results can be checked in Table 12.

As we can see, the different objective functions contribute to the overall performance of the algorithm in different ways and in different levels, however there is a definite increase in performance the more fitness functions we combine that goes beyond the individual sum of the performance of each function.

# 9 Discussion

Throughtout the second part of our work we have demonstrated how Genetic Algorithms can be used as a tool for integrating several sources of biological data into a clustering process used for protein function prediction, and the advantages this have when giving more significance to the functional predictions our algorithm makes. We have described how easy it is to add new information to the genetic engine with minimal changes to the existing search process, and the immediate effect this can have on the validity of the results we obtain.

It is important to always have in mind that in bioinformatics the best clustering is not the one obtained by the most complicated clustering techniques, but the one that makes the most biological sense. As such, providing our algorithms with as much contextual information as we can becomes a vital task. Genetic Algorithms (in specific its Multi Objective version) presents a framework that adheres to this philosophy by being extremely easy to introduce new optimization information into an existing searching method. Further study on the applications of this type of algorithms in relationship to the data mining of biological data should be of great interest to the community.

However, even with these many advantages there are some limitations that should be taken into account if one is interested into implementing these kind of algorithms. Because of the inherent exploratory nature of genetic algorithms, the convergence time (in other words, the time it takes for the algorithm to find an answer) is highly variable, between 5 and 15 minutes. This time is clearly greater than that of traditional clustering algorithms, specially when compared for $k$-means, whose convergence time is in the range of 0.5-1 minutes. As such, our algorithm is better suited for applications when the generation of a backend database is preferable to generating the relationships under demand.

From the Biology side, despite the Multi Objective system working significantly better than its Single Objective and Classical counterparts, there is still a big number of operons that our algorithm fails to classify (including the widely known *lac* operon) the algorithm could only classify about 20% of the total operon database we had at our disposal. It is our belief that introducing more biological data (like other non-homology data) would help in the algorithm being able to discover all these relationships eventually, all the while its ability to detect false positives would also increase, leading to more biological coherent clusters.

**Acknowledgements.** This work was supported by the Consejo Nacional de Ciencia y Tecnología (CONACYT) under SEP-CONACYT award No. SEP-2004-C01-47434. The authors would like to thank Ricardo Ciria for his invaluable contribution to this chapter.

# References

1. Bork, P., Dandeker, T., et al.: Predicting function: from genes to genomes and back. J. Mol. Biol. 283, 707–725 (1998)
2. Coello-Coello, C., et al.: Evolutionary Algorithms for Solving Multi Objective Problems. Kluwer Academic Publishers, Dordrecht (2002)
3. Curteanu, S., Leon, F., Galea, D.: Alternatives for Multi-objective Optimization of a Polymerization Process. J. Applied Polymer Science (2006)
4. Deb, K., Reddy, A.R.: Reliable Classification of Two-Class Cancer Data Using Evolutionary Algorithms. Biosystems 72(1-2), 111–129 (2003)
5. Duester, G., Campen, R., et al.: Nucleotide sequence of an Escherichia coli tRNA (Leu 1) operon and identification of the transcription promoter signal. Nucleic Acids Research 9, 2121–2139 (1981)
6. Eisenberg, D., Marcotte, E., et al.: Protein function in the post-genomic era. Nature 405, 823–826 (2000)
7. Facelli, K., de Souto, M.: Multi-objective Clustering Ensemble. In: Proceedings of the Sixth International Conference on Hybrid Intelligent Systems (2006)
8. Fernández, J.C., Vallejo, E.E., Morett, E.: Fuzzy-C means for inferring functional coupling of proteins from their phylogenetic profiles. In: Ashlock, D., et al. (eds.) IEEE Computational Intelligence in Bioinformatics and Computational biology. IEEE Computer Society Press, Los Alamitos (2006)
9. Goldberg, D.E., Korb, B., Deb, K.: Messy genetic algorithms: Motivation, analysis, and first results. In: Complex Systems, pp. 493–530 (1989)
10. Holland, J.H.: Adaptation in Natural and Artificial Systems. An introduction. University of Michigan Press (1975)
11. Horn, J., et al.: Fitness Sharing and Niching Methods Revisited. IEEE Transactions on Evolutionary Computation, 82–87 (1994)
12. Huynen, M., Snel, B., et al.: Predicting Protein Function by genomic context: quantitative evaluation and qualitative inferences. Genomic Research 10(8), 1204–1210 (2000)
13. Jang, J.S.R., Sun, C.T., et al.: Neuro-fuzzy and soft-computing. Prentice Hall, Englewood Cliffs (1997)
14. Jensen, L.J., Kuhn, M., et al.: STRING 8–a global view on proteins and their functional interactions in 630 organisms. In: Ashlock, D., et al. (eds.) Pubmed (2009)
15. Lin, C., Wang, M.: Genetic-clustering algorithm for intrusion detection system. International Journal of Information and Computer Security 2(2), 218–234 (2008)
16. Marcotte, E.: Computational genetics: finding protein function by nonhomology methods. Current Option in Structural Biology 10, 359–365 (2000)
17. Mandal, C., Gudi, R.D., Suraishkumar, G.K.: Multi-Objective Optimization in Aspergillus Niger Fermentation for Selective Product Enhancement. Bioprocess and Biosystems Eng. 28, 149–164 (2005)
18. Marcotte, E., Xenarios, I., et al.: Localizing proteins in the cell from their phylogenetic profiles. In: PNAS, vol. 97, pp. 12115–12120 (2000)
19. von Mering, C., et al.: STRING 7–recent developments in the integration and prediction of protein interactions. Nuclear Acid Res., D358–D362 (January 2007)
20. Milano., M., et al.: A Clustering Genetic Algorithm for Actuator Optimization in Flow Control. In: Proceedings of the 2nd NASA/DoD workshop on Evolvable Hardware (2000)

21. Ciria, R., Abreu-Goodger, C., Morett, E., Merino, E.: GeConT: gene context analysis. Bioinformatics 20, 2307–2308 (2004)
22. Pellegrini, M., Marcotte, E., et al.: Assigning protein function by comparative genome analysis: Protein phylogenetic profiles. In: PNAS, vol. 96, pp. 4285–4288 (1999)
23. Sali, A.: Funtional links between proteins. Nature 402, 23–26 (1999)
24. Sareni, B., Laurent, K.: Fitness Sharing and Niching Methods Revisited. IEEE Transactions on Evolutionary Computation, 97–108 (1998)
25. Sun, J., Xu, J., et al.: Refined phylogenetic profiles method for predicting protein-protein interactions. Bioinformatics 21, 3409–3415 (2005)
26. Vert, J.F.: A tree kernel to analyze phylogenetic profiles. Bioinformatics 18, S276–S284 (2002)
27. Marcotte, E., Pellegrini, M., et al.: A combined algorithm for genomewide prediction of protein function. Nature 402, 83–86 (1999)
28. Karp, P.D., Keseler, I.M., et al.: Multidimensional annotation of the Escherichia coli K-12 genome. Nucleic Acids Research (2007)
29. Salwinski, L., Miller, C.S., et al.: The Database of Interacting Proteins: update. NAR 32(Database issue), D449–D451 (2004)
30. Falkenaeur, E.: Genetic algorithms and grouping problems. Wiley, Chichester (c1998)
31. Sammon, J.: A Nonlinear Mapping for Data Structure Analysis. IEEE Transactions on Computers c18, 401–409 (1969)
32. Someren, E.P., et al.: Multi-Criterion Optimization for Genetic Network Modeling. Signal Processing 83, 763–775 (2003)
33. Tapia, J.J., Vallejo, E.E.: A Clustering Genetic Algorithm for Inferring Protein-Protein Functional Interactions from Phylogenetic Profiles. In: 2008 IEEE World Congress on Computational Intelligence (2008)
34. Tatusov, R.L., Natale, D.A., et al.: The COG database: new developments in phylogenetic classification of protein from complete genomes. Nucletic Acids Research 29(1), 22–28 (2001)
35. Tatusov, R.L., Koonin, E.V., et al.: A genomic perspective on protein families. Science 278, 631–637 (1997)
36. Tatusov, R.L., Fedorova, N.D., et al.: The COG database: an updated version includes eukaryotes. BMC Bioinformatics 4, 41–54 (2003)
37. Watanabe, R.L.A., Morett, E., Vallejo, E.E.: Inferring modules of functionally interacting proteins using the Bond Energy Algorithm. BMC Bioinformatics 9, 285 (2008)
38. Wren, J.: The emerging in silico scientist how text-based bioinformatics is bridging biology and artificial intelligence. IEEE Engineering in Medicine and Biology Magazine, 87–93 (2004)
39. Wu, J., Kasif, S., et al.: Identification of functional links between genes using phylogenetic profiles. Bioinformatics 19, 1524–1530 (2003)
40. Wu, F.X.: Genetic weighted k-means algorithm for clustering large-scale gene expression data. BMC Bioinformatics 28(suppl. 6), S12 (2008)

# Detection of Remote Protein Homologs Using Social Programming

Gerard Ramstein, Nicolas Beaume, and Yannick Jacques

**Summary.** We present a Grammatical Swarm (GS) for the optimization of an aggregation operator. This combines the results of several classifiers into a unique score, producing an optimal ranking of the individuals. We apply our method to the identification of new members of a protein family. Support Vector Machine and Naive Bayes classifiers exploit complementary features to compute probability estimates. A great advantage of the GS is that it produces an understandable algorithm revealing the interest of the classifiers. Due to the large volume of candidate sequences, ranking quality is of crucial importance. Consequently, our fitness criterion is based on the Area Under the ROC Curve rather than on classification error rate. We discuss the performances obtained for a particular family, the cytokines and show that this technique is an efficient means of ranking the protein sequences.

## 1 Introduction

Remote homology detection is a challenging problem in bioinformatics. The identification of an unknown protein is generally based on sequence analysis, a method that is not well adapted to remote homologs presenting poor sequence similarities. Notable advances have been made in this area during the last decades and several protein classifiers inspired from machine learning algorithms have been proposed. The study of their performances seems to indicate that the limits of individual methods have been reached. Another strategy consists in combining existing well performing methods, the fusion of classifiers being a common way of improving the

Gerard Ramstein and Nicolas Beaume
LINA, Polytech'Nantes, rue Christian Pauc BP 50609 44306 Nantes cedex 3, France
e-mail: gerard.ramstein@univ-nantes.fr,
nicolas.beaume@univ-nantes.fr

Yannick Jacques
INSERM U601, 9 Quai Moncousu, F-44035 Nantes Cedex
e-mail: yjacques@nantes.inserm.fr

A. Abraham et al. (Eds.): Foundations of Comput. Intel. Vol. 4, SCI 204, pp. 277–296.
springerlink.com                    © Springer-Verlag Berlin Heidelberg 2009

reliability of the prediction. We apply a fusion method that produces a classification score based on the combination of the classifier outputs. The function performing this calculation will be called an aggregation operator. The aim of this work is to discover efficient aggregation operators for protein classification.

We propose to construct aggregation operators using a biogically-inspired technique. Biocomputation is a field of Artificial Intelligence that explores biological processes as metaphors for developing original and powerful methods. This approach has fertilized a variety of new developments in computational intelligence, including evolutionary algorithms, artificial neural networks and artificial immune systems. Many researchers have investigated the social behavior of animals, a source of inspiration that has produced many efficient algorithms. For instance, ant colony optimization is a class of metaheuristics that has been successfully applied to solve complex combinatorial problems. Bird flocking is another form of collective behavior that has been exploited, leading to a new optimization technique.

We propose to find efficient aggregation operators using social programming. A Grammatical Swarm [1] has been designed for the optimization of these functions: the metaphor of swarm refers to a collection of programs exploring the solution space. Each individual of this swarm computes a global score from a set of inputs given by the classifiers. The input values are probability estimates used to guarrantee a commensurability of the classifier results. In our experiment, these are given by Support Vector Machine (SVM) and Naive Bayes classifiers. We apply our method to the identification of remote homologs. More precisely, we search new members of a protein family based on the similarity of a candidate sequence with a target set of already known sequences. As the number of candidates is very large, it is necessary to optimize the quality of the ranking. We therefore use a dedicated objective function called the Area Under ROC Curve (AUC) to evaluate the aggregation operators. The remainder of the paper is structured as follows: we first review the related work, then we introduce the concept of Grammatical Swarm. In the next section we describe the application domain and the classifiers involved. Finally we discuss the performances of our technique and the information that can be extracted from an analysis of the generated programs.

## 2 Related Work

Grammatical Evolution (GE [2]) is an evolutionary method whose aim is to determine a computer program with an acceptable fitness value. The term fitness defines an objective function to be minimized. This recent technique has been mainly used in financial applications such as predicting corporate bankruptcy, and in bioinformatics [3]. In comparison with traditional Genetic Programming techniques, GE is based on a grammar defining the algorithmic structure of the outcome. Hence GE guarantees that the program is syntactically correct. It also permits to constrain the search space by biasing the grammar. GE can thus produce solutions compatible with the domain knowledge. The power of GE relies on the distinction between genotype and phenotype. This offers a great flexibility, since the user can easily modify the language defining the programs by merely changing the specification of the grammar. This feature involves that GE can use any evolutionary technique, the most widely-used approach being based on Genetic Algorithms. Grammatical Swarm (GS) is an emerging variant based on Particle Swarm Optimization [4]. This method is a stochastic optimization technique inspired by social behavior of bird flocking or fish schooling. Compared to Genetic Algorithms that use computationally expensive operators (i.e. crossover and mutation), Particle Swarm Optimization appears to be a competitive method. It has been demonstrated that GS needs smaller populations while having a fixed-length vector representation.

Fusion of classifiers is an important topic of research implying many data mining applications. Some authors suggest to apply majority voting scheme [5], boosting techniques [6] or team evolution [7]. Our approach differs from these works in that we want to integrate the results of base classifiers by considering a specific fitness criterion. Indeed, the classification error rate is not an adequate measure in our case. As we focus on the quality of ranking, the $AUC$ measure represents a finer evaluation model.

## 3 Grammatical Swarm

To construct a program, a GS needs two kinds of information. First, a training data set and a fitness function are given to define the optimization criterion. Second, a set of grammar rules specifies the syntax of the program. Fig.1 presents the overall structure of a GS. Two essential building blocks are shown: the search engine and the programming tools. The search engine block explores the search space and generates new candidate programs that are next translated and interpreted in the second block. One can note that the two blocks are independent; in section 5.4, we will compare two different search engines using the same programming tools. The process starts with a random intitialization of the search engine. Arbitrary source codes are generated. A source code is a program coded as a sequence of integers. The translator generates a function using this code and the grammar rules, as explained later. The interpreter takes the inputs provided by the user and computes the result as specified by the program. The output is evaluated using a given fitness function. The system checks if the termination criterion is met, otherwise a new iteration cycle is performed.

We will first describe the programming tools and explain the grammar, the coding scheme and the principle of the translator. Section 3.3 will define the algorithm used for the search engine, called Particle Swarm Optimization.

As illustrated in Fig.2, the principle of a GS relies on a genotype-phenotype mapping. The genotype is defined by an ordered list of integers coding the program. These values determine which production rules (expressed in Backus-Naur form) will be employed. This mapping produces a derivation sequence, which in turn provides the genotype, that is to say the program represented by a tree structure as in standard genetic programming.

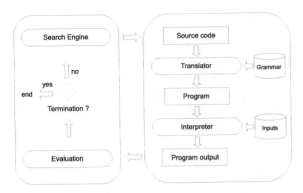

**Fig. 1** Architecture of a Grammatical Swarm. The left block represents the search engine and the right block the programming tools

**Fig. 2** Example of genotype-phenotype mapping as described in [1]. The derivation sequence details the transformation of production rules that determine the derivation tree. This can be simplified and represented by a parse tree

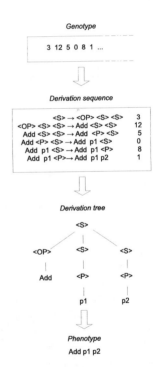

## 3.1 Context-Free Grammar

Backus-Naur form is a notation defining a language in terms of production rules. Fig.3 presents the grammar designed for our application. The symbol $\langle S \rangle$ denotes the start symbol of the grammar. In addition to the four arithmetic operators, we introduce two operators often used for aggregating classifiers: the operators $Min$(a,b) and $Max$(a,b) (these operators return the minimum (resp. maximum) value of a and b). We also design a C-like operator named *CND* that allows us to introduce conditional computations. The operator $(cnd\ ?\ i_1{:}\ i_2)$ evaluates the condition $cnd$. If it is true, then instruction $i_1$ is used, otherwise $i_2$ is used. The only condition that we have implemented is based on the logical operator $Less$, but more complex operators can be generated through nested conditional operators. Note that GS can either choose the minimum operator or define it through the following expression: $((Less\ x\ y)\ ?\ x : y\ )$. This logical redundancy also occurs in programming languages: a Java programmer for instance can either select the predefined function Math.min(a,b) or write it using an if-else statement. Terminal symbols are associated to the inputs, i.e. the probability estimates $(p_1, \ldots, p_{nc})$. An additional terminal $\tau$ has been designed to express generated thresholds randomly. In the next section, we describe how to create these values.

| $\langle S \rangle$ | :: = | $\langle Op \rangle\ \langle S \rangle\ \langle S \rangle$ | (1.0) |
| | | | $((\langle CND \rangle\ ?\ \langle S \rangle : \langle S \rangle))$ | (1.1) |
| | | | $\langle P \rangle$ | (1.2) |
| | | | | |
| $\langle Op \rangle$ | :: = | $Add$ | (2.0) |
| | | | $Sub$ | (2.1) |
| | | | $Mul$ | (2.2) |
| | | | $Div$ | (2.3) |
| | | | $Min$ | (2.4) |
| | | | $Max$ | (2.5) |
| | | | | |
| $\langle CND \rangle$ | :: = | $Less\ \langle P \rangle\ \langle P \rangle$ | (3.0) |
| | | | $Less\ \langle P \rangle\ \langle T \rangle$ | (3.1) |
| | | | $Less\ \langle T \rangle\ \langle P \rangle$ | (3.2) |
| | | | | |
| $\langle T \rangle$ | :: = | $\tau$ | (4.0) |
| | | | | |
| $\langle P \rangle$ | :: = | $p_1$ | (5.0) |
| | | | $p_2$ | (5.1) |
| | | | $\ldots$ | (5.$i$) |
| | | | $p_{nc}$ | (5.$nc$-1) |

**Fig. 3** The generative grammar used for the construction of aggregation operators. The expressions $(p.q)$ in brackets denote the $q^{th}$ right-hand side branch of the $p^{th}$ rule

## 3.2 Program Generation

The mapping process transforms an expression by applying production rules. The initial expression contains only the start symbol of the grammar $\langle S \rangle$. This symbol can be replaced by using the three right-hand side branches of the rule (1.0), (1.1) and (1.2) defined in Fig.3. The selected rule depends on the ordered list of positive integers $(c_1, c_2, \ldots, c_n)$ representing the genome of the individual.

Let $r$ be the number of right-hand side branches of the current rule. In basic GEs, the alternative branch $a$ to be applied is given by the following formula:

$$a = c_i \% r \tag{1}$$

where $c_i$ is the $i^{th}$ codon of the genome and $\%$ the modulo operator. In Fig.2, the mapping process uses the branch (1.0) for the first rule applied, since $c_1 = 3$ and $r = 3$. The expression $\langle S \rangle$ is then replaced by the expression $\langle Op \rangle \langle S \rangle \langle S \rangle$. The same procedure is performed, the first symbol to be replaced being now $\langle Op \rangle$. The corresponding rule has 6 alternative branches: (2.0) to (2.5). As $c_2 = 12$, the symbol is replaced by the terminal *Add*. Fig.2 indicates the whole derivation sequence that can be extracted from the genome. Two situations may arise: either all non-terminals have been replaced and the program is valid, or the process has used the last codon before completion. In the latter case, a wrapping operator can be invoked (for example, the genome can be read again from the beginning). In our implementation, we prefer to generate a new random genome. If the mapping still does not succeed, we set the fitness of the individual to the lowest possible value (i.e. 0 in our case). The introduction of new individuals presents the advantage of exploring new regions of the search space during all the evolutionary process.

Rule 4 is a dummy rule that presents no alternative branches. Actually, this rule has been inserted to determine a threshold value as a function of the current codon value $c_i$. For instance, let us consider the instruction $(0.9 < p_1 ? p_1 : p_2)$ which returns $p_1$ if its value is greater than 0.9 and $p_2$ otherwise. The value of the threshold has been assigned using the formula $\tau = c_i / cmax$ where $cmax$ is the maximum value of the codons so that $\tau$ has the same range as a probability.

We have implemented a variant of GS in which the selection of $a$ is not given by eq.1. We instead associate a different probability for each alternative branch. Fig.4 shows an example of such a rule. This variant allows us to parameter the program structure produced without modifying the grammar.

**Fig. 4** Probabilistic production rule with unequal probabilities. The probability of rule $X ::= Y$ is 25%. The selection procedure is based on the roulette-wheel algorithm

$$
\begin{array}{llll}
X & ::= & Y & c_i \in [0, 63] & \text{(a0)} \\
 & | & Z & c_i \in [64, 255] & \text{(a1)}
\end{array}
$$

## 3.3 Particle Swarm Optimization

Particle Swarm Optimization (PSO) is an algorithm inspired by the model of social learning that simulates the flocking behavior of birds and fish. A PSO is based on a population of interacting elements (particles). A swarm of particles explores an n-dimensional space. The location of a particle determines a potential solution. Each particle knows the most promising position (*pbest*) encountered so far, as well as the best location (*gbest*) found by its neighbors (the neighborhood in our implementation corresponds to the whole swarm). Particles adjust their flight according to their own experience but are also influenced by their peers. The displacement of

a particle is determined by its current position and a velocity vector that is updated
at each iteration of the algorithm. The algorithm starts with a population composed
of random particles. Then, the fitness value is computed for each particle and *pbest*
revised if necessary, as well as *gbest*.

The velocity vector $v_i$ at iteration $t + 1$ is determined as follows:

$$v_i(t + 1) = w * v_i(t) + \gamma_1 * R_1 * (pbest - x_i)$$
$$+ \gamma_2 * R_2 * (gbest - x_i) \tag{2}$$

where

$$w = wmax - ((wmax - wmin)/itermax) * iter \tag{3}$$

In eq.2, $\gamma_1$ and $\gamma_2$ are weights associated to the *pbest* and *gbest* terms, $R_1$ and $R_2$
random values in $U(0, 1)$ and $x_i$ is the $i^{th}$ component of the location vector of the
particle. The parameter $w$ defined in eq.3 is a momentum coefficient that linearly
decreases from $wmax$ to $wmin$. The next step of the algorithm updates the location
vector as follows:

$$x_i(t + 1) = x_i(t) + v_i(t + 1) \tag{4}$$

The PSO algorithm can be expressed as follows:

Step1.   Generation of an initial swarm composed of $N$ particles.
The location of the particles are uniform random values bound to a predefined
range $[dmin, dmax]$. The initial velocities are set to 0 for all the particles.
Step2.   Estimation of the fitness of all the particles.
If *pbest* is undefined or if the current fitness of a particle is better than the fitness
of *pbest*, set *pbest* to the current location. In our application, the fitness quantifies
the discrimination power of the aggregation operator encoded by the particle
location. The fitness function estimates the ranking quality of a training set of
proteins. If the particle achieving the best fitness value among the swarm obtains
a better fitness than *gbest* or if *gbest* is undefined, set *gbest* to the location of
this particle.
Step3.   Displacement of the swarm.
For all the particles, compute the new velocity according to eq.2 and update the
location according to eq.4. The velocity cannot exceed a maximum value $vmax$
and the particles are bound to the dimension of the search space, defined by the
range $[dmin, dmax]$.
Step4.   If the termination criterion is not met, return to Step2.
The termination criterion can be a fixed fitness value, a stagnation of *gbest* over
a fixed number of iterations or a predefined number of iterations.

## 3.4   Area under ROC Curve

An objective function $\varphi$ is associated to a PSO for the quantification of a candidate solution. As our criterion is the quality of the ranking, the commonly-used

**Fig. 5** ROC curves. A binary classifier defines a membership estimate and uses a decision value to separate two classes. This cut point determines the individuals considered as positives. Among these, actual positives are called true positives (TPs) whereas actual negatives are called false positives (FPs). Each point of the ROC curve depicts a cut point which determines the rate of TPs and FPs. The closer the curve follows the left-hand border and then the top border of the ROC space, the more accurate the classifier is. For instance, the GS curve indicates a better result than the BEST one. Therefore, the area under the ROC curve gives an efficient measure of the quality of the classifier

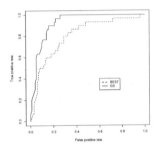

classification error rate is not an appropriate evaluation model. A more precise measure is the Area Under ROC Curve ($AUC$) which has the advantage of being independent from a particular decision cut point.

$AUC$ is defined as:

$$AUC = \frac{\sum_{i=1}^{n^+} \sum_{j=1}^{n^-} \mathbf{1}_{f(x_i^+) > f(x_j^-)}}{n^+ n^-} \tag{5}$$

where $f(.)$ is the scoring function used for the ranking, $x^+$ (resp. $x^-$) represents a positive (resp. negative) example, $n^+$ (resp. $n^-$) their number, and $\mathbf{1}_\pi$ is the indicator function (equals to 1 if the predicate $\pi$ holds and 0 otherwise). $AUC$ is equivalent to the Wilcoxon-Mann-Whitney statistic [8] and can be interpreted as the probability that a positive example will achieve a higher score than a negative one, when both examples are selected at random. Thus, a solution having an $AUC$ fitness equals to 0.5 would do no better than a random guess, whereas a fitness of 1 would indicate an error-free aggregation operator. Fig.5 shows the graphical interpretation of the ROC curve.

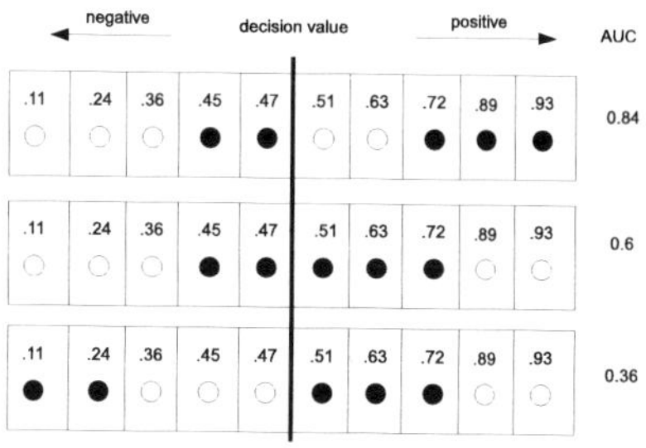

**Fig. 6** Examples of $AUC$ scores. We consider three data sets containing 10 individuals (5 positive examples indicated by the symbol • and 5 negative ones indicated by the symbol ○ ) presented in increasing order of probabilities. All these data sets are related to the same decision value: if the probability estimate is greater than 0.5, the example is predicted as positive. Thus, the classification error rate is common to the three data sets and equals 40%. Note that the $AUC$ score is sensible to the actual ranking of the examples and can better discriminate between the three classifications

## 3.5 Objective Function

As the $AUC$ represents a pertinent measure of the ranking quality, the fitness $\varphi$ of a solution can be computed using eq.5. We also have introduced an adjustment factor $\phi$ to control the program structure produced by the GS algorithm. The objective function $\varphi$ associated to a program $\pi$ is defined by $\varphi(\pi) = \phi(\pi) \times AUC(\pi)$ where $\phi$ is a factor returning a real value in $[0, 1]$. Its aim is to force the swarm to explore particular regions of the search space. The complexity of $\pi$ can be modeled by the number $c$ of distinct classifiers that are used as operands and the number $o$ of operations involved: $\phi(\pi) = \phi(c, o)$. For instance $\pi$=( Add (Min $p_1$ $p_2$) $p_1$))) presents a structure $(c = 2, o = 2)$. As *CND* is a complex instruction involving 4 operands, we have associated a weight of 2 for this particular operator. Thus $\pi$ =((Less $p_1$ $p_2$) ? Min($p_1$, $p_3$) : $p_2$) possesses a structure $(c = 3, o = 3)$. In this paper we propose the three following factors :

$$\phi(c, o) = 1 \quad \forall(c, o) \in [1, nc] \times \mathbb{N}^* \tag{6}$$

$$\phi(c, o) = \begin{cases} 1 & \text{if } (c, o) = (x_c, y_c) \\ 0 & \text{otherwise} \end{cases} \tag{7}$$

$$\phi(c, o) = \frac{1}{1 + e^{dist(c,o) - \delta}} \tag{8}$$

**Fig. 7** Algorithm of method
*dist* used in eq.8

$dist(c, o)$ : returns a real value
    parameters:
        $x_c, y_c$       : coordinates of the central point $C$
        $\sigma_x, \sigma_y$     : distance weights
        $x := c - x_c;$
        $y := o - y_c;$
        $x' := \frac{\sqrt{2}}{2}(x - y);$
        $y' := \frac{\sqrt{2}}{2}(x + y);$
        $dist := \sqrt{\frac{x'^2}{2\sigma_x^2} + \frac{y'^2}{2\sigma_y^2}};$

**Fig. 8** Perspective view
of the adjustment factor.
The parameters used in this
paper are the following:
$x_c = 4, y_c = 4, \sigma_x = 1,$
$\sigma_y = 2, \delta = 3$

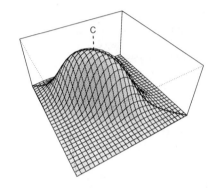

Eq.6 provides an objective function that accepts all programs regardless of their structures, contrary to eq.7 that focuses on a particular program structure presenting exactly $x_c$ classifiers and $y_c$ operations. Eq.8 corresponds to a radial basis function which decreases the fitness according to a function *dist* described in Fig.7 ($\delta$ is an offset parameter introduced to shift the classical sigmoid function). This adjustment factor follows two objectives. The first one is a parsimony criterion which guarantees that the generated programs keep a reasonable complexity expressed by the number $o$ of operations performed. The parsimony criterion also concerns the ratio between the number of operations and the number $c$ of distinct classifiers involved. These two parameters must proportionally evolve to produce concise aggregation operators. The second objective focuses on the parameter $c$ which must remain in a median range. The underlying idea is that the reliability of a decision depends to some extent on the number of deciders. On the one hand, a fusion based on one or two classifiers defines a poor strategy (probably due to an over-estimation of the performances obtained in the learning phase). On the other hand, it seems advantageous for computing cost depletion to remove some classifiers that do not influence the global performance. Indeed, as some SVM predictions are based on similar feature spaces, they may be redundant.

In eq.8 the $AUC$ score is not penalized if the concerned program shows a median structure represented by the central point $C = (c = x_c, o = y_c)$. Thereby we want $\phi$ to remain near 1 in the neighborhood of $C$ and to progressively decrease as the couple $(c, o)$ moves away from the central region (we did not use the gaussian model but instead applied a sigmoid function that has a less abrupt decrease near $C$). The distance of the point $(c, o)$ from $C$ is a weighted euclidian metrics described in Fig.7. The function $dist$ takes into account the fact that the variance of $c$ and $o$ are different in essence. We also perform a rotation of 45 degrees to orientate the function along the bisector axis. This transformation corresponds to the postulate that $c$ and $o$ evolve proportionally. Fig.8 shows the shape of the resulting curve.

## 4 Classification of Protein Sequences

Proteins are defined by a unique sequence of amino acid residues, called the primary structure of the protein. A set of 20 amino acids forms the alphabet from which all the variety of proteins are drawn. We apply our GS to the identification of a particular protein family called the cytokines. These proteins are involved in the immune system and constitute one of the largest and most heterogeneous protein family of the human genome. There are 45 identified members but some observations have suggested the existence of still unknown homologs. The discovery of new cytokines is decisive for biomedical research, notably the emergence of anti-cancer drug therapies. In this section, we detail the classifiers coming from two different strategies.

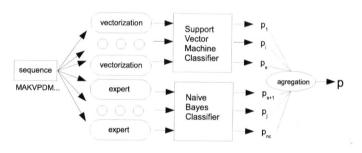

**Fig. 9** Classification of protein sequences

We first use SVM classifiers that have proven very efficient for the identification of remote homologs. Secondly, we apply Naive Bayes classifiers for exploiting relevant biological characteristics. The first classifier category involves large feature vectors (up to several thousand components) whereas the other one only focuses on a unique feature. This study will show that this additional information significantly improves the quality of the final ranking. Fig.9 shows the classification scheme that has been designed. A candidate protein sequence is analyzed by a set of classifiers and experts. The classifiers use different strategies of vectorization whereas the experts exploit particular features of the sequence. All the tools provide a probability estimate. The agregation operator delivers a unique score from these individual values.

## *4.1   SVM Classifiers*

Support Vector Machine (SVM)[9] is a method based on Statistical Learning Theory that addresses the problem of learning to discriminate between two classes (usually positive and negative members of a given class). SVMs have been successfully applied to a wide range of real-world applications, such as face recognition [10] and text categorization [11]. This technique has shown to perform well in computational biology in a wide range of problems, including gene functional classification [12], tumor identification [13], protein secondary structure prediction [14], intron detection [15], signal peptide cleavage site localization [16] and protein fold recognition [17]. Moreover, SVM tends to give the best performances in our application, namely the protein classification task [18].

The use of SVMs in bioinformatics is motivated by the fact that many biological applications involve high-dimensional data, for which SVMs achieve remarkable performances compared to other classification methods. Another advantage of SVM is that it relies on strong theoretical foundations. SVM is based on the Structural Risk Minimization principle consisting in minimizing the expected risk, rather than the error of the training data. Therefore, SVMs have proven to have a greater ability to generalize than other techniques such as traditional neural networks [19].

SVMs operate a mapping of a training set into a high dimensional feature space. A hyperplane boundary is defined between both classes so that the separation plane maximizes a margin from any point of the training set. An unlabeled point can then be predicted by simply considering the space region where it lies. As an SVM requires that the input be fixed-length numeric vectors, a feature extraction technique is generally used to transcribe the variable-length strings representing the sequences into real vectors. Four methods of vectorization have been developed, stemming from the literature and dedicated to protein classification. The first two perform a direct transformation whereas the other two use a similarity measure between the candidate sequence and a set of known sequences.

- *Spectrum.* Developed by Leslie *et al* [18], this classifier is based on n-grams, i.e. the set of sequences of size $n$ that can be built from the amino-acid alphabet. The vectorization of a sequence $s$ consists in computing the frequency of all n-grams in $s$.
- *Mismatch.* Like Spectrum, Mismatch [20] is based on n-grams but this method tolerates inexact string matching: an n-gram is recognized even if at most $k$ errors (mismatches) occur.
- *Pairwise.* This classifier [21] compares a sequence $s$ with a collection of known proteins. This learning set is composed of positive and negative examples. The sequence is transformed into a numeric vector for which the $i^{th}$ component represents a similarity measure between $s$ and the $i^{th}$ sequence of the learning set. This value is given by the Smith-Waterman (SW) algorithm [22], which is a widely-used local alignment method.
- *SVM-BLAST.* This technique is an alternative version of Pairwise proposed by the same authors. The Smith-Waterman algorithm is replaced by BLAST, which provides a standard measure for sequence similarity [23].

- *Local Alignment Kernel.* The LA kernel [24] follows the same sequence-versus-family comparisons as Pairwise but instead of a local alignment score, this classifier introduces a string kernel function that is mathematically valid. Contrary to the SW algorithm that only considers the best local alignment, the kernel performs a summation over all the possible ones.

In conformity with the results found in the literature, our own experiments have shown that the last two algorithms clearly outperform the other methods in ten-fold cross-validation. However, we have observed that the first two SVMs achieve performances comparable to the two others when the training set decreases in size. This phenomenon has motivated us to consider all five SVMs.

## 4.2 Experts

To improve performances, we define a set of Naive Bayes classifiers denoted experts and based on specific biological clues. Three experts have been currently selected to complement the SVM classifiers:

- *the length of the sequence.* Although this criterion seems trivial, the length of the sequence can partially discriminate the cytokines. Most cytokines have a restricted range of lengths in comparison with the sizes of the human genome proteins.
- *the isoelectrical point.* The isoelectrical point is the pH where a given molecule is in a zwitterionic state, *i.e.* it carries no net electrical charge. Cytokines have similar ways of functioning: they use the same cellular components suggesting that they share physico-chemical features.
- *the Segment OVerlap (SOV) criterion.* More than the protein sequence, the protein structure is a well conserved feature in protein families, especially for the cytokines. The secondary structure is defined as the sequential chaining of local structures of amino acids. Proteins are made of three canonical local structures: alpha helix, beta sheets and random foils. We use PSIpred [25] to predict the secondary structure from the protein sequence. The SOV criterion [26] quantifies the secondary structure similarity between a candidate and each cytokine: the best SOV is retained to compute the bayesian probability.

## 5 Experiments and Results

In this section we first present the data sets involved, then the parameters of the GS used for all experiments and finally we discuss the results obtained.

## 5.1 Data Sets

Our cytokine protein set is composed of 45 sequences. Negative examples are extracted from the SCOP database [27], excluding the cytokine family. This collection

classifies known proteins according to their structural characteristics. Sequences
have been randomly selected so that each protein superfamily is equally represented.

As shown in Fig.10, three different data sets have been prepared for our experiment. A first learning set $S_1$ is used to train the classifiers, a second learning set
$S_2$ is applied to train the evolutionary operator and finally a test set $S_3$ is needed to
evaluate the performance of the ranking. The cytokine set has been split into 3 subsets of 15 sequences corresponding to the positive examples of $S_1$, $S_2$ and $S_3$. One
hundred negative examples have been randomly extracted from the SCOP database
and added to $S_1$. The set $S_1$ remains unique so that the SVM learning capacity does
not interfere during our experiment. The two other sets $S_2$ and $S_3$ are used to define
a two-fold cross-validation and are also composed of 15 positive and 100 negative
examples. Positive examples are randomly selected from the remaining cytokines.
As we want to investigate the usefulness of the experts, we have intentionally selected the negative examples that are the most difficult to classify. A pool of 200
sequences has been extracted. These examples present the highest prediction for
each SVM classifier. We retain the 40 top ranked negative sequences for each of the
five classifiers (duplicates have been removed). This pool is randomly partitioned
into two equal subsets to complete $S_2$ and $S_3$.

**Fig. 10** Construction of the
data sets. *Random split* first
defines the learning set $S_1$
for the individual classifiers (*Learning* methods).
*Classifier prediction* then
computes the probability
estimates of the sequences
used for testing. The *Selection* method retains 200
negative examples. The last
*Random split* defines the
learning and test sets for
cross-validation

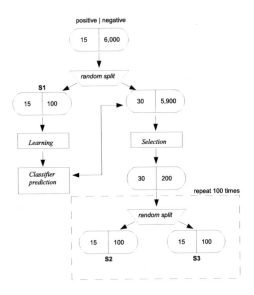

## 5.2  GS Parameters

The parameters used across the experiments for GS are $\gamma_1 = \gamma_2 = 1.0$, $wmin =$
$0.4$, $wmax = 0.9$. The location dimension values are bound to the range $[0, 255]$
and the velocities are bound to $\pm 255$. The swarm is composed of 50 particles, the
length of the genome is set to 100. The termination criterion is set to a maximum
number of iterations: $itermax = 1000$.

## 5.3   Results

Figure 11 represents the $AUC$ scores obtained from the cross-validation. For comparison purpose, we implement a basic aggregation rule denoted BEST. Like the GS method, BEST needs a learning phase. This method retains the best classifier among the $nc = 8$ individual classifiers that can be observed on the learning set $S_2$. The selected classifier is afterwards applied to predict the test set $S_3$. Concerning GS, we first discuss the results obtained when the search space of the swarm is not restricted by the adjustment factor. Using eq.6, we consider any program, regardless of its complexity. We called GS this version of our grammatical swarm, which is based on the grammar of Fig.3 with equiprobable rule probabilities. As shown later, the $CND$ instruction plays a paramount role in the obtained solutions. We restrain the language to this unique operator by applying the grammar variant described in Fig.4. We set all the rule probabilities to 0, except for the conditional operator (the probability of alternative branch 1.1 has been set to 1) and its derived rules (the alternatives of rules 3 to 5 have equal probabilities). The generated programs are then only based on if-else statements handling input parameters or threshold values. In spite of the simplicity of its grammar, this variant outperforms BEST and achieves comparable results to GS. A study of the outcomes of GS reveals that a common trend can be observed. The most representative profile consists in a short program performing few operations (3 operations involving 3 distinct classifiers). Moreover, the programs are most often based on very few distinct operators, as shown in Fig.12. Most of the time, the program only applies a unique operator type. One also observes that in 19 percent of the cases, the GS does not perform any operation at all but directly selects a classifier. The performances of BEST have shown that this strategy is not optimal. To avoid this pitfall, we introduced the adjustment factor of eq.8.

Fig.13 presents the operators that have been selected by GS. The most frequent one is the $CND$ operator. Table 14 indicates that it has been retained by the GS in 13 percent of the cases: a remarkable fact that explains why we performed a GS based on this unique operator in Fig.11. The second most used operator is $Min$: this shows that the best strategy is to compare the probability estimates and to retain the score of the most pessimistic classifier. It is interesting to note that conversely the optimistic point of view, represented by the $Max$ operator, has rarely been

**Fig. 11** Boxplot representation of the $AUC$ scores for BEST and variants of GS. CND only uses the conditional operators, whereas GS and AGS accept any operator. The fitness of GS is equal to the $AUC$; AGS multiplies this score by the adjustment factor $\phi$

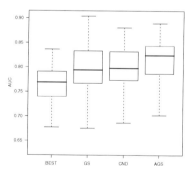

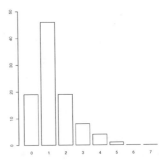

considered. The third most efficient operator is the product, which corresponds to the mutual independence hypothesis in terms of probability ($Div$ is less pertinent and is also the less used operator). As we anticipated, the $Add$ operator is also indicated, but $Sub$ is more frequent than expected. These results should be viewed with circumspection because of the possible presence of useless operations. So far, we have not implemented a post-processing for optimizing the code (e.g. ($Add$ $p_1$ ($Sub$ $p_1$ $p_1$)) should be simplified into $p_1$).

Table 14 indicates the most frequent combination of operators. One third of the programs is only based on conditional operations if one considers that the $Min$ operator is a particular case of *CND*. However one observes a great diversity in the solutions: 32 combinations in total have been found.

All these results somehow confirm the Occam's Razor principle that can be re-stated as follows: operators should not be multiplied beyond necessity. However, the GS does not systematically adopt an adequate profile, as discussed in section III.E. Although GS generally finds good solutions, it sometimes produces a pro-gram whose structure is inappropriate. To observe this phenomenon, we force the GS to find solutions having a fixed program structure. We therefore apply the ad-justment factor defined in eq.7. The contour map shown in Fig.15 indicates that the program structure significantly influences the performance of the ranking, justifying the weighting scheme of eq.8.

Fig.11 indicates the performances of the variant AGS relying on eq.8. One ob-serves a significant gain on the $AUC$ score and the construction of concise aggre-gation operators. Indeed a great advantage of evolutionary programming is that the algorithms produced are easily understandable. Thus, the analysis of the generated

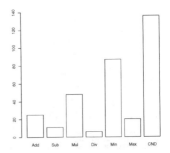

**Fig. 13** Operations involved
during the cross-validation.
The bars represent the number of operations over 100
validation samples (we
count 1 for each operation
that occurs in a program)

**Fig. 14** Conjunction of operators in the computer programs. This table gives the most representative outcomes (in percentage) and their structures

| % | Operators |
|---|-----------|
| 17% | *CND Min* |
| 13% | *CND* |
| 9% | *CND Mul* |
| 7% | *CND Min Mul* |
| 7% | *Min Mul* |
| 5% | *Min* |

**Fig. 15** Contour map of the ranking quality as a function of the program structure. The region yielding the best results is roughly centered around the point ($x_c = 4, y_c = 4$)

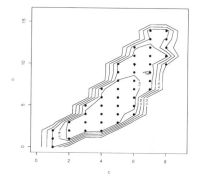

programs gives useful insights into the aggregation procedure that can be employed. The examples given in Fig.16 reveal that only half of the classifiers are necessary to achieve good performances. Moreover only two SVM classifiers are needed if they are associated to experts. These results show that our experts can efficiently improve the performances of the SVM classifiers.

| | |
|---|---|
| $(Mul\ (Min\ p_6\ p_8)\ (Min\ p_1\ p_4))$ | $(\pi_1\colon 0.86)$ |
| $(Min\ (Less\ p_7\ p_6)\ ?\ p_4\colon p_7)\ p_1))$ | $(\pi_2\colon 0.87)$ |
| $(Min\ p_8\ (Mul\ (p_6,$ | |
| $((Less\ 0.83\ p_1)\ ?\ \colon p_1\ \colon p_2))))$ | $(\pi_3\colon 0.87)$ |
| $((Less\ 0.70\ p_4)\ ?\ p_5\colon (Min\ p_8\ p_6))$ | $(\pi_4\colon 0.88)$ |
| $(Add\ (Add\ p_6\ p_7)$ | |
| $(Less\ 0.61\ p_1)\ ?\ p_4\colon p_1))$ | $(\pi_5\colon 0.90)$ |
| $((Less\ p_4\ p_2)\ ?\ p_1\ \colon$ | |
| $(Add\ p_6\ ((Less\ 0.84\ p_4)\ ?\ p_1\colon p_7)))$ | $(\pi_6\colon 0.91)$ |

**Fig. 16** Examples of aggregation operators. The values in brackets give the $AUC$ scores obtained. The input parameters $p_i$ refer to either SVM classifiers ($i \leq 5$) or experts ($i \geq 6$). Fig.5 shows the ROC of the operator $\pi_6$ compared to the ROC of the BEST method

## 5.4 Comparison with Genetic Evolution

As described in Fig.1, the structure of our system comprises two independent blocks. This means that any optimization algorithm can be used to generate candidate aggregation operators. It seems interesting to compare the Particle Swarm Optimization with more conventional search engines. We focus our study on a classical search engine, namely the Genetic Algorithm. We applied standard parameters: tournament selection, crossover probability of 0.8 and mutation probability of 0.01. These settings have produced the best convergence speed for our experiments. Table 17 shows the performances relative to the grammar of Fig.3 without any adjustment factor. To achieve a fair comparison between GS and GE, we generated the same number of individuals for both methods (GS: 1,000 iterations over a swarm size of 50, GE: 100 iterations over a population size of 500). A t-test showed that GS slightly but significantly outperforms GE ($p < 10^{-15}$).

**Fig. 17** Comparison of the fitness obtained for GS and GE over 100 runs

| method | mean fitness | std dev. | median |
|--------|--------------|----------|--------|
| GS     | 0.843        | 0.008    | 0.842  |
| GE     | 0.823        | 0.009    | 0.822  |

## 6 Conclusion

In this work, we have addressed the problem of identifying remote protein homologs. This challenging task consists in determining the protein family associated to an unknown sequence, given its primary structure. We have designed a hybrid system comprising 5 classifiers and 3 experts to improve the prediction performances. The classifiers use different strategies to transform the primary sequence into a real-valued vector. After this preprocessing step, a common classification process is performed, based on Support Vector Machines. The experts are biological clues introducing complementary information. The outputs of the classifiers and experts are probability estimates that are combined to form a unique score. The function that performs this calculation, called an aggregation operator, has been automatically constructed by using social programming. We have shown that a Grammatical Swarm is capable of providing compact and efficient aggregation algorithms. Our study indicates that a search engine based on a Particle Swarm optimization gives remarkable performances, compared with a genetic approach using computationally expensive operators. From a practical point of view, a great advantage of this technique is to produce understandable formulas that can be analyzed by a human expert. The experiments on the cytokine family, an important protein family involved in the immune system, have shown that the fusion of classifiers significantly improves the ranking quality of the candidate sequences. We particularly demonstrate the usefulness of biological clues that have proven to enrich commonly used SVM classifiers. Since, for instance, the genomic structure is relatively well conserved among the cytokines, we are currently working at implementing new experts based on particular features of our protein family.

# References

1. O'Neill, M., Brabazon, A.: Grammatical swarm: The generation of programs by social programming. Natural Computing: an international journal 5(4), 443–462 (2006)
2. O'Neill, M., Ryan, C.: Grammatical Evolution: Evolutionary Automatic Programming in an Arbitrary Language. Kluwer Academic Publishers, Hingham (2003)
3. O'Neill, M., Adley, C., Brabazon, A.: A grammatical evolution approach to eukaryotic promoter recognition. In: Bioinformatics Inform Workshop and Symposium, Dublin, Ireland (2005)
4. Kennedy, J., Eberhart, R.C.: Particle swarm optimization. In: Proceedings of the 1995 IEEE International Conference on Neural Networks, Perth, Australia, vol. 4, pp. 1942–1948. IEEE Service Center, Piscataway (1995)
5. Handstad, T., Hestnes, A.J.H., Saetrom, P.: Motif kernel generated by genetic programming improves remote homology and fold detection. BMC Bioinformatics 8, 23 (2007) (Evaluation Studies)
6. Paris, G., Robilliard, D., Fonlupt, C.: Applying boosting techniques to genetic programming. In: Selected Papers from the 5th European Conference on Artificial Evolution, pp. 267–280. Springer, London (2002)
7. Brameier, M., Banzhaf, W.: Evolving teams of predictors with linear genetic programming. Genetic Programming and Evolvable Machines 2(4), 381–407 (2001)
8. Yan, L., Dodier, R.H., Mozer, M., Wolniewicz, R.H.: Optimizing classifier performance via an approximation to the wilcoxon-mann-whitney statistic. In: ICML, pp. 848–855 (2003)
9. Vapnik, V.N.: The nature of statistical learning theory. Springer, Heidelberg (1998)
10. Guo, G., Li, S., Chan, K.: Face recognition by support vector machines (2000)
11. Joachims, T.: Text categorization with support vector machines: learning with many relevant features. In: Nédellec, C., Rouveirol, C. (eds.) Proceedings of ECML-98, 10th European Conference on Machine Learning, Chemnitz, pp. 137–142. Springer, Heidelberg (1998)
12. Brown, M., Grundy, W., Lin, D., Cristianini, N., Sugnet, C., Furey, T., Ares Jr., M., Haussler, D.: Knowledge-based analysis of microarray gene expression data by using suport vector machines. In: Proc. Natl. Acad. Sci., vol. 97, pp. 262–267 (2000)
13. Segal, N.H., Pavlidis, P., Antonescu, C.R., Maki, R.G., Noble, W.S., DeSantis, D., Woodruff, J.M., Lewis, J.J., Brennan, M.F., Houghton, A.N., Cordon-Cardo, C.: Classification and subtype prediction of adult soft tissue sarcoma by functional genomics. Am. J. Pathol. 163(2), 691–700 (2003)
14. Hua, S., Sun, Z.: A novel method of protein secondary structure prediction with high segment overlap measure: Svm approach (2001)
15. Saeys, Y., Degroeve, S., Aeyels, D., Rouze, P., Van de Peer, Y.: Feature selection for splice site prediction: a new method using EDA-based feature ranking. BMC Bioinformatics 5, 64 (2004) (Comparative Study)
16. Vert, J.: Support vector machine prediction of signal peptide cleavage site using a new class of kernels for strings (2002)
17. Ding, C., Dubchak, I.: Multi-class protein fold recognition using support vector machines and neural networks (2001)
18. Leslie, C., Eskin, E., Noble, W.S.: The spectrum kernel: a string kernel for svm protein classification. In: Pac. Symp. Biocomput., pp. 564–575 (2002)
19. Gunn, S.: Support vector machines for classification and regression (1998)
20. Leslie, C.S., Eskin, E., Cohen, A., Weston, J., Noble, W.S.: Mismatch string kernels for discriminative protein classification. Bioinformatics 20(4), 467–476 (2004)

21. Liao, L., Noble, W.S.: Combining pairwise sequence similarity and support vector machines for detecting remote protein evolutionary and structural relationships. J. Comput. Biol. 10(6), 857–868 (2003)
22. Smith, T.F., Waterman, M.S.: Identification of common molecular subsequences. Journal of Molecular Biology 147, 195–197 (1981)
23. Altschul, S.F., Madden, T.L., Schaffer, A.A., Zhang, J., Zhang, Z., Miller, W., Lipman, D.J.: Gapped blast and psi-blast: a new generation of protein database search programs. Nucleic Acids Res. 25(17), 3389–3402 (1997)
24. Saigo, H., Vert, J.P., Ueda, N., Akutsu, T.: Protein homology detection using string alignment kernels. Bioinformatics 20(11), 1682–1689 (2004)
25. Jones, D.T.: Protein secondary structure prediction based on position-specific scoring matrices. J. Mol. Biol. 292(2), 195–202 (1999)
26. Zemla, A., Venclovas, C., Fidelis, K., Rost, B.: A modified definition of sov, a segment-based measure for protein secondary structure prediction assessment. Proteins 34(2), 220–223 (1999)
27. Conte, L., Ailey, L., Hubbard, B., Brenner, T., Murzin, S., Chothia, A.: Scop: a structural classification of proteins database (2000)

# Part V
# Bio-Inspired Approaches in Information Retrieval and Visualization

# Optimizing Information Retrieval Using Evolutionary Algorithms and Fuzzy Inference System

Václav Snášel, Ajith Abraham, Suhail Owais, Jan Platoš, and Pavel Krömer

**Summary.** With the rapid growth of the amount of data available in electronic libraries, through Internet and enterprise network mediums, advanced methods of search and information retrieval are in demand. Information retrieval systems, designed for storing, maintaining and searching large-scale sets of unstructured documents, are the subject of intensive investigation. An information retrieval system, a sophisticated application managing underlying documentary databases, is at the core of every search engine, including Internet search services. There is a clear demand for fine-tuning the performance of information retrieval systems. One step in optimizing the information retrieval experience is the deployment of Genetic Algorithms, a widely used subclass of Evolutionary Algorithms that have proved to be a successful optimization tool in many areas. In this paper, we revise and extend genetic approaches to information retrieval leverage via the optimization of search queries. As the next trend in improving search effectiveness and user-friendliness, system interaction will use fuzzy concepts in information retrieval systems. Deployment of fuzzy technology allows stating flexible, smooth and vague search criteria and retrieving a rich set of relevance ranked documents aiming to supply the inquirer with more satisfactory answers.

Václav Snášel, Jan Platoš, and Pavel Krömer
Department of Computer Science, Faculty of Electrical Engineering and
Computer Science, VŠB - Technical University of Ostrava, 17. listopadu 15, 708 33
Ostrava - Poruba, Czech Republic
e-mail: {vaclav.snasel, jan.platos, pavel.kromer.fei}@vsb.cz

Ajith Abraham
Center of Excellence for Quantifiable Quality of Service, Norwegian University of
Science and Technology, O.S. Bragstads plass 2E, N-7491 Trondheim, Norway
e-mail: ajith.abraham@ieee.org

Suhail Owais
Information Technology, Al-Balqa' Applied University - Ajloun University College,
P.O. Box 6, JO 26810 Ajloun, Jordan
e-mail: suhailowais@yahoo.com

A. Abraham et al. (Eds.): Foundations of Comput. Intel. Vol. 4, SCI 204, pp. 299–324.
springerlink.com                                   © Springer-Verlag Berlin Heidelberg 2009

# 1  Introduction

Information retrieval activity is a derivation of real-world human communication. An information or knowledge, stored in a data repository by one person, is desired to be retrieved by another. Data repositories, emphasizing the Internet as the ultimate one, are used for persisting information in both, time and space. Data available on the Internet might be accessed by users distant in time and place. Unfortunately, the omnipresence of data is not equal to instant availability of information. In general, data can be seen as a state of information used for storage purposes, encapsulating the information content itself. A speech is not information, it contains information. An article is not information, it contains information. An electronic document can be seen similarly. To exploit stored data, it is desired to access the contained information in an efficient way. Such *information retrieval* activity is not an easy task and its complexity depends specially on the dimension of searched data basis. Moreover, when we are trying to automate information search process, the requirement to understand is becomes crucial. To retrieve the information in document, its content should be understood. To present required information to inquirer, the requests must be understood and correctly interpreted. Advanced techniques of information retrieval are under investigation to provide both - better content representation and better query apprehension.

There is fuzziness in human mind. It involves the means of communication. Estimations and intuition are present. Vagueness, imprecision and mistakes occur. These facts influence both - information content of documents and search request formulations. Contrariwise, any automated search tool has rather crisp and rough picture (i.e. model) of the information content of data, providing satisfactory search service for data collections up to certain size. Inevitably, the enormous growth of data repositories and especially of the Internet brings up more and more problems when performing information retrieval tasks. The amount of regular users of search services is growing as well. One approach to improve information retrieval in such conditions is approximating reality better than before. To improve the efficiency of information retrieval, soft computing techniques with special emphasis on fuzzy technology are being intensively investigated. When modelling information and requests containing vagueness or imprecision, fuzzy set theory providing formal background to deal with imprecision, vagueness, uncertainty and similar concepts might be used, introducing significant improvements to the search results.

User profiles, personalization of web search tasks and soft information retrieval are current challenges. Information retrieval optimization based on knowledge of previous user search activities and fuzzy softening of both, search criteria and information models, aims at enriching document sets retrieved in response to user requests and helping user when she or he has no clear picture of searched information. In this chapter we introduce genetic and fuzzy oriented approach to these tasks with the goal to determine useful search queries describing documents relevant to users area of interest as deducted from previous searches as a tool helping user to fetch the most relevant information in his or her current context.

The rest of this paper is organized as follows: In Section 2, some background on information retrieval and fundamentals of information retrieval systems are provided. Section 3 introduces evolutionary computation, genetic algorithms, genetic programming and its application to information retrieval. In Section 4, we present our contribution extending the usage of genetic algorithms for search optimization in both, crisp and fuzzy information retrieval systems. Experimental results are presented in Section 5 and the work is concluded in Section 6.

## 2 Information Retrieval

The area of *Information Retrieval* (IR) is a branch of Computer Science dealing with data storage, maintenance and information search. The data could be all - textual, visual, audio or multimedia documents [7]. The rest of this article is devoted to information retrieval dealing with extensive collections of unstructured textual documents.

An *Information Retrieval System* (IRS) is a software tool for data representation, storage and information search. The amount of documents contained in data collections managed by IRS is usually very large and the task of easy, efficient and accurate information search is specially highlighted. General architecture of an information retrieval system is shown in Fig. 1 [7].

Document collection is for the search purposes analyzed and transformed into suitable internal representation in a process called indexing. The real world information need of an IRS user must be for the use with particular IRS expressed by the means of query language understandable to that system. A search query is evaluated against the internal document representation and

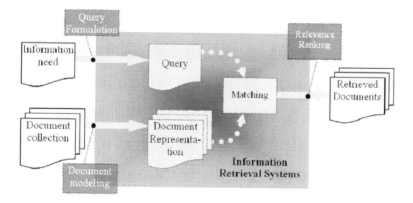

**Fig. 1** An Information Retrieval System

the system decides whether and how much are particular documents relevant to the query. The way of document indexing, structure of internal document representation, query language and document-query matching mechanism depends on certain IRS model which is a theoretical background below particular information retrieval system [7]. For regular users provides an IRS two main functions: data storage and information retrieval.

An *information need* is a state in which is ones own knowledge insufficient for satisfying her or his demands. If an IRS is to be used for information search, the demanded information need must be expressed in query language of the particular IRS in a process called querying. The search system attempts to find in managed documentary collection entries relevant to the query. Ordered set of *retrieved* documents is then offered to the user. Retrieved documents are such subset of documentary collection that is considered by the information retrieval application to be *relevant* to the user query. Retrieved documents are presented in certain ordering as a source of information to satisfy information need stated in the query. The document ordering is based on particular ranking strategy which is realized by certain ranking function.

The typical allocation of documents within the collection in response to a query is illustrated in Fig. 2. We can see that not all relevant documents are allways retrieved and moreover, some non-relevant documents could be included in the set of retrieved documents. We may also legitimately consider different documents to be relevant to the query in a certain degree. One of the main goals in the research of IR systems is to improve the accuracy of retrieved document set. It means to maximize the subset of retrieved relevant documents and minimize the subset of retrieved non-relevant documents.

**Fig. 2** Documents in collection classified in response to a query

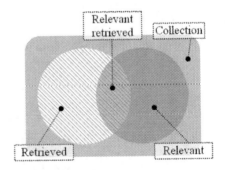

In the previous paragraphs were documents classified against a query as relevant and non-relevant, though the entire concept of relevance is a subject of discussion with no universal convergence yet. Objective relevance is an algorithmic measure of the degree of similarity between the query representation and the document representation. It is also referred to as a topicality measure, referring to the degree to which the topic of the retrieved information matches the topic of the request [12]. Subjective relevance is user-centric and deals with fitness for use of the retrieved information [27]. Subjective relevance involves intellectual interpretation by human assessors or users [5] and should be seen as a cognitive, dynamic process involving interaction between the information user and the information source. A general high-level relevance criterion is whether or not (and alternatively how much) the particular document contributes to the saturation of user's information need. Different inquirers might be satisfied with different response to the same question. Among the most important factors having impact on the user request is long and short term context of the particular inquirer. When evaluating a search expression, the knowledge of user's area of interest, abilities, language capabilities, current needs etc., can be important contribution to the search efficiency improvement. These are among the most fundamental reasons for personalized search research, user modelling and user profiling.

## 2.1 Information Retrieval Models

An *IR model* is a formal background defining internal document representation, query language and document-query matching mechanism. Consequently, the model determines document indexing procedure, result ordering and other aspects of particular information retrieval system. In the following, we will present two influential IR models - Boolean IR model and vector space IR model [7, 15].

### Boolean IR model

Boolean IR model belongs to the oldest but till nowadays widely used information retrieval models [7, 2]. It is based on set theory, Boolean logic and exact document-query match principle. The name Boolean originates in the fact that the query language uses as search expressions Boolean logic formulas composed of search terms and standard Boolean operators AND, OR and NOT [2]. The documents are represented as sets of indexed terms. The document indexing procedure distinguishes only whether a term is contained in the document or not and assigns to the term indexing weight 1 if the term is contained in the document or 0 if not. The inner representation of a document collection is a binary matrix composed of vectors (rows) representing documents with term weights as coordinates. Therefore every column represents weight of certain term in all documents in the whole collection. Formally, an index of documentary collection containing $n$ terms and $m$ documents in Boolean IR model is described as shown in Equations 1 and 2, where $d_i$ represents $i$-th document, $t_{ij}$ the weight of $j$-th term in $i$-th document and $D$ denotes the index matrix.

$$d_i = (t_{i1}, t_{i2}, \ldots, t_{in}), \forall t_{ij} \in \{0, 1\} \tag{1}$$

$$D = \begin{pmatrix} t_{11} & t_{12} & \cdots & t_{1n} \\ t_{21} & t_{22} & \cdots & t_{2n} \\ \vdots & \vdots & \ddots & \vdots \\ t_{m1} & t_{m2} & \cdots & t_{mn} \end{pmatrix} \tag{2}$$

The document-query matching procedure is based on the exact match principle. Only documents utterly satisfying all conditions stated by particular search query are considered to be relevant and thus retrieved in response to the query. When a document fully conforms to the search request, the query is against it evaluated, according to the Boolean algebra rules, as true. In the contrary case, when the document is in conflict with at least one of the clauses in the search request, the query is evaluated as false. In that way, the set of all documents in the collection is divided into two disjunctive subsets - retrieved and non-retrieved documents. There is no consideration of different degrees of document-query relevancy. All retrieved documents are supposed to be equally (fully) relevant to the query and all non-relevant documents are expected to be equally non-relevant. The ordering, in which are the results presented to the user, does not depend on the relevancy but on other factors such as date of last modification, document length, number of citations and so on [2, 7, 11]. There are numerous variations of basic Boolean IR model. Frequent modification consists in addition of advanced query operators - XOR implementing the logical exclusive OR operation, operator OF simplifying the notion of search formulas or, among others, operator NEAR expressing the requirement to retrieve documents having several terms near each other [11].

Apparently, the greatest advantage of Boolean IR model lies in exuberance and flexibility of its query language, allowing expressing very sophisticated and complex search requirements. On the other hand, to formulate such powerful search queries appropriately, the user should have at least minimal knowledge of Boolean algebra. Remarkable disadvantage of Boolean IR model is the crisp differentiation of documentary collection in response to query and therefore impossibility to use some relevance ranking technique to present retrieved documents sorted in relevance order. Because of this, a too restrictive query could cause denial of useful documents and contrariwise a too general query might retrieve additional non-relevant documents [14]. The Boolean IR model provides the basis for extended Boolean IR model introducing the principles of fuzzy set techniques and fuzzy logic to the area of information retrieval.

**Vector space model**

Vector space model (VSM) is based on interpretation of both, documents and queries, as points in a multidimensional document space [7, 11]. The dimension of the document space is given by the number of indexed terms in the documentary collection. Every term has in every document assigned a weight representing the coordinate in multidimensional space. The weight is based on the importance of corresponding term in the document and in the scope of whole collection respectively. Greater weight means greater importance of particular term [7, 2, 15]. Formal description of VSM is almost identical to the description of Boolean model as provided in Equations 1 and 2. The domain of $t_{ij}$ in VSM is the set of real numbers $R$. Query $q$ is formalized as a vector of searched terms (Equation 3).

$$q = (t_{q1}, t_{q2}, \ldots, t_{qn}), \forall t_{qj} \in R \tag{3}$$

In Boolean IR, indexing procedure was due to the simplicity of internal document representation trivial task. In VSM is the matrix representing document collection composed of real values - the weights of terms in documents. The weight assessment can be done manually (this is too expensive and inefficient) or automatically [11]. Several automatic indexing approaches were proposed. They assign real weights to the terms in documents. The weighting algorithms are usually based on statistical distribution of the terms in particular document with respect to their distribution among all documents in the collection. Among the most popular and widely deployed indexing techniques takes significant place Gerard Salton's $TFIDF_t$ introduced in [24]. Consider normalised term frequency of term $t$ in document $d$ shown in Equation 4 as the ratio of frequency of each term in the document to the maximum term frequency in that document. Therefore, the greater the frequency of particular term in the document, the greater the normalized frequency of such term in the document.

$$f_{dt} = \frac{freq(t, d)}{\max(freq(t_i, d))} \tag{4}$$

Normalized inverse document frequency, defined as shown in Equation 5, reflects the distribution of given term among all documents in the collection. The rarer is the term in the scope of whole collection, the greater is its inverse document frequency. $N$ stands in Equation 5 for the number of all documents in the collection, $N_t$ is number of documents containing at least one occurrence of the term $t$ and $g$ is some normalizing function. Finally, the weight of term $t$ in document $d$ according to $TFIDF_t$ is defined in Equation 6.

$$IDF_t = g(\log \frac{N}{N_t}) \tag{5}$$

$$F(d, t) = f_{dt} \cdot IDF_t \tag{6}$$

Summarizing previous definitions, high weight will be assessed to the terms frequent in given document and rare in the scope of whole collection. It is obvious that such terms are good significant marks distinguishing current document from other documents. More indexing functions for VSM can be found i.e. in [11]. Also queries have in VSM the form of documents (term vectors) and a term weighting function should be deployed. Query term weighting function example is shown in Equation 7.

$$F(q,t) = (\frac{1}{2} + \frac{f_{qt}}{2}) \tag{7}$$

The document-query matching procedure is in VSM based on the best match principle. Both, document and query are interpreted as points in multidimensional space and we can evaluate similarity between them. Several formulas expressing numerically the similarity between points in the document space have been introduced [11]. Among the most popular are scalar product (8) and cosine measure (9) that can be interpreted as an angle between the query vector and document vector in $m$-dimensional document space.

$$Sim(q, d_i) = \sum_{j=1}^{m} t_{qj} \cdot t_{ij} \tag{8}$$

$$Sim(q, d_i) = \frac{\sum_{j=1}^{m} t_{qj} \cdot t_{ij}}{\sqrt{\sum_{j=1}^{m} t_{qj}^2 \cdot \sum_{j=1}^{m} t_{ij}^2}} \tag{9}$$

The similarity measure does not directly predicate document's relevance to the query. It is supposed that among documents similar to the query should be many relevant documents whereas among dissimilar documents is only few relevant ones [11]. Querying is in VSM based on the best match principle. All documents are during the query evaluation process sorted according their distance to the query and presented to the user. Omitting the vague relationship between point distance and document relevance, we can consider this ordering as relevance ranking.

VSM is more recent and advanced than Boolean IR model. Its great advantage lies in relevance based ordering of retrieved documents allowing easy deployment of advanced IR techniques such as document clustering, relevance feedback, query reformulation and topic evolution. Disadvantages are vague relationship between relevance and similarity and unclear query term explication. From the interpretation of query as a searched document prescription originates another significant disadvantage of VSM - the query language allows specifying only what should be searched and there are no natural means on how to point out what should not be contained in retrieved documents.

## 2.2  *IR Effectiveness Evaluation*

When evaluating an information retrieval system, we are interested in the speed of search processing, user comfort, the possibilities of querying, result presentation and especially in the ability of *retrieving relevant documents*. As mentioned before, the concept of relevance is vague and uncertain. Though, it is useful to measure IR effectiveness by the means of query-document relevance. Precision $P$ and recall $R$ are among the most used IR effectiveness measures (10). In the precision and recall definition, REL stands for the set of all relevant documents and RET for the set of all retrieved documents. Precision can be then understood as the probability of retrieved document to be relevant and recall as the probability of retrieving relevant document. For easier effectiveness evaluation were developed measures combining precision and recall into one scalar value. Among most popular of these measures are effectiveness $E$ and $F$-score $F$ [18] as shown in Equation 11.

$$P = \frac{|REL \cap RET|}{|RET|} \qquad R = \frac{|REL \cap RET|}{|REL|} \tag{10}$$

$$E = 1 - \frac{2}{\frac{1}{P} + \frac{1}{R}} \qquad F = 1 - E = \frac{2PR}{P + R} \tag{11}$$

## 2.3 User Profiles in IR Systems

In previous section was shown that the concept of document-query relevance is highly subjective matter. Information need of particular user can be satisfied better if there is some knowledge about ones specific needs, abilities, long and short term context. That is the field of personalized IR systems exploiting user profiles. An *user profile* (or user model) is a stored knowledge about particular user. Simple profile consists usually of keywords describing user's area of long time interest. Extended profile is replenished with information about the user such as name, location, mother tongue and so on. Advanced user profiles contain rather than set of keywords a list of queries (persistent queries) characterizing user's behavior and habits [22, 9].

User profile can be exploited to make the search task more personalized. Information retrieval system equipped with user profiles can utilize user-specific information from the profile for retrieving documents satisfying stated query with special respect to individual user, her or his preferences, needs, abilities, history, knowledge and context. User profile information might be evaluated when improving search process. Keywords from the profile can be used for query extension, query reformulation for other techniques improving the search results. Such IR improvement techniques aim at retrieving information that satisfy users needs rather than information that was explicitly asked by potentially imprecise query [11]. User profile can be also exploited for document re-ranking according to individual preferences [22].

Explicit profiles, created by users or system administrators, are imprecise, not enough flexible and do not reflect dynamic changes of user preferences. Instead, various techniques for automated creation and maintenance of user profiles are being investigated [6]. Automatically created and updated user profiles are referred to as implicit user profiles. From the perspective of user profiling, IR systems can be divided into two categories: personalized IR systems providing personalized search services and consensual search system not aware of individual users [10].

# 3 Evolutionary Computation

Evolutionary algorithms (EA) belongs to a family of iterative stochastic search and optimization methods based on mimicking successful optimization strategies observed in nature [8, 13, 19, 3]. The essence of EAs lies in the emulation of Darwinian evolution utilizing the concepts of Mendelian inheritance for the use in computer science and applications [3]. Together with fuzzy sets, neural net-works and fractals, evolutionary algorithms are among the fundamental members of the class of soft computing methods.

EA operate with population (also known as pool) of artificial individuals (referred often as items or chromosomes) encoding possible problem solutions. Encoded individuals are evaluated using objective function which assigns a fitness value to each individual. Fitness value represents the quality (ranking) of each individual as solution of given problem. Competing individuals search the problem domain towards optimal solution [13]. In the following sections will be introduced general principles common for all methods belonging to the class of evolutionary algorithms.

## 3.1 Evolutionary Search Process

For the purpose of EAs, a proper encoding representing solutions of given problem as en-coded chromosomes suitable for evolutionary search process, is necessary. Finding proper en-coding is non-trivial problem dependent task affecting the performance and results of evolutionary search while solving given problem. The solutions might be encoded into binary strings, real vectors or more complex, often tree-like, hierarchical structures, depending on the needs of particular application.

The iterative phase of evolutionary search process starts with an initial population of individuals that can be generated randomly or seeded with potentially good solutions. Artificial evolution consists of iterative application of genetic operators, introducing to the algorithm evolutionary principles such as inheritance, survival of the fittest and random perturbations. Current population of problem solutions is modified with the aim to form new and hopefully better population to be used in next generation. Iterative evolution of problem solutions ends after satisfying specified termination criteria and especially the criterion of finding optimal solution. After terminating the search process, evolution winner is decoded and presented as the most optimal solution found.

## 3.2  Genetic Operators

Genetic operators and termination criteria are the most influential parameters of every evolutionary algorithm. All bellow presented operators have several implementations performing differently in various application areas. Selection operator is used for selecting chromosomes from population. Through this operator, selection pressure is applied on the population of solutions with the aim to pick more promising solutions to form following generation. Selected chromosomes are usually called parents. Crossover operator modifies the selected chromosomes from one population to the next by exchanging one or more of their subparts. Crossover is used for emulating sexual reproduction of diploid organisms with the aim to inherit and increase the good properties of parents for offspring chromosomes. Mutation operator introduces random perturbation in chromosome structure; it is used for changing chromosomes randomly and introducing new genetic material into the population.

Besides genetic operators, termination criteria are important factor affecting the search process. Widely used termination criteria are i.e.:

- Reaching optimal solution (which is often hard, if not impossible, to recognize)
- Processing certain number of generations
- Processing certain number of generations without improvement in population

EAs are successful general adaptable concept with good results in many areas. The class of evolutionary techniques consists of more particular algorithms, each having numerous variants. They are forged and tuned for specific problem domains. The family of evolutionary algorithms consists of genetic algorithms, genetic programming, evolutionary strategies and evolutionary programming.

## 3.3  Genetic Algorithms

Genetic algorithms (GA) introduced by John Holland and extended by David Goldberg are wide applied and highly successful EA variant. Basic workflow of original (standard) generational GA (GGA) is:

1. Define objective function
2. Encode initial population of possible solutions as fixed length binary strings and evaluate chromosomes in initial population using objective function
3. Create new population (evolutionary search for better solutions)
   a. Select suitable chromosomes for reproduction (parents)
   b. Apply crossover operator on parents with respect to crossover probability to produce new chromosomes (known as offspring)

    c. Apply mutation operator on offspring chromosomes with respect to mutation probability. Add newly constituted chromosomes to new population

    d. Until the size of new population is smaller than size of current population go back to (a).

    e. Replace current population by new population

4. Evaluate current population using objective function

5. Check termination criteria; if not satisfied go back to (3).

Many variants of standard generational GA have been proposed. The differences are mostly in particular selection, crossover, mutation and replacement strategy [13]. Different high-level approach is represented by steady-state Genetic Algorithms (SSGA). In GGA, in one iteration is replaced whole population [8] or fundamental part of population [26] while SSGA replace only few individuals at time and never whole population. This method is more accurate model of what happens in the nature and allows exploiting promising individuals as soon as they are created. However, no evidence that SSGA are fundamentally better than GGA was found [26].

### 3.4 Genetic Programming

Genetic programming by John Koza is referred to as special case [26] or an extension [17] to GA. Encoded individuals (chromosomes) have hierarchical

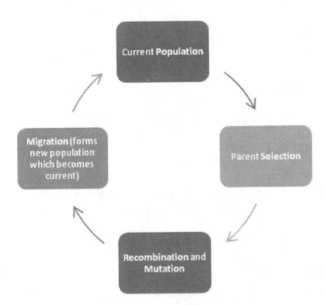

**Fig. 3** Iterative phase of Genetic Algorithm

structure, unlimited size and they are often modelled as tree structures. So can be modelled mathematical formulas, logical expressions or even whole computer programs (i.e. Lisp programs). Genetic programming is a native tool for modelling and artificial evolution of search queries.

# 4 Evolutionary Techniques and Fuzzy Logic Principles in IRS

Fuzzy theory, as a framework describing formally the concepts of vagueness, imprecision, uncertainty and inconsistency provides interesting extensions to the area of information retrieval. Imprecision and vagueness are present in natural language and take part in real-world human communication. User friendly and flexible advanced IRS should be able to offer user interface for non experienced users allowing natural deployment of these concepts in user-system interaction for more efficient information retrieval.

IR models exploiting fuzzy techniques can overcome some of the limitations pointed out in first part of this article [14]. They support different grades of document-query relevance, cut inaccuracies and oversimplifications happening during document indexing and introduce the concepts of vagueness and imprecision in query language.

## 4.1 Genetic Algorithms in Information Retrieval

Multiple works have been recently published in the area of IR and search query optimization as this topic becomes increasingly challenging. The use of various evolutionary algorithms was proposed at multiple stages of the information retrieval process.

Yeh et al. [28] described in 2007 a genetic programming for IR ranking function discovery and introduced a set of numerical experiments demonstrating the contribution of proposed method to IR efficiency. Yeh et al. innovatively combined different types of evidences including classical IR content features, structure features, and query independent features to the ranking function.

Several contributions towards evolutionary optimization of search queries were introduced. Kraft et al. [14] used genetic programming to optimize Boolean search queries over a documentary database with an emphasis on the comparison of several IR effectiveness measures as objective functions. Cordn et al. [6] introduced MOGA-P, an algorithm to deal with search query optimization as a multi-objective optimization problem and compared their approach with several other methods including Kraft's. Yoshioka and Haraguchi [29] introduced query reformulation interface to transform Boolean search queries into more efficient search expressions.

In one of the recent contributions to evolutionary query optimization, A. Aly [1] introduced a genetic algorithm for vector query reformulation based

on vector space model of an IR system. The method is based on evolutionary learning of significant terms from search results to modify user queries.
Snášel, Nyongesa et al.[25] used similar approach to learn user profiles in IRS
based on

This work aims to evaluate evolutionary learning of Boolean search queries
in both, traditional crisp Information Retrieval frameworks and advanced
fuzzy Information retrieval systems.

## 4.2   Fuzzy Principles in Information Retrieval

Fuzzy concepts affect most phases of IR process. They are deployed during
document indexing, query formulation and search request evaluation. Information retrieval is seen as fuzzy multi-criteria decision making in the presence
of vagueness. In general, document is interpreted as a fuzzy set of document
descriptors and queries as a composite of soft search constraints to be applied on documents. Document-query evaluation process is based on fuzzy
ranking of the documents in documentary collection according to the level
of their conformity to the soft search criteria specified via user queries. The
document-query matching has to deal with the uncertainty arising from the
nature of fuzzy decision making and from the fact that user information needs
can be recognized, interpreted and understood only partially. Moreover, the
document content is described only in a rough, imperfect way [4].

In the fuzzy enabled IR frameworks, soft search criteria could be specified
using linguistic variables. User search queries can contain elements declaring
level of partial importance of the search statement elements. Linguistic variables such as "probably" or "it is possible that", can be used to declare the
partial preference about the truth of the stated information. The interpretation of linguistic variables is then among the key phases of query evaluation
process. Term relevance is considered as a gradual (vague) concept. The decision process performed by the query evaluation mechanism computes the
degree of satisfaction of the query by the representation of each document.
This degree, called retrieval status value (RSV), is considered as an estimate
of the relevance of the document with respect to the query. $RSV = 1$ corresponds to maximum relevance and $RSV = 0$ denotes no relevance. The values
within the range $(0, 1)$ correspond to particular level of document relevance
between the two extremes 0 and 1 [4].

Possibility theory together with the concept of linguistic variable defined
within fuzzy set theory provides a unifying formal framework to formalize the
processing of imperfect information [4]. Inaccurate information is inevitably
present in information retrieval systems and textual databases applications.
The automatically created document representation based on a selection
of index terms is invariably incomplete and far worse than document representations created manually by human experts who utilize their subjective theme knowledge when performing the indexing task. Automated text

indexing deals with imprecision since the terms are not all fully significant to characterise the document content and their statistical distribution does not reflect their relevance to the information included in the document necessarily. Their significance depends also on the context in which they appear and on the unique personality of the inquirer. During query formulation, users might have only a vague idea of the information they are looking for therefore face difficulties when formulating their information needs by the means of query language of particular IR system. A flexible IRS should be designed to provide detailed and rich representation of documents, sensibly interpret and evaluate soft queries and hence offer efficient information retrieval service in the conditions of vagueness and imprecision [4].

In the following, Extended Boolean IR model as the representative of fuzzy IR models will be discussed in details. Some other recent fuzzy IR models will be briefly presented.

## 4.3  Extended Boolean IR Model

Fuzzy generalizations of the Boolean model have been defined to extend existing Boolean IRSs without the need to redesign them. Classic Boolean model of IR represents documents as sets of indexed terms. Therefore we can for every term say whether it belongs to the set repre-senting the document (then a weight 1 is assigned to the term for the particular document representation) or not (a weight 0 is assigned). The term weight is either 0 or 1 and multiple occur-rences of the term in the document do not affect its internal representation.

Extended Boolean model of IR is based on fuzzy set theory and fuzzy logic. Documents are interpreted as fuzzy sets of indexed terms, assigning to every term contained in the document particular weight from the range of $[0, 1]$ expressing the degree of significance of the term for document representation. Hence documents are modelled more accurately than in classic Boolean IR model. Formal collection description in extended Boolean IR model is shown in Equations 12 and 13.

$$d_i = (t_{i1}, t_{i2}, \ldots, t_{in}), \forall t_{ij} \in \{0, 1\} \tag{12}$$

$$D = \begin{pmatrix} t_{11} & t_{12} & \cdots & t_{1n} \\ t_{21} & t_{22} & \cdots & t_{2n} \\ \vdots & \vdots & \ddots & \vdots \\ t_{m1} & t_{m2} & \cdots & t_{mn} \end{pmatrix} \tag{13}$$

Next new feature of extended Boolean IR model is fuzzy extension of query language aiming at providing apparatus to express more flexible and accurate search requests. Two techniques are being used for query enhancement query term weighting using numeric weights or linguistic variables and Boolean conjunction parameterization for expressing relationships among the extremes of AND, OR, NOT etc. [14]. Choosing appropriate indexing procedure is essential for exploitation of extended Boolean IR model benefits. Internal documentary collection model should be as accurate as possible snapshot of the collection of textual documents in natural language and at the same time a basis for efficient and practical search. Fuzzy indexing function is defined as shown in Equation 14, where D stands for the set of all documents and T for set of all indexed terms.

$$F : D \times T \to [0, 1] \tag{14}$$

Kraft in [14] used Salton's $TFIDF_t$ indexing formula introduced for VSM as textual document indexing mechanism in extended Boolean IR model. Query language is in extended Boolean model of IR upgraded by the possibility of weighting query terms in order to express different importance of those in search request and by weighting (parameterizing) aggregation operators to soften their impact on query evaluation [7, 15]. Consider $Q$ to be the set of user queries over a collection then the weight of term $t$ in query $q$ is denoted as $a(q, t)$ satisfying $a : Q \times T \to [0, 1]$. To evaluate atomic query of one term, stating therefore only one search criterion, will be used function $g : [0, 1] \times [0, 1] \to [0, 1]$. $q(F(q, t), a)$ is the retrieval status value (RSV). For RSV enumeration is crucial the interpretation of query term weight $a$. The most used interpretations are to see query term weight as importance weight, threshold or ideal document description [7, 15]. The theorems for RSV evaluation in the case of importance weight interpretation and threshold interpretation are shown in Equations 15 and 16 respectively [15, 7], where $P(a)$ and $Q(a)$ are coefficients used for tuning the threshold curve. An example of $P(a)$ and $Q(a)$ could be as follows: $P(a) = \frac{1+a}{2}$ and $Q(a) = \frac{1+a^2}{4}$. The RSV formula in Equation 16 is illustrated in Fig. 4a. Adopting the threshold interpretation, an atomic query containing term $t$ of the weight $a$ is a request to retrieve documents having $F(d, t)$ equal or greater to $a$. For documents satisfying this condition will be rated with high RSV and contrariwise documents having $F(d, t)$ smaller than $a$ will be rated with small RSV.

$$RSV = \begin{cases} \min(a, F(d, t)) & \text{if t is operand of OR} \\ \max(1 - a, F(d, t)) & \text{if t is operand of AND} \end{cases} \tag{15}$$

$$RSV = \begin{cases} P(a)\frac{F(d,t)}{a} & \text{for } F(d, t) < a \\ P(a) + Q(a)\frac{F(d,t)-a}{1-a} & \text{for } F(d, t) \geq a \end{cases} \tag{16}$$

Query term weight $a$ can be understood as an ideal document term weight prescription. In that case, RSV will be evaluated according to Equation 17, enumerating the distance between $F(d, t)$ and $a$ in a symmetric manner as shown in Fig. 4b. This means that a document with lower term weight will be rated with the same RSV as document with higher term weight, considering

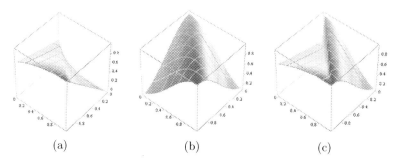

(a)                              (b)                              (c)

**Fig. 4** Graphic representation of the three RSV functions

the same differences. Asymmetric version of Equation 17 is shown in Equation 18 and illustrated in Fig. 4c.

$$RSV = e^{K \cdot (F(d,t)-a)^2} \tag{17}$$

$$RSV = \begin{cases} e^{K \cdot (F(d,t)-a)^2} & \text{for } F(d,t) < a \\ P(a) + Q(a)\frac{F(d,t)-a}{1-a} & \text{for } F(d,t) \geq a \end{cases} \tag{18}$$

Single weighted term is basic element of fuzzified Boolean query. Aggregation operators concatenating query elements into more flexible and powerful search expressions might be weighted as well. The operator weight interpretation is another key part of fuzzy Boolean query evaluation. In general, various T-norm and T-conorm pairs might be used for fuzzy generalization of AND and OR operators while evaluating NOT as fuzzy complement. Operator weights are in these cases handled in the same manner as query term weight achieving higher flexibility and expressiveness of search expressions. Nevertheless, such approach does not reduce the complexity of Boolean logic needed to use the queries efficiently [15]. Alternatively, new definitions of aggregation operators for fuzzy queries have been introduced. Vague relationship among selection criteria is expressed using linguistic quantifiers such as all, most of, at least $n$, introducing blurred behaviour between AND and OR and allowing easier query formulation [7, 15].

## 4.4 Fuzzy IR Effectiveness Evaluation

When evaluating effectiveness of an IR system, precision and recall are among the most popular performance measures serving as a basis for numerous derived indicators such as effectiveness $E$ or F-score $F$. For the enumeration of precision and recall in the framework of fuzzy IR systems cannot be used crisp precision and recall as specified in Equation 10. New definitions were proposed on the basis of Zadehs cardinality (see Equation 19; the function

$\mu_A(x)$ assigns to every item $x$ its fuzzy weight in fuzzy set $A$) as shown in Equations 20 and 21 [16].

$$card(A) = \|A\| = \sum_{x \in A} \mu_A(x) \tag{19}$$

$$\rho(X|Y) = \begin{cases} \frac{\|X \cap Y\|}{\|Y\|} & \|Y\| \neq 0 \\ 1 & \|Y\| = 0 \end{cases} \tag{20}$$

$$P = \rho(REL|RET) \text{ a } R = \rho(RET|REL) \tag{21}$$

## 5   Experimental Evaluation

A series of computer experiments was conducted in order to evaluate proposed GA enabled IR framework in both, crisp Boolean IR model and fuzzified Extended Boolean IR model[21, 20, 23, 25]. Experiments were executed using data taken from the LISA[1] collection. The collection was indexed for both, Boolean IR and Extended Boolean IR system.Indexed collection contained 5999 documents and 18442 unique indexed terms.

Genetic programming was used to evolve Boolean search queries. Boolean expressions were parsed and encoded into tree like chromosomes, as shown in Figure 5. Genetic operators were applied on nodes of the tree chromosomes. Several parameters were fixed for all experiments:

- mutation probability $P_M = 0.2$
- crossover probability $P_C = 0.8$
- maximum number of 1000 generations
- population of 70 individuals (queries)

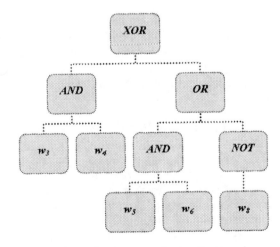

**Fig. 5** Search query *(w3 and w4) xor ((w5 and w6) or not w8)* encoded for GP

[1] Available at: http://www.dcs.gla.ac.uk/idom/ir_resources/test_collections/

We have used two scenarios for initial population. In the first case, all queries in initial population were generated randomly. In the second scenario, three good ranked queries, created by the experiment administrators, were added to the initial population. Two selection strategies were investigated: elitary selection choosing parents among the best ranked individuals and probabilistic selection implementing the roulette wheel selection algorithm. Two mutation strategies were under investigation. Single point mutation performing random perturbation of one gene (i.e. one node) of the query chromosome and each point mutation applying mutation operator on

**Table 1** Summary for experiments results

| IR Model | User Query | Initial Population Enhancement |
|---|---|---|
| BIRM | (("EXTREMELY" AND "POOR") OR "FUNDS") | "FUNDS" OR "BIBLIOGRAPHIC" "EXTREMELY" AND "INNOVATORS" NOT ("POOR" XOR "FUNDS") |
| EBIRM | (("EXTREMELY":0.94 AND "POOR":0.50) OR:0.50 "FUNDS":0.90 ) | "FUNDS":0.9 OR "BIBLIOGRAPHIC":0.8 "EXTREMELY":0.3 AND "INNOVATORS" NOT ("POOR" XOR:0.03 "FUNDS":0.5 ) |

**Table 2** Summary of experimental results in BIRM

| Scenario | Precision | Recall | F-Score |
|---|---|---|---|
| REI | 0.04699 | 0.089552 | 0.0486915 |
| REC | 0.040411 | 0.11194 | 0.0621065 |
| RPI | 0.064519 | 0.074627 | 0.069205 |
| RPC | 0.053471 | 0.119403 | 0.0689775 |
| SEI | 1 | 0.985075 | 0.992481 |
| SEC | 1 | 0.985075 | 0.992481 |
| SPI | 1 | 0.985075 | 0.992481 |
| SPC | 1 | 0.985075 | 0.992481 |

**Table 3** Summary of experimental results in EBIRM

| Scenario | Precision | Recall | F-Score |
|---|---|---|---|
| REI | 0.078706 | 0.027165 | 0.04039 |
| REC | 0.078706 | 0.027165 | 0.04039 |
| RPI | 0.0765365 | 0.0760845 | 0.0754315 |
| RPC | 0.163975 | 0.0389625 | 0.060813 |
| SEI | 0.9933375 | 0.9045225 | 0.9454495 |
| SEC | 0.993873 | 0.968469 | 0.9810005 |
| SPI | 0.9138465 | 0.9696315 | 0.940731 |
| SPC | 0.9965815 | 0.968436 | 0.9823045 |

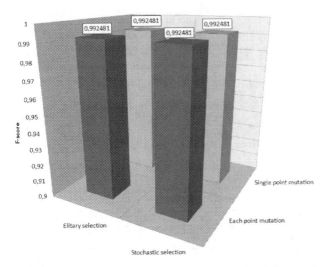

**Fig. 6** The comparison of achieved F-score for different algorithm setups in BIRM with seeded initial population

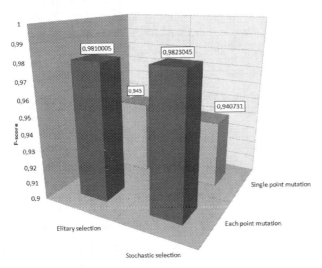

**Fig. 7** The comparison of achieved F-score for different algorithm setups in EBIRM with seeded initial population

every gene in the chromosome. Mutation was implemented as replacement of one node by an equivalent (i.e. another node with the same arity). This means that OR might be replaced by XOR and AND. NOT operator might be inserted or removed. A term might be replaced by another term.

An *user query* was used to assign some initial relevance degree to documents in the collection. The user query (or its equivalent) represents in

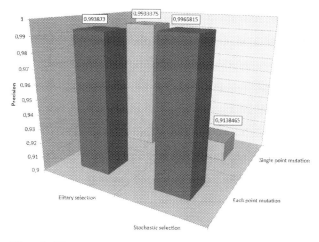

**Fig. 8** The comparison of achieved Precision for different algorithm setups in EBIRM with seeded initial population

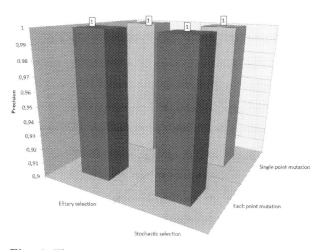

**Fig. 9** The comparison of achieved Precision for different algorithm setups in BIRM with seeded initial population

laboratory conditions desired output of the optimization algorithm. The experiments were conducted in crisp and fuzzy laboratory information retrieval framework. The crisp IR framework was labeled as Boolean Information Retrieval Model (BIRM) and the fuzzy IR framework was denoted as Extended Boolean Information Retrieval Model (EBIRM). Due to the stochastic character of GP process, all experiments were executed several times and mean experimental results evaluated.

Table 1 lists the user query and better ranked queries injected into initial population in some experiments.

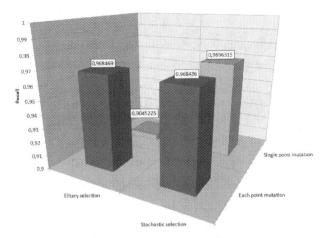

**Fig. 10** The comparison of achieved Recall for different algorithm setups in EBIRM with seeded initial population

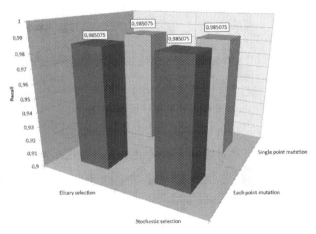

**Fig. 11** The comparison of achieved Recall for different algorithm setups in BIRM with seeded initial population

Tables 2 and 3 summarize the experimental results obtained for different test scenarios. Experiments were labeled with the following letters: single point mutation **I**, each point mutation **C**, elitism **E**, probabilistic selection **P**, seeded queries **S** and random initial population **R**. The presented results are averages of the fitness values of precision, recall and $F - score$.

From the experiments with Boolean queries we conclude the following results: Genetic algorithms succeeded in optimization of Boolean and extended Boolean search queries. Crucial for the result of the optimization process was the quality of initial population. For successful optimization, initial population must contain at least some quality queries pointing to documents related

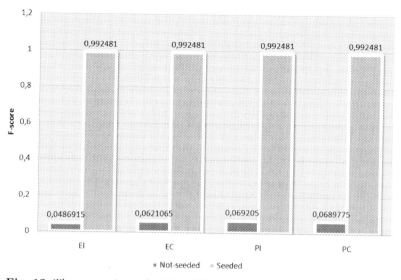

**Fig. 12** The comparison of achieved F-score for used algorithm setups with seeded (S) and random (R) initial population in BIRM

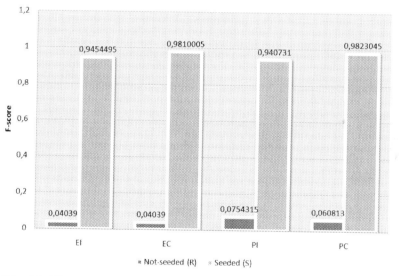

**Fig. 13** The comparison of achieved F-score for used algorithm setups with seeded (S) and random (R) initial population in EBIRM

to user needs. This fact was especially significant when optimizing extended queries with weighted terms and operators. Weight assessment rapidly increases search domain of the problem.

$F - score$ fitness was preferred as a measure combining precision and recall into one value by the means of information retrieval and therefore simplifying query optimization from multi-objective to a single-objective task. Figures 6 to 11 illustrate the improvements of $F - Score$, $Precision$ and $Recall$ of the optimized queries in different experimental cases. The figures 12 and 13 respectively show the significant differences of optimization results when using random initial population and seeded initial population.

# 6   Conclusion

The area of information retrieval faces today enormous challenges. The information society in the age of Internet excels in producing huge amounts of data and it is often complicated to retrieve information contained in such data sources. Decades ago, sophisticated techniques and algorithms forming information retrieval systems were designed to handle document collections available at that time. Information retrieval systems have gone over an intensive evolution to satisfy increasing needs of growing data bases. In their mature form, they are still present in the heart of Internet search engines, the key communication hubs of our society.

Internet search allows exploitation of large amount of knowledge available in the ubiquitous multitude of data. Information search is one of the most important e-activities. The IR systems, despite their advanced features, need revision and improvement in order to achieve better performance and provide inquirer with more satisfactory answers. Aiming to achieve better performance, more flexible models and techniques are requested. Fuzzy set framework has been proved as suitable formalism for modelling and handling vagueness and imprecision, the hot topics of information retrieval. Numerous researches considering various applications of fuzzy set technology have been initiated and conducted, some recent summarized in this article. The deployment of fuzzy techniques in all phases of IR has brought improvement of IR results and therefore increases user satisfaction. Lotfi Zadeh once called fuzzy technology computing with words. Information retrieval performs real world computation with words for decades. The symbiosis of these two progressive areas promises exciting results for the coming years.

Evolutionary techniques are an excellent tool to extract non-explicit information from data. Their unique ability to estimate, evolve and improve can be used to model Internet search user. Implicit data, such as the click-stream, produced during the web browsing activities could be exploited to keep track of the preferences of every single user. Such model is accurate, flexible, and can be well exploited for query optimization. Simultaneous deployment of fuzzy set techniques for better document modelling and genetic algorithms for query optimization brings a significant contribution to the ultimate goal of web search: bringing knowledge to man.

# References

1. Aly, A.A.: Applying genetic algorithm in query improvement problem. Int. Journal on Information Technologies and Knowledge 1(12), 309–316 (2007)
2. Belkin, N.J., Croft, W.B.: Information filtering and information retrieval: two sides of the same coin? Communications of the ACM 35(12), 29–38 (1992)
3. Bodenhofer, U.: Genetic Algorithms: Theory and Applications. Lecture notes, Fuzzy Logic Laboratorium Linz-Hagenberg (Winter, 2003/2004)
4. Bordogna, G., Pasi, G.: Modeling vagueness in information retrieval, pp. 207–241 (2001)
5. Borlund, P., Ingwersen, P.: Measures of relative relevance and ranked half-life: performance indicators for interactive IR. In: SIGIR 1998, Melbourne, Australia, August 1998, pp. 324–331 (1998)
6. Cordón, O., de Moya, F., Zarco, C.: Fuzzy logic and multiobjective evolutionary algorithms as soft computing tools for persistent query learning in text retrieval environments. In: IEEE International Conference on Fuzzy Systems 2004, Budapest, Hungary, pp. 571–576 (2004)
7. Crestani, F., Pasi, G.: Soft information retrieval: Applications of fuzzy set theory and neural networks. In: Kasabov, N., Kozma, R. (eds.) Neuro-Fuzzy Techniques for Intelligent Information Systems, pp. 287–315. Springer, Heidelberg (1999)
8. Dianati, M., Song, I., Treiber, M.: An introduction to genetic algorithms and evolution strategies. Technical report, University of Waterloo, Ontario, N2L 3G1, Canada (July 2002)
9. Fan, W., Gordon, M.D., Pathak, P.: A generic ranking function discovery framework by genetic programming for information retrieval. Inf. Process. Manage 40(4), 587–602 (2004)
10. Fan, W., Gordon, M.D., Pathak, P., Xi, W., Fox, E.A.: Ranking function optimization for effective web search by genetic programming: An empirical study. In: HICSS (2004)
11. Greengrass, E.: Information retrieval: A survey. DOD Technical Report TR-R52-008-001 (2001)
12. Harter, S.P.: Psychological relevance and information science. JASIS 43(9), 602–615 (1992)
13. Jones, G.: Genetic and evolutionary algorithms. In: von Rague, P. (ed.) Encyclopedia of Computational Chemistry. John Wiley and Sons, Chichester (1998)
14. Kraft, D.H., Petry, F.E., Buckles, B.P., Sadasivan, T.: Genetic Algorithms for Query Optimization in Information Retrieval: Relevance Feedback. In: Sanchez, E., Shibata, T., Zadeh, L.A. (eds.) Genetic Algorithms and Fuzzy Logic Systems. World Scientific, Singapore (1997)
15. Kraft, D.H., Bordogna, G., Pasi, G.: Fuzzy set techniques in information retrieval. In: Bezdek, J.C., Didier, D., Prade, H. (eds.) Fuzzy Sets in Approximate Reasoning and Information Systems, MA. The Handbook of Fuzzy Sets Series, vol. 3, pp. 469–500. Kluwer Academic Publishers, Dordrecht (1999)
16. Larsen, H.L.: Retrieval evaluation. In: Modern Information Retrieval course. Aalborg University Esbjerg (2004)
17. Leroy, G., Lally, A.M., Chen, H.: The use of dynamic contexts to improve casual internet searching. ACM Transactions on Information Systems 21(3), 229–253 (2003)

18. Losee, R.M.: When information retrieval measures agree about the relative quality of document rankings. Journal of the American Society of Information Science 51(9), 834–840 (2000)

19. Mitchell, M.: An Introduction to Genetic Algorithms. MIT Press, Cambridge (1996)

20. Owais, S., Kromer, P., Snasel, V., Husek, D., Neruda, R.: Implementing GP on optimizing both boolean and extended boolean queries in IR and fuzzy IR systems with respect to the users profiles. In: Yen, G.G., Wang, L., Bonissone, P., Lucas, S.M. (eds.) Proceedings of the 2006 IEEE Congress on Evolutionary Computation, Vancouver, BC, Canada, July 6-21, pp. 5648–5654. IEEE Computer Society, Los Alamitos (2006)

21. Owais, S.S.J., Krömer, P., Snášel, V.: Evolutionary Learning of Boolean Queries by Genetic Programming. In: ADBIS Research Communications, pp. 54–65 (2005)

22. Owais, S.S.J., Krömer, P., Snášel, V.: Query Optimization by Genetic Algorithms. In: DATESO, pp. 125–137 (2005)

23. Owais, S.S.J., Krömer, P., Snášel, V.: Implementing gp on optimizing boolean and extended boolean queries in irs with respect to users profiles. In: Fahmy, H.M.A., Salem, A.M., El-Kharashi, M.W., El-Din, A.M.B. (eds.) Proceedings of the 2006 International Conference on Computer Engineering & Systems (IC-CES 2006), Cairo, Egypt, pp. 412–417. IEEE Computer Society, Los Alamitos (2006)

24. Salton, G., Buckley, C.: Term-weighting approaches in automatic text retrieval. Information Processing and Management 24(5), 513–523 (1988)

25. Snášel, V., Krömer, P., Owais, S.S.J., Nyongesa, H.O., Maleki-Dizaji, S.: Evolving web search expressions. In: Third International Conference on Natural Computation (ICNC 2007), Haikou, Hainan, China, vol. 4, pp. 532–538. IEEE Computer Society Press, Los Alamitos (2007)

26. Townsend, H.A.R.: Genetic Algorithms - A Tutorial (2003)

27. Vakkari, P., Hakala, N.: Changes in relevance criteria and problem stages in task performance. Journal of Documentation 5(56), 540–562 (2000)

28. Yeh, J.-Y., Lin, J.-Y., Ke, H.-R., Yang, W.-P.: Learning to rank for information retrieval using genetic programming. In: SIGIR (2007)

29. Yoshioka, M., Haraguchi, M.: An Appropriate Boolean Query Reformulation Interface for Information Retrieval Based on Adaptive Generalization. In: WIRI, pp. 145–150 (2005)

# Web Data Clustering

Dušan Húsek, Jaroslav Pokorný, Hana Řezanková, and Václav Snášel

**Abstract.** This chapter provides a survey of some clustering methods relevant to clustering Web elements for better information access. We start with classical methods of cluster analysis that seems to be relevant in approaching the clustering of Web data. Graph clustering is also described since its methods contribute significantly to clustering Web data. The use of artificial neural networks for clustering has the same motivation. Based on previously presented material, the core of the chapter provides an overview of approaches to clustering in the Web environment. Particularly, we focus on clustering Web search results, in which clustering search engines arrange the search results into groups around a common theme. We conclude with some general considerations concerning the justification of so many clustering algorithms and their application in the Web environment.

## 1 Introduction

Efficient content mining is an important task for Web communities. Most of existing information is stored on the Web. Its electronic format makes it easy searchable by means of informational technologies. Such a way Web becomes an electronics repository of human knowledge.

Dušan Húsek
Institute of Computer Science, Academy of the Sciences of the Czech Republic
Pod vodárenskou věží 2, 182 07 Praha 8, Czech Republic
e-mail: dusan@cs.cas.cz

Jaroslav Pokorný
Department of Software Engineering, Charles University
Malostranské náměstí 25, 118 00 Praha 1, Czech Republic
e-mail: pokorny@ksi.ms.mff.cuni.cz

Hana Řezanková
Department of Statistics and Probability, University of Economics, Prague
nám. W. Churchilla 4, 130 67 Praha 3, Czech Republic
e-mail: rezanka@vse.cz

Václav Snášel
Department of Computer Science, Technical University of Ostrava
17. listopadu 15, 708 33 Ostrava-Poruba, Czech Republic
e-mail: vaclav.snasel@vsb.cz

A. Abraham et al. (Eds.): Foundations of Comput. Intel. Vol. 4, SCI 204, pp. 325–353.
springerlink.com                    © Springer-Verlag Berlin Heidelberg 2009

As the information on the Web has no fixed structure the web users has to rely on dynamic, learning based methods to get efficient access to needed information. The most popular approach to learning is "Learning by example". Clustering is another setting for learning, which does not use labeled objects and therefore is unsupervised. The objective of clustering is finding common patterns, grouping similar objects, or organizing them in hierarchies. In the context of the Web, the objects manipulated by the learning system are web documents, and the class labels are usually topics or user preferences. Thus, a supervised web learning system would build a mapping between documents and topics, while a clustering system would group web documents or organize them in hierarchies according to their topics.

A clustering system can be useful in web search for grouping search results into closely related sets of documents. Clustering can improve similarity search by focusing on sets of relevant documents. Hierarchical clustering methods can be used to create topic directories automatically, directories that can then be labeled manually. On the other hand, manually created topic directories can be matched against clustering hierarchies and thus the accuracy of the topic directories can be evaluated. Clustering is also a valuable technique for analyzing the Web.

## 2   Cluster Analysis

*Cluster analysis* is a name for the group of the methods, techniques and algorithms, by which the object set is divided into subsets (groups) of similar objects. Each group, called a *cluster*, consists of objects that are more similar between themselves and less similar to objects of other groups.

Clustering can be realized by means of such techniques as multivariate statistical methods, neural networks, genetic algorithms, and formal concept analysis. In the terminology of machine learning, we can talk about *unsupervised learning*. Statistical methods for clustering can be classified to the groups like cluster analysis, multidimensional scaling ([42,68]), factor analysis and correspondence analysis.

The following terms and notation will be used throughout this chapter.

- An *object* (vector of term weights) $\mathbf{x}$ is a single data item. It typically consists of a vector of $m$ components $\mathbf{x} = (x_1, ..., x_m)$, where $m$ is the dimensionality of the objects or of the feature space.

- The individual scalar components $x_i$ of an object $\mathbf{x}$ are called values of *features* (or values of variables or attributes).

- An object set will be denoted $X = \{\mathbf{x}_1, ..., \mathbf{x}_n\}$. The $i$-th object in $X$ will be enoted $\mathbf{x}_i = (x_{i1}, ..., x_{im})$. In many cases an object set to is viewed as an $n \times m$ object matrix.

The base of the most of methods is dissimilarity or similarity measure. A *dissimilarity* (or *distance*) between object $\mathbf{x}$ and $\mathbf{y}$ (or distance measure) is function $d(\mathbf{x}, \mathbf{y}) : X \times X \rightarrow R$ which satisfied the following conditions:

$$d(\mathbf{x}, \mathbf{x}) = 0$$
$$d(\mathbf{x}, \mathbf{y}) \geq 0$$
$$d(\mathbf{x}, \mathbf{y}) = d(\mathbf{y}, \mathbf{x})$$

For distance we require triangle inequality to satisfy, i.e. for any objects $\mathbf{x}$, $\mathbf{y}$, and $\mathbf{z}$

$$d(\mathbf{x}, \mathbf{z}) \leq d(\mathbf{x}, \mathbf{y}) + d(\mathbf{y}, \mathbf{z}).$$

A *similarity* $s(\mathbf{x}, \mathbf{y})$ between object $\mathbf{x}$ and $\mathbf{y}$ is function $s(\mathbf{x}, \mathbf{y}) : X \times X \to R$ which satisfied the following conditions:

$$s(\mathbf{x}, \mathbf{x}) = 1$$
$$s(\mathbf{x}, \mathbf{y}) \geq 0$$
$$s(\mathbf{x}, \mathbf{y}) = s(\mathbf{y}, \mathbf{x})$$

Both dissimilarity and similarity functions are often defined by a matrix.

Some clustering algorithms operate on a dissimilarity matrix (they are called distance-space methods in [89]). How the dissimilarity between two objects is computed depends on the type of the original objects.

Here are some most frequently used dissimilarity measures for continuous data.

- *Minkowski* $L_q$ distance (for $1 \geq q$ )

$$d(\mathbf{x}_i, \mathbf{x}_j) = \sqrt[q]{\sum\nolimits_{l=1}^{p} |x_{il} - x_{jl}|^q} \ .$$

- *City*-block (or Manhattan distance or $L_1$ )

$$d(\mathbf{x}_i, \mathbf{x}_j) = \sum\nolimits_{l=1}^{p} |x_{il} - x_{jl}| \ .$$

- *Euclidean distance* (aliases $L_2$ )

$$d(\mathbf{x}_i, \mathbf{x}_j) = \sqrt{\sum\nolimits_{l=1}^{p} (x_{il} - x_{jl})^2} \ .$$

- *Chebychev distance metric* (or maximum or $L_\infty$ )

$$d(\mathbf{x}_i, \mathbf{x}_j) = \max_{l=1,\dots,p} (|x_{il} - x_{jl}|) \ .$$

A relation between two objects can be expressed also as a similarity [11]. It can be measured as a correlation between feature vectors. For interval-scaled data, Pearson correlation coefficient is used (but values are from the interval $<-1, 1>$). Further possibility is a *cosine measure*. Cosine of feature vectors is calculated according the following formula:

$$s(\mathbf{x}_i,\mathbf{x}_j) = \frac{\sum_{l=1}^{P} x_{il} x_{jl}}{\sqrt{\sum_{l=1}^{P} x_{il}^2}\sqrt{\sum_{l=1}^{P} x_{jl}^2}}.$$

Further, we can use *Jaccard coefficient* or *Dice coefficient*. The former can be expressed as

$$s(\mathbf{x}_i,\mathbf{x}_j) = \frac{\sum_{l=1}^{P} x_{il} x_{jl}}{\sum_{l=1}^{P} x_{il}^2 + \sum_{l=1}^{P} x_{jl}^2 - \sum_{l=1}^{P} x_{il} x_{jl}},$$

and the latter as

$$s(\mathbf{x}_i,\mathbf{x}_j) = \frac{2\times\sum_{l=1}^{P} x_{il} x_{jl}}{\sum_{l=1}^{P} x_{il}^2 + \sum_{l=1}^{P} x_{jl}^2}.$$

As concerning as *binary* variables, we distinguish symmetric ones (both categories are equally important) and asymmetric ones (one category carries more importance than the other). For document clustering, the latter has to be considered. Let us consider the following contingency table with frequencies $a$, $b$, $c$ and $d$.

**Table 1** Scheme of contingency table

| $\mathbf{x}_i/\mathbf{x}_j$ | 1 | 0 |
|---|---|---|
| 1 | $a$ | $b$ |
| 0 | $c$ | $d$ |

For asymmetric variables, we can use for example Jaccard coefficient or Dice coefficient. The former can be expressed as

$$s(\mathbf{x}_i,\mathbf{x}_j) = \frac{a}{a+b+c},$$

and the latter as

$$s(\mathbf{x}_i,\mathbf{x}_j) = \frac{2a}{2a+b+c}.$$

We can also use cosine of feature vectors, i.e. Ochiai coefficient

$$s(\mathbf{x}_i,\mathbf{x}_j) = \sqrt{\frac{a}{a+b}\times\frac{a}{a+c}}.$$

There are a lot of measures for clustering. We will mention a way how a distance between clusters can be measured. *Log-likelihood* distance between clusters $h$ and $h'$ is

$$D_{hh'} = \xi_h + \xi_{h'} - \xi_{\langle h, h' \rangle}$$

where $\langle h, h' \rangle$ denotes a cluster created by joining objects from clusters $h$ and $h'$, and

$$\zeta_v = -n_v \sum_{l=1}^{p} \frac{1}{2} \log(s_l^2 + s_{vl}^2)$$

where $n_v$ is the number of objects in the $v$-th cluster, $p$ is the number of variables, $s_l^2$ is a sample variance of the $l$-th continuous variable, and $s_{vl}^2$ is a sample variance of the $l$-th continuous variable in the $v$-th cluster.

## 2.1 Algorithms

As different communities use clustering, the associated terminology varies widely. There are a lot of ways how to classify clustering methods. One of them is *hard vs. fuzzy*. In hard clustering, each object belongs to the certain cluster unambiguously. In fuzzy clustering, membership grades indicate the degree to which the objects belong to the different clusters.

In hard clustering, the assignment of the $i$-th object to the $v$-th cluster is expressed by the value $u_{iv}$ which satisfies $u_{iv} \in \{0, 1\}$, $i = 1, \ldots, n$, $v = 1, \ldots, k$.

In fuzzy clustering ([1]), $u_{iv}$ satisfies the following conditions

$$u_{iv} \geq 0 \text{ for all } i = 1, \ldots, n \text{ and all } v = 1, \ldots, k, \quad \sum_{v=1}^{k} u_{iv} = 1 \text{ for all } i = 1, \ldots, n$$

Further, hard clustering can be regarded as *non-overlapping and overlapping*. In the former, each object belongs to one and only one cluster, i.e.

$$u_{iv} \in \{0, 1\}, i = 1, \ldots, n, v = 1, \ldots, k,$$

$$\sum_{v=1}^{k} u_{iv} = 1 \text{ for all } i = 1, \ldots, n.$$

In the latter, one object can be assigned to one or more clusters.

In the further text, we shall consider mainly hard clustering. The classification of the hard clustering methods can be the following (see [37]): hierarchical clustering techniques, centre-based algorithms, graph-based algorithms, grid-based algorithms, density-based algorithms, model-based algorithms, search-based algorithms and subspace clustering. However, some clustering algorithms are hybrid, which combine different approaches, e.g. hierarchical and graph-based.

## Hierarchical Clustering Techniques

Hierarchical algorithms divide a set of the objects into a sequence of nested partitions. There are some more detail types in this group of techniques. We can mention *agglomerative vs. divisive*. Agglomerative methods start with each object in a separation group. Then it merges clusters until only one large cluster remains which is the whole data set. It is sometimes ([74]) called AGNES (AGlomerative NESting). Divisive methods start with all objects in a single group and proceed until each object is in a separate group. The algorithm is sometimes ([74]) called DIANA (DIvisive ANAlysis). Further, we can mention division to *monothetic vs. polythetic* algorithms. Monothetic methods use single-feature based assignment into clusters. Divisive monothetic approach for binary data is called MONA (MONothetic Analysis). Polythetic algorithms consider multiple-features based assignment.

For polythetic agglomerative clustering, the user must select variables, choose dissimilarity or similarity measure and agglomerative procedure. At the first step, when each object represents its own cluster, the dissimilarity $d(\mathbf{x}_i, \mathbf{x}_j)$ between objects $\mathbf{x}_i$ and $\mathbf{x}_j$ is defined by the chosen dissimilarity measure. However, once several objects have been linked together, we need a linkage or amalgamation rule to determine when two clusters are sufficiently similar to be linked together. Numerous linkage rules have been proposed.

The distance between two different clusters can be determined by the distance of the two closest objects in the clusters (single linkage method), the greatest distance between two objects in the clusters (complete linkage method), or average distance between all pairs of objects in the two clusters (unweighted pair-group average method). Further, this distance can be determined by weighted average distance between all pairs of objects in the two clusters (the number of objects in a cluster is used as a weight), or distance between centroids (unweighted or weighted). Moreover, we can use the method that attempts to minimize the sum of squares of differences of individual values from their average in the cluster (Ward's method).

The hierarchical approach is used in some algorithms proposed for clustering large datasets. We can mention the BIRCH (Balanced Iterative Reducing and Clustering using Hierarchies) method [117] as an example. Objects in the data set are arranged into subclusters, known as "cluster-features". These cluster-features are then clustered into $k$ groups, using a traditional hierarchical clustering procedure. A cluster feature (CF) represents a set of summary statistics on a subset of the data. The algorithm consists of two phases. In the first one, an initial CF tree is built (a multi-level compression of the data that tries to preserve the inherent clustering structure of the data). In the second one, an arbitrary clustering algorithm is used to cluster the leaf nodes of the CF tree. Disadvantage of this method is its sensitivity to the order of the objects.

Another method for the clustering large datasets is the CURE (Clustering Using Representatives) algorithm [45] which combines a random sampling and partitioning. It consists of six main steps: drawing a random sample, partitioning the sample, partially clustering the partitions, elimination of the outliers, clustering the partial clusters, and labelling the data vectors.

## Centre-based algorithms

This type of algorithms is denoted as partitioning. It divides a whole set of the objects into single partitions. These techniques are very efficient for clustering large and high-dimensional datasets. The goal of centre-based algorithms is to minimize an objective function. Final clusters have convex shapes and each cluster is represented by a centre. This centre is either one object (corresponding vector) from the cluster ($k$-medoids algorithm) or a vector of the certain characteristics of individual variables (centroid).

The former technique is called PAM (Partitioning Around Medoids) algorithm. This algorithm was extended to the CLARA (Clustering LARge Applications) method [74]. CLARA clusters a sample from the data set and then it assigns all objects in the data set to these clusters. The process is repeated several times and then the clustering with the smallest average distance is selected.

The improvement of CLARA algorithm is CLARANS (Clustering Large Applications based on a RANdomized Search), see [93]. It proceeds by searching a random subset of the neighbours of a particular solution. Thus the search for the best representation is not confined to a local area of the data.

From the several $k$-centroid techniques, we can mention $k$-means algorithm (and its modifications), in which the average is used, and $k$-median algorithm, in which the median is applied. For categorical variables, the $k$-modes [24], and $k$-histogram [55] algorithms were proposed, for mixed-type datasets, $k$-probabilities and $k$-prototypes algorithms were designed, see [37].

## Graph-based algorithms

These types of algorithms apply a clustering to partition of the graph or hypergraph. The similar principal is used in *link-based clustering algorithms*, in which the individual vectors are represented by points in the space. The links between data points can be consider as links between graph nodes. We can mention the algorithms Chameleon, CACTUS and ROCK.

The Chameleon algorithm [71] is based on a similarity matrix. It constructs a sparse graph representation of the data items based on the commonly used $k$-nearest neighbour graph approach. Then agglomerative hierarchical clustering algorithm is used to merge the most similar subclusters by taking into account the relative interconnectivity and closeness of the clusters.

The CACTUS and ROCK algorithms are suitable for clustering categorical data. In the CACTUS (Categorial Clustering Using Summaries) algorithm [38], the inter-attribute and intra-attribute summaries of the dataset are calculated and the similarity graph is constructed according to these summaries. Then the clusters are found with respect to this graph. For the measurement of the similarity between two categorical variables, the support is applied.

ROCK (Robust Clustering using links) algorithm [46] is based on agglomerative clustering. It uses the link-based similarity measures to measure the similarity between two data vectors and also between two clusters. The link of two vectors is a number of common neighbours (a neighbour is a vector with the similarity above the given threshold). This method uses a random sampling just like the CURE algorithm (see above).

## Grid-based algorithms

These algorithms are suitable for large high-dimensional datasets, because their great advantage is the significant reduction of the computational complexity. The grid-based algorithms differ from the traditional clustering approaches in that they are concerned not with the data points but with the value space that surrounds the data points. A typical grid-based clustering algorithm consists of the five steps: creating the grid structure (partitioning the data space into a finite number of cells), calculating the cell density for each cell, sorting the cells according to their densities, identifying centres of clusters, and traversal of neighbour cells. We can mention the methods STING [109] and OptiGrid [61] as examples.

## Density-based algorithms

The density-based clustering is a technique which is capable of finding arbitrarily shaped clusters, where clusters are defined as dense regions separated by low-density regions. It can detect the outliers and the number of clusters. We can mention the methods DBSCAN [31] and DENCLUE [60] as examples.

## Model-based algorithms

Clustering algorithms can be also developed based on probability models, such as the finite mixture model for probability densities. It is assumed that the data are generated by a mixture of probability distributions in which each component represents a different cluster. The certain models are used for clusters and it tries to optimize the fit between the data and the models. There are two approaches to formulating the model for the composite of the clusters: the classification likelihood approach and the mixture likelihood approach.

For clustering categorical variables, the COOLCAT [8] algorithm was proposed. It uses entropy for this purpose. This algorithm consists of two steps. The first step is initialization, which finds a suitable set of clusters from a small sample of the entire data set, and the second step is incremental, which assigns the remaining records to suitable clusters.

## Search-based algorithms

To this group of algorithms, some special approaches and techniques are assigned. First, there are genetic algorithms which are inspired by the analogy of evolution and population genetics. Further, heuristic or metaheuristic algorithms can be used for clustering. In the problem of clustering $n$ objects into $k$ clusters such that the distance between feature vectors within a cluster and its centre is minimized, many local optimal solutions exist. In order to find the global optimum, Al-Sultan in [4] proposed a clustering algorithm based on a tabu search technique which is used to solve combinatorial optimization problems.

**Subspace clustering**

In high dimensional spaces, clusters often lie in a *subspace*. To handle this situation, some algorithms were suggested. Instead of creation of reduced matrix based on new features (obtained for example by linear combination of original features), subspaces of the original data space are investigated. The task is based on the original features which have a real meaning while linear combination of many dimensions may be sometimes hard to interpret. Subspace clustering is based on density based approach. The aim is to find subsets of features that projections of the input data include high density regions. The principle is partitioning of each dimension into the same number of equal length intervals. The clusters are unions of connected high density units within a subspace.

CLIQUE (CLustering In QUEst) suggested for numerical variables by Agrawal et al. in [3] is a clustering algorithm that finds high-density regions by partitioning the data space into cells (hyper-rectangles) and finding the dense cells. Clusters are found by taking the union of all high-density cells. For simplicity, clusters are described by expressing the cluster as a DNF (disjunctive normal form) expression and then simplifying the expression.

MAFIA (Merging of Adaptive Finite Intervals (And more than a CLIQUE)) is a modification of CLIQUE that runs faster and finds better quality clusters. pMAFIA is the parallel version. MAFIA was presented by Goil et al. in [40,90]. The main modification is the use of an adaptive grid. Initially, each dimension is partitioned into a fixed number of cells.

Moreover, we can mention the algorithm ENCLUS (ENntropy-based CLUStering) suggested by Cheng et al. in [25]. In comparison with CLIQUE, it uses a different criterion for subspace selection.

# 3  Graph Clustering

Networks arising from real life are concerned with relations between real objects and are important part of modern life. Important examples include links between Web pages, citations of references in scientific papers, social networks of acquaintance or other connections between individuals, electric power grids, etc. Word "network" is usually used for what mathematicians and a few computer scientists calls graphs [91]. A *graph* (*network*) is a set of items called *nodes* (*vertices*) with connections between them, called *edges* (*links*). The study of graph theory is one of the fundamental pillars of discrete mathematics.

A *social network* is a set of people or groups of people with some pattern of contacts or interactions between them. Social networks have been studied extensively since the beginning of 20 th century, when sociologists realized the importance of the understanding how the human society is functioned. The traditional way to analyze a graph is to look at its picture, but for large networks this is unusable. A new approach to examine properties of graphs has been driven largely by the availability of computers and communication networks, that allow us to analyze data on a scale far larger than before now [47,92].

Interesting source of reliable data about personal connections between people is communication records of certain kinds. For example, one could construct a network in which each vertice represents an email address and directed edges represent messages passing from one address to another.

Complex networks such as the Web or social networks or emails often do not have an engineered architecture but instead are self-organized by the actions of a large number of individuals. From these local interactions nontrivial global phenomena can emerge as, for example, small-world properties or a scale-free distribution of the degree [91]. In *small-world networks* short paths between almost any two sites exist even though nodes are highly clustered. *Scale-free networks* are characterized by a power-law distribution of a node's degree, defined as the number of its next neighbours, meaning that structure and dynamics of the network are strongly affected by nodes with a great number of connections. There is reported in [30] that networks composed of persons connected by exchanged emails show both the characteristics of small-world networks and scale-free networks.

The Web can be considered as a graph where nodes are HTML pages and edges are hyperlinks between these pages. This graph is called the *Web graph*. It has been the subject of a variety of recent works aimed at understanding the structure of the Web [65].

A *directed graph* $G = (V, E)$ consists of a set of nodes, denoted $V$ and a set of edges, denoted $E$. Each edge is an ordered pair of nodes $(u, v)$ representing a directed connection from $u$ to $v$. The graph $G = (V, E)$ is often represented by the *adjacency matrix* $W$ by $|V| \times |V|$, where $w_{ij} = 1$ if $(v_i, v_j) \in E$ and $w_{ij} = 0$ in other cases. The *out-degree* of a node $u$ is the number of distinct edges $(u, v_1)...(u, v_k)$ (i.e., the number of links from $u$), and the *in-degree* is the number of distinct edges $(v_1, u)...(v_k, u)$ (i.e., the number of links to $u$). A *path* from node $u$ to node $v$ is a sequence of edges $(u, u_1), (u_1, u_2),...(u_k, v)$. One can follow such a sequence of edges to "walk" through the graph from $u$ to $v$. Note that a path from $u$ to $v$ does not imply a path from $v$ to $u$. The *distance* from $u$ to $v$ is one more than the smallest $k$ for which such a path exists. If no path exists, the distance from $u$ to $v$ is defined to be infinity. If $(u, v)$ is an edge, then the distance from $u$ to $v$ is 1.

Given a directed graph, a *strongly connected* component (*strong component* for brevity) of this graph is a set of nodes such that for any pair of nodes u and v in the set there is a path from u to v. In general, a directed graph may have one or many strong components. The strong components of a graph consist of disjoint sets of nodes. One focus of our studies will be in understanding the distribution of the sizes of strong components on the web graph.

An *undirected graph* consists of a set of nodes and a set of edges, each of which is an unordered pair $\{u, v\}$ of nodes. In our context, we say there is an edge between $u$ and $v$ if there is a link between $u$ and $v$, without regard to whether the link points from $u$ to $v$ or the other way around. The *degree* $deg(u)$ of a node $u$ is the number of edges incident to $u$. A path is defined as for directed graphs, except that now the existence of a path from $u$ to $v$ implies a path from $v$ to $u$. A *component* of an undirected graph is a set of nodes such that for any pair of nodes $u$ and $v$ in the set there is a path from $u$ to $v$. We refer to the components of the undirected graph obtained from a directed graph by ignoring the directions of its edges as the weak components of the directed graph. Thus two nodes on the web may be in the same weak component even though there is no directed path between them (consider, for instance, a node u that points to two other nodes v and $w$; then v and $w$ are in the same weak component even though there may be no sequence of links leading from $v$ to $w$ or vice versa). The interplay of strong and weak components on the (directed) web graph turns out to reveal some unexpected properties of the Web's connectivity.

Informally we can say that two nodes are considered *similar* if there are many short paths connecting them. On the contrary, the "shortest path" distance does not necessarily decrease when connections between nodes are added, and thus it does not capture the fact that strongly connected nodes are at a smaller distance than weakly connected nodes.

The main findings about the Web structure are as follows:

- A power-law distribution of degrees [81]: in-degree and out-degree distribution of the nodes of the Web graph follows the power law.
- A bow-tie shape [13]: the Web's macroscopic structure.
- The average path length between two Web pages: 16 [13] and 19 [7].
- Small world phenomenon [2]: Six degrees of separation between any two Web pages.
- Cyber-communities [81]: groups of individuals who share a common interest, together with the most popular Web pages among them.
- Self-similarity structure [27]: the Web shows a fractal structure in many different ways.

Link analysis plays an import role in understanding of the Web structure. There are three well known algorithms for ranking pages, such as, HITS, PageRank, and SALSA [101].

The book [101] describes exciting new opportunities for utilizing robust graph representations of data with common machine learning algorithms. Graphs can model additional information which is often not present in commonly used data representations, such as vectors. Through the use of graph distance a relatively new approach for determining graph similarity the authors show how well-known algorithms, such as $k$-means clustering and $k$-nearest neighbours classification, can be easily extended to work with graphs instead of vectors. This allows for the utilization of additional information found in graph representations, while at the same time employing well-known, proven algorithms.

## 3.1 Linear Algebra Background

Any $m \times n$ matrix $A$ can be expressed as

$$A = \sum_{t=1}^{r} \sigma_t(A)u(t)v(t)^T,$$

where $r$ is the rank of $A$, $\sigma_1(A) \geq \sigma_2(A) \geq ... \geq \sigma_r(A) > 0$ are its *singular values* and $u(t) \in \mathbb{R}^m$, $v(t) \in \mathbb{R}^n$, $t = 1,...,r$ are its left and right singular vectors, respectively. The $u(t)$'s and the $v(t)$'s are orthonormal sets of vectors; namely, $u(i)^T u(j)$ is one if $i = j$ and zero otherwise. We also remind the reader that

$$\|A\|_F^2 = \sum_{i,j} A_{i,j}^2 = \sum_{i=1}^{r} \sigma_i^2(A)$$

$$\|A\|_2 = \max_{x \in \mathbb{R}^n : \|x\|=1} \|Ax\| = \max_{x \in \mathbb{R}^m : \|x\|=1} \|x^T A\| = \sigma_1(A)$$

In matrix notation, SVD is defined as $A = U\Sigma V^T$ where $U$ and $V$ are orthogonal (thus $U^T U = I$ and $V^T V = I$, an $I$ matrix is the identity matrix $I = \{\delta_{ij}\}$ where $\delta_{ij}$ is the Kronecker symbol) matrices of dimensions $m \times r$ and $n \times r$ respectively, containing the left and right singular vectors of $A$. $\Sigma = diag(\sigma_1(A),...,\sigma_r(A))$ is an $r \times r$ diagonal matrix containing the singular values of $A$. If we define $A_l = \sum_{t=1}^{l} \sigma_t(A)u(t)v(t)^T$, then $A_l$ is the best rank $l$ approximation to $A$ with respect to the 2-norm and the Frobenius norm. Thus, for any matrix $D$ of rank at most $l$, $\|A - A_k\|_2 \leq \|A - D\|_2$ and $\|A - A_k\|_F \leq \|A - D\|_F$. A matrix $A$ has a "good" rank $l$ approximation if $A - A_l$ is small with respect to the 2-norm and the Frobenius norm. It is well known that $\|A - A_k\|_F^2 = \sum_{t=l+1}^{r} \sigma_t^2(A)$ and $\|A - A_k\|_2 = \sigma_{l+1}(A)$. From basic Linear Algebra, $A_l = U_l \Sigma_l V_l^T = A V_l V_l^T = U_l U_l^T A$, where $U_l$ and $V_l$ are sub-matrices of $U$ and $V$, containing only the top $k$ left or right singular vectors of $A$ respectively; for a detailed treatment of SVD see Golub and Van Loan [41].

## 3.2 Eigenvector Clustering of Graphs

Donath and Hoffman [28] introduced the use of eigenvectors for the purpose of partitioning an undirected graph in a balanced way. Since then, there has been a lot of work on spectral approaches for graph partitioning. See Chung [26] for an excellent overview of the field. Shi and Malik [103] showed that the eigenvectors of different matrices based on the adjacency matrix of a graph are related to different kinds of balanced cuts in a graph. Let $W$ be the adjacency matrix of an undirected graph $G = (V, E)$ with nodes $1, 2, ..., n$ and let $D$ be a diagonal matrix with $d_i = deg(i)$. Let $A$ and $B$ be sets of nodes and let $E(A, B)$ be the set of edges $(u, v)$ with $u \in A$ and $v \in B$. Two subsets $A$ and $B$ of $V$, such that $A \cup B = V$ and $A \cap B = \emptyset$, define a *cut* in $G$, which we denote as $(A, B)$.

The *average association* of a set $A$ is

$$| E(A, A) | / | A |.$$

The *average cut* of a set A is

$$| E(A, V - A) | / | A | + | E(A, V - A) | / | V - A |.$$

The *normalized cut* of a set A is

$$| E(A, V - A) | / | E(A, V) | + | E(A, V - A) | / | E(V - A, V) |.$$

Then Shi and Malik in [103] show that

- the second largest eigenvector of $W$ is related to a set that maximizes the average association;
- the second smallest eigenvector of $D - W$ (also known as the algebraic connectivity or Fiedler value [35]) is related to a set that minimizes the average cut; and
- the second smallest eigenvector of the generalized eigenvector problem $(D - W)x = \lambda Dx$ gives an approximation of the smallest normalized cut.

These results hold for undirected graphs, but the Web graph is a directed graph. Thus, it would be interesting to understand what the above relationships are for directed graphs, i.e., whether the eigenvectors of the corresponding matrices of a directed graph are also related to balanced decompositions of the directed graph. It is possible that this would lead to an interesting clustering of the Web graph or for a topic-specific subgraph. The first step in this direction was taken by Gibson et al. [39]. They used the eigenvectors of the matrix $AA^T$ and the matrix $A^TA$, where $A$ is the adjacency matrix of a topic-specific subgraph, to decompose topic-specific subgraphs. They show that the principal eigenvector and the top few nonprincipal eigenvectors decompose the topic graphs into multiple "hyperlinked communities," i.e., clusters of pages on the same subtopic [56]. Lot of examples of eigenvector computations we can found in the survey paper [82].

Roughly speaking, from spectral analysis we obtain decomposition of graph to "high order" connected component [34,35]. The work [54] compares clustering based on Fiedler vector with $k$-means clustering method and founds the results of spectral partitioning usually much better.

## 3.3 Connectivity Clustering of Graphs

Although there are numerous algorithms for cluster analysis in the literature, we briefly review the approaches that are closely related to the structure of a graph.

Matula [85,86,87] uses a high connectivity in similarity graphs to cluster analysis, which is based on the cohesiveness function. The function defines every node and edge of a graph to be the maximum edge-connectivity of any subgraph containing that element. The *k-connected subgraphs* of the graph are obtained by deleting all elements with cohesiveness less than $k$ in the graph, where $k$ is a constant value. It is hard to determine the connectivity values in real clustering applications with this approach.

There are approaches using biconnected components (maximal $2-$connected subgraphs). The work [15] introduces a new algorithm for protein structure prediction based on biconnected components. In [56] Henzinger presents fully dynamic algorithms for maintaining the biconnected components.

There is a recent work related to clustering of a graph. The HCS algorithms [51] use a similarity graph as the input data. The algorithm recursively partitions a current set of elements into two subsets. It then identifies highly connected subgraphs, in which the number of edges exceeds half the number of their corresponding nodes, as kernels among them. A kernel is considered as a cluster. Unfortunately, the result of the clustering is not uniquely determined.

The CLICK algorithm [102] builds on a statistical model. It uses the same basic scheme as HCS to form kernels, and includes the following processing: singleton adoption, recursive clustering process on the set of remaining singletons, and an iterative merging step.

The CAST [9] uses a single parameter $t$, and starts with a single object. Objects are added or removed from the cluster if their affinity is larger or lower than $t$, respectively, until the process stabilizes.

In [65] there are introduced definitions of homogeneity and separation to measure the quality of a graph clustering.

In [92] Newman's $Q$ function is used for graph embedding into Euclidean space. This representation is used for fast geometric clustering.

## 3.4 Combined Methods

Deng Cai et al. in [14] described a method to organize Web image search results. Based on the Web context, they proposed three representations for Web images, i.e. representation based on a visual feature, representation based on a textual feature and representation induced from the image link graph. Spectral techniques were applied to cluster the search results into different semantic categories. They show that the combination of textual feature based representation and graph based representation actually reflects the semantic relationships between Web images.

In [83] the algorithm S-GRACE is presented. S-GRACE is a hierarchical clustering algorithm on XML documents, which applies categorical clustering algorithm (ROCK [46]) on the $s$-graphs (structure graph) extracted from the XML documents.

For two XML documents $\mathbf{x}_i$ and $\mathbf{x}_j$, the distance between them is defined by

$$d(\mathbf{x}_i, \mathbf{x}_j) = 1 - \frac{|sg(\mathbf{x}_i) \cap sg(\mathbf{x}_j)|}{\max(|sg(\mathbf{x}_i)|, |sg(\mathbf{x}_j)|)}$$

where $sg(\mathbf{x}_i)$ is a structure graph ($i = 1, 2$), $|sg(\mathbf{x}_i)|$ is the number of edges in $sg(\mathbf{x}_i)$; and $sg(\mathbf{x}_i) \cap sg(\mathbf{x}_j)$ is the set of common edges of $sg(\mathbf{x}_i)$ and $sg(\mathbf{x}_j)$.

## 4 Artificial Neural Networks

Artificial Neural Networks (ANNs) belong to the adaptive class of techniques in the machine learning area. ANNs try to mimic the biological neural network, the brain to solve basic computationally hard problems of AI.

There are three important, and attractive, features of ANNs:
- it is their capability of learning from example (extracting knowledge from data),
- there are natural parallel, and thus should be computationally effective, and
- they should work incrementally - not whole data set is necessary at once.

This feature makes ANNs a very interesting and promising clustering choice for large datasets including multimedia and text files.

Most models of ANNs are organized in the form of a number of processing units (also called artificial neurons, or simply neurons [88]), and a number of weighted connections (artificial synapses) between the neurons. The process of building an ANN, similar to its biological inspiration, involves a learning episode (also called training). During learning episode, the network observes a sequence of recorded data, and adjusts the strength of its synapses according to a learning algorithm and on the observed data. The process of adjusting the synaptic strengths in order to be able to accomplish a certain task, much like the brain, is called learning. Learning algorithms are generally divided into two types, supervised and unsupervised. The supervised algorithms require labelled training data. In other words, they require more a priori knowledge about the training set.

There is a very large body of research that has resulted in a large number of ANN designs. For a more complete review of the various ANN types see [52,98]. In this chapter, we discuss only some of the types that have been used data mining area.

## 4.1  Layered, Feed-Forward, Backpropagation Neural Networks

These are a class of ANNs whose neurons are organized in layers. The layers are normally fully connected, meaning that each element (neuron) of a layer is connected to each element of the next layer. However, self-organizing varieties also exist in which a network starts either with a minimal number of synaptic connections between the layers and adds new ones as training progresses (*constructive*), or starts as a fully connected network and prunes connections based on the data observed in training (*destructive*) [52,98].

Backpropagation [98] is a learning algorithm that, in its original version, belongs to the gradient descent optimization methods [111]. The combination of backpropagation learning algorithm and the feed-forward, layered networks provide the most popular type of ANNs. These ANNs have been applied to virtually all pattern recognition problems, and are typically the first networks tried on a new problem. The reason for this is the simplicity of the algorithm, and the vast body of research that has studied these networks. As such, in sequencing, many researchers have also used this type of network as a first line of attack. Examples can be mentioned in [111,112]. In [112] Wu has developed a system called gene classification artificial neural system (GenCANS), which is based on a three layered, feed-forward backpropagation network.

## 4.2  Self-organizing Neural Networks

These networks are a very large class of neural networks whose structure (number of neurons, number of synaptic connections, number of modules, or number of layers) changes during learning based on the observed data. There are two classes of this type of networks: destructive and constructive. Destructive networks are initially a fully connected topology and the learning algorithm prunes synapses (sometime entire neurons, modules, or layers) based on the observed data. The final remaining network after learning is complete, usually is a sparsely connected network. Constructive algorithms start with a minimally connected network, and gradually add synapses (neurons, modules, or layers) as training progresses, in order to accommodate for the complexity of the task at hand.

**Self-Organizing Map.** A self-organizing map (SOM) [78] is a neural network paradigm first proposed by Kohonen [77]. SOMs have been used as a divisive clustering approach in many areas. Several groups have used SOMs to discover patterns clusters in Web pages or in textual documents [5,6]. Special version of this paradigm WEBSOM was developed for Web pages clustering [72,79]. With the WEBSOM method a textual document collection is organized onto a graphical map display that provides an overview of the collection and facilitates interactive browsing. Interesting documents can be located on the map using a content-directed search. Each document is encoded as a histogram of term categories which are formed by the SOM algorithm based on the similarities in the contexts of the terms. The encoded documents are organized on another self-organizing map, a document map, on which nearby locations contain similar documents.

Special consideration is given to the computation of very large document maps which is possible with general-purpose computers if the dimensionality of the term category histograms is first reduced with a random mapping method and if computationally efficient algorithms are used in computing the SOMs.

SOM as a clustering method has some disadvantages. One of them is necessity of introduction of decay coefficient that stops the learning (clustering) phase. If the map is allowed to grow indefinitely, the size of SOM is gradually increased to a point when clearly different sets of expression patterns are identified. Therefore, as with $k$-means clustering, the user has to rely on some other source of information, such as PCA, to determine the number of clusters that best represents the available data. For this reason, Sásik [100] and his colleagues believe that SOM, as implemented by Tamayo et al. [104], is essentially a restricted version of $k$-means: Here, the $k$ clusters are linked by some arbitrary user-imposed topological constraints (e.g. a $3 \times 2$ grid), and as such suffers from all of the problems mentioned above for $k$-means (and more), except that the constraints expedites the optimization process. There are many varieties to SOM, among which the self-organizing feature maps (SOFM) should be mentioned [77,78]. The *growing cell structure* (GCS) [36] is another derivative of SOFM. It is a selforganizing and incremental (constructive) neural learning approach.

**Self-organizing trees.** Self-organizing trees are normally constructive neural network methods that develop into a tree (usually binary tree) topology during learning. Among examples of these networks the work of Dopazo et al. [29], Wang et al. [107], and Herrero et al. [59] can be mentioned. Dopazo and Carazo introduce the self-organizing tree algorithm (SOTA) [29]. SOTA is a hierarchical neural network that grows into a binary tree topology. For this reason SOTA can be considered a hierarchical clustering algorithm. SOTA is based on Kohonen's SOM discussed above and Fritzke's growing cell [36]. The SOTA's performance is superior to that of classical hierarchical clustering techniques. Among the advantages of SOTA as compared to hierarchical cluster algorithms are its lower time complexity, and its top-to-bottom hierarchical approach. SOTA's runtimes are approximately linear with the number of items to be classified, making it suitable for large datasets. Also, because SOTA forms higher clusters in the hierarchy before forming the lower clusters, it can be stopped at any level of hierarchy and still produces meaningful intermediate results. There are many other types of self-organizing trees.

## 4.3  Recurrent ANNs

**ART and its derivatives.** *Adaptive Resonance Theory* was introduced by Stephen Grossberg [43,44] in 1976. Networks based on ART are unsupervised and self-organizing, and only learn in the so called "resonant" state. ART can form (stable) clusters of arbitrary sequences of input patterns by learning (entering resonant states) and self-organizing. Since the inception, many derivatives of ART have emerged. Among these ART-1 (the binary version of ART; forms clusters of binary input data) [17], ART-2 (analog version of ART) [16], ART-2A (fast version of ART-2) [22], ART-3 (includes "chemical transmitters" to control the search process in a hierarchical ART structure) [18], ARTMAP (supervised version of ART) [20] can be mentioned. Many hybrid varieties such as Fuzzy-ART [23], Fuzzy-ARTMAP (supervised Fuzzy-ART) [19,21] and simplified Fuzzy-ARTMAP (SFAM) [73] have also been developed.

**The ART family of networks.** These networks have a broad application in virtually all areas of clustering. In general, in problem settings when the number of clusters is not previously known apriori, researchers tend to use unsupervised ART, where when the number of clusters is known a priori, usually the supervised version, ARTMAP, is used. Among the unsupervised implementations, the work of Tomida et al. [105] should be mentioned. Here the authors used Fuzzy ART for expression level data analysis. Fuzzy ART incorporates the basic features of all ART systems, notably, pattern matching between bottom-up input and top-down learned prototype vectors. This matching process leads either to a resonant state that focuses attention and triggers stable prototype learning or to a self-regulating parallel memory search. If the search ends by selecting an established category, then the category's prototype may be refined to incorporate new information in the input pattern. If the search ends by selecting a previously untrained node, then learning of a new category takes place. Fuzzy ART performs best in noisy data. Although ART has been used in several research works as a text clustering tool, the level of quality of the resulting document clusters has not been clearly established. In [84] the author presents experimental results with binary ART that address this issue by determining how close clustering quality is to an upper bound on clustering quality.

**Associative Clustering Neural Networks.** Since the introduction of the concept of auto-associative memory by Hopfield [63], there have been many associative memory models built with neural networks [75,80]. Most of them can be considered into store-recall models and the correlation between any two $D$-bit bipolar patterns $s(\mathbf{x}_i, \mathbf{x}_j)$, $x_{id} \in \{-1,1\}$ for all $l = 1, \ldots, p$ is often determined by a static measurement such as

$$s(\mathbf{x}_i, \mathbf{x}_j) = \frac{1}{p} \sum_{l=1}^{p} x_{il} x_{jl}.$$

The human mind, however, associates one pattern in memory to others in a much more sophisticated way than merely attempting to homogeneously link vectors. Such associations would interfere with each other [62]. To mimic the formation of such associations in cybernetics, Yao at al. [113] build a recurrent neural network to dynamically evaluate the association of any pairwise patterns through the interaction among a group patterns and incorporate the results of interaction into data clustering. The novel rule based on the characteristic of clusters has been proposed to determine the number of clusters with a reject option. Such a hybrid model they named Associative Clustering Neural Network (ACNN). The performance of ACNN has been studied by authors on simulated data only, but and the results have demonstrated that ACNN has the feasibility to cluster data with a reject option and label the data robustly.

**Bayesian Neural Networks.** There are a number of recent networks that have been suggested as solutions clustering. For instance, Bayesian neural networks (BNNs), are another technique that has been recently used for Web clustering. Her et al. [58] have used the BNNs for clustering Web query results. Their BNN is based on SOM and it differs in the last step when $n$ documents are assigned under each cluster by Bayesian rule. The BNNs are an important addition to the host of ANN solutions that have been offered to the problem at hand, as they represent a large group of hybrid ANNs that combine classical ANNs with statistical classification and prediction theories.

# 5  Web Clustering

The Web has undergone exponential growth since its birth, which is the cause of a number of problems with its usage. Particularly, the quality of Web search and corresponding interpretation of search results are often far from satisfying due to various reasons like huge volume of information or diverse requirements for search results.

The lack of a central structure and freedom from a strict syntax allow the availability of a vast amount of information on the Web, but they often cause that its retrieval is not easy and meaningful. Although ranked lists of search results returned by a search engine are still popular, this method is highly inefficient since the number of retrieved search results can be high for a typical query. Most users just view the top ten results and therefore might miss relevant information. Moreover, the criteria used for ranking may not reflect the needs of the user. A majority of the queries tend to be short and thus, consequently, non-specific or imprecise. Moreover, as terms or phrases are ambiguous in the absence of their context, a large amount of search results is irrelevant to the user.

In an effort to keep up with the tremendous growth of the Web, many research projects were targeted on how to deal its content and structure to make it easier for the users to find the information they want more efficiently and accurately. In last years mainly data mining methods applied in the Web environment create new possibilities and challenges.

Methods of Web data mining can be divided into a number of categories according to kind of mined information and goals that particular categories set. In [94], three categories are distinguished: *Web structure mining* (WSM), *Web usage mining* (WUM), and *Web Content Mining* (WCM). Particularly, WCM refers broadly to the process of uncovering interesting and potentially useful knowledge from Web documents.

WCM shares many concepts with traditional text mining techniques. One of these, *clustering*, groups similar documents together to make information retrieval more effective. When applied to Web pages, clustering methods try to identify inherent groupings of pages so that a set of clusters is produced in which clusters contain relevant pages (to a specific topic) and irrelevant pages are separated. Generally, text document clustering methods attempt to collect the documents into groups where each group represents some topic that is different than those topic represented by the other groups. Such clustering is expected to be helpful for discrimination, summarization, organization, and navigation for unstructured Web pages.

In a more general approach, we can consider Web documents as collections of Web pages including not only HTML files but also XML files, images, etc. An important research direction in Web clustering is Web XML data clustering stating the clustering problem with two dimensions: content and structure [106].

WUM techniques use the Web-log data coming from users' sessions. In this framework, Web-log data provide information about activities performed by a user from the moment the user enters a Web site to the moment the same user leaves it. In WUM, the clustering tries to group together a set of users' navigation sessions having similar characteristics [106]. Concerning WSM techniques, graph-oriented methods described in Section 3 can be used.

Considering Web clustering techniques, it is important to be aware of two main categories of approaches:

- clustering Web pages in a space of resources to facilitate some search services and
- clustering Web search results.

In [12], these categories are called *offline clustering* and *online clustering*, respectively. We mention approaches of both categories although the main accent is put on the latter.

## 5.1   *Application of Web Clustering*

Web clustering is currently one of the crucial IR problems related to Web. It is used by many intelligent software agents in order to retrieve, filter, and categorize Web documents. Various forms of clustering are required in a wide range of applications: efficient information retrieval by focusing on relevant subsets (clusters) rather than whole collections, clustering documents in collections of digital libraries, clustering of search results to present them in an organized and understandable form, finding mirrored Web pages, and detecting copyright violations, among others.

Clustering techniques are immensely important for Web applications to assist the automated (or semiautomated) generation of proper categories of documents and organize repositories of search engines. Hierarchical categorization of documents is often used (see Google, Yahoo, Open Directory, and LookSmart as examples). The reason is that the search results are not summarized in terms of topics; they are not well suited for browsing tasks. One possible solution is to create manually a static hierarchical categorization of a reasonable part of the Web and use these categories to organize the search results of a particular query. However, this solution is feasible only for small collections. To categorize the entire Web either manually or automatically is, unfortunately, not real.

In [95], document clustering and a WUM technique are used for construction of Web Community Directories, as a means of personalizing Web services. Also effective summarization of Web page collections becomes more and more critical as the amount of information continues to grow on the Web. The significance of Web collection clustering for automatic Web collection summarization is investigated in [115].

Clustering is also useful in extracting salient features of related Web documents to automatically formulate queries and search for other similar documents on the Web.

## 5.2 Principles of Web Clustering Methods

Most of the documents clustering methods that are in use today are based on the VSM. A similarity between documents is measured using one of several similarity measures that are based on relations of feature vectors, e.g. cosine of feature vectors (see Section 2). Many of traditional algorithms based on VSM, however, falter when the dimensionality of the feature space becomes high relative to the size of the document space. In a high dimensional space, the distance between any two documents tends to be constant, making clustering on the basis of distance ill-defined. This phenomenon is called a *curse of dimensionality*. Therefore the issue of reducing the dimensionality of the space is critical. The methods presented in Section 2 are often used.

Traditional clustering algorithms either use a priori knowledge of document structures to define a distance or similarity among these documents, or use probabilistic techniques such as Bayesian classification.

Taxonomies are generated using document clustering algorithms which typically result in topic or concept hierarchies. This classification and clustering techniques are combined. Concept hierarchies expose the different concepts presented in the Web pages (or search result) collection. The user can choose the concept he/she is interested in and can browse it in detail.

## 5.3 Classification of Web Clustering Methods

Generally, clustering approaches could be classified in two broad categories [92]: *term-based clustering* and *link-based clustering*. Recent work in online clustering has included both link-based and term-based methods.

**Term-based clustering.** We start with methods where each term is a single word. Zamir et al. mention in [116] very simple *word-intersection clustering* method, where words that are shared by documents are used to produce clusters. Let $n$ denote the number of documents to be clustered. The method runs in $O(n^2)$ and produces good results for Web documents originating rather from on a corpus of texts. We point out that standard methods such as $k$-means, are also in this category since they usually exploit single words as features. Most of methods based on VSM belong to this category. They do not make use of any word proximity or phrase-based approach.

Word-based clustering that is used on common words shared among documents does not adapt well to Web environment since it ignores the availability of hyperlinks between Web pages and is susceptible to spam. Also the curse of dimensionality restricts a usability of these methods. A more successful clustering in this case (also ignoring links among documents) is based on multi-word terms (phrases, sentences). Then we speak about *term-based clustering* [115]. Extracting terms significantly reduces the high dimensionality. Authors of [115] show that this reduction is almost an order of magnitude while maintaining comparable performance with word-based model.

Among first works using phrases in clustering we find approach [114] based on Suffix Tree Clustering (STC). STC firstly transforms the string of text representing each document to a sequence of stems. Secondly, it identifies the sets of documents that shared a common phrase as base clusters by a suffix tree. Finally, these base clusters are combined into clusters. Tree building often requires $O(n \log n)$ time and produces high quality clusters. On the other hand, the suffix tree model can have a high number of redundancies in terms of the suffixes stored in the tree. However, the STC clustering based on phrases shared between documents generates inferior results to those based on the full text of the document.

In [48], a system for Web clustering is based on two key concepts. The first is the use of weighted phrases as an essential constituent of documents. Similarity between documents will be based on matching phrases and their weights. The similarity calculation between documents combines single-word similarity and phrase-based similarity. The latter is proven to have a more significant effect on the clustering quality due to its insensitivity to noisy terms that could lead to incorrect similarity measure. The second concept is the incremental clustering of documents using a histogram-based method to maximize the tightness of clusters by carefully watching the similarity distribution inside each cluster. In the system a novel phrase-based document index model is used, the Document Index Graph (DIG), that captures the structure of sentences in the document set, rather than single words only. The DIG model is based on graph theory and utilizes graph properties to match any-length phrase from a document to any number of previously seen documents in a time nearly proportional to the number of words of the document. Improvement over traditional clustering methods was 10 to 29 percent.

**Link-based clustering.** Links among Web pages could provide valuable information to determine the related page since they give objective opinions for the topic of the pages they point to. Many works tried to explore link analysis to improve the term-based methods. In general these methods belong to the category of graph clustering (Section 3). Kleinberg in [76] suggested that there are two kinds of pages on the Web for a specific query topic: hub and authority and they reinforce each other. HITS algorithm, which was used to locate hubs and authorities from the search results given a query topic, provided a possible way to alleviate the problems. However, sometimes one's "most authoritative" pages are not useful for others. It is also observable that many "authority" pages contain very little text. The work [108] combines successfully link-based features (co-citations and bibliographic coupling) and contents information in clustering. Co-citation measures the number of citations (out-links) in common between two documents and coupling measures the number of document (in-links) that cites both of two documents under consideration.

**Structure of clusters.** Two clustering algorithms that can effectively cluster documents, even in the presence of a very high dimensional feature space are described in [53]. These clustering techniques, which are based on generalizations of graph partitioning, do not require pre-specified ad hoc distance functions, and are capable of automatically discovering document similarities or associations.

As we mentioned in Introduction, most of clustering methods can be divided into two categories: hierarchical clusters and flat clusters. Hierarchical clustering is exceedingly slow when it is used for online for very high $n$. Its implementing time can be from $O(n^2)$ up to $O(n^3)$.

The flat clustering algorithms are model-based algorithms that search for the model parameters given the data and prior expectation. For example, $k$-means is $O(nkT)$ algorithm, where $T$ is the number of iterations, but the task to determine model describing data complicates its use for large collections, particularly in a Web environment.

## 5.4 Clustering with Snippets

Today search engines return with a ranked list of search results also some contextual information, in the form of a Web page excerpt, the so called *snippet*. *Web-snippet clustering* is an innovative approach to help users in searching the Web. It consists of clustering the snippets returned by a (meta-) search engine into a hierarchy of folders which are labelled with a term. The term expresses latent semantics of the folder and of the corresponding Web pages contained in the folder. The folder labels vary from a bag of words to variable-length sentences.

Web-snippet clustering methods are classified in [33] according to two dimensions: words vs. terms and flat vs. hierarchical. Four categories of approaches are distinguished.

**Word-based and flat clustering.** This category includes systems like SCATTER-GATHER and WEBCAT. Other systems use e.g. fuzzy relations [70] or take into account in-linking and out-linking pages to improve precision.

**Term-based and flat clustering.** Authors of [115] used sentences of variable length to label the folders, but these sentences were drawn as contiguous portions of the snippets by means of a Suffix Tree data structure. Other systems use SVD on a term-document matrix to find meaningful long labels. This approach is restricted by the time complexity of SVD applied to a large number of snippets. In addition, the similar snippets can lead to very high overlap, means of a STC.

**Word-based and hierarchical clustering.** There are approaches based on the Frequent Itemsets Problem and a concept lattice [97] on single words in order to construct the folder hierarchy.

**Term-based and hierarchical clustering.** This class includes the best meta-search engines of the years 2000-2003 Vivisimo and Dogpile. These tools add to the flat list of search results a hierarchy of clusters built on-the-fly over snippets. It improves precision over recall by using a snippet representation made of pair of words (not necessarily contiguous) linked by a lexical affinity, i.e. a correlation of their common appearance. Among older approaches there is a simple extension of Grouper [114] to hierarchical clustering based on the size of folders overlap. A hierarchical engine SNAKET introduced in [33] organizes on-the-fly the search results from 16 commodity search engines and offers folder labelling with variable-length sentences. Hierarchies are overlapping because snippet might cover multiple themes.

# 6 Conclusion

Clustering is currently one of the most crucial techniques for dealing with massive amount of heterogeneous information on the Web as well as organizing Web search results. Unlike clustering in other fields, Web clustering separates unrelated pages and clusters related pages (to a specific topic) into semantically meaningful groups, which is useful for discrimination, summarization, organization and navigation of unstructured Web pages. In this chapter we have presented a lot of general approaches to clustering as well as a lot of various classifications of clustering algorithms. Consequently, two important questions arise: (1) why so many clustering algorithms, and (2) which of them are usable for Web clustering?

In his paper [32] Estivill-Castro tries to answer the first question in terms of the model of data to be clustered and the cluster model (inductive principle in his terminology). For a single model of data and a cluster model there are many clustering algorithms. As there are cluster models and many algorithms for each cluster models, there are many clustering algorithms. And why are here so many clustering models? Because clustering is in part beholder dependent. Cluster models are just formal models of what researchers believe is a definition of cluster. Thus, it is very hard to compare particular approaches.

To answer the second question, we can first consider the techniques that are not usable for Web clustering. Observe that clustering in a Web environment eliminates naturally a use of some general clustering techniques. The reason is easy. Since clustering translates into optimization problem, its computational complexity is typically intractable in the case of huge Web data collections.

Another reason for inapplicability of some classical techniques is associated with usability of the clustering achieved. Given a large document collection, it is difficult to provide the number of real categories for users when they attempt to categorize the documents. Organizing Web search results into a hierarchy of topics and subtopics facilitates browsing the collection and locating results of interest. Traditional clustering techniques are inadequate for Web since they do not generate clusters with highly readable names. It seems that Web-snippet clustering methods deal successfully with this issue. We have also mentioned how link information can be used to improve classification results for Web collections. In practice, it is desirable to combine term-based clustering and link-based clustering.

This survey represents only a small part of the research being conducted in the area. Furthermore, as new techniques and algorithms are being proposed for Web datasets, it makes survey such as this highly time dependent.

**Acknowledgement.** The work was partly supported by the project 1ET100300419 of the Program Information Society of the Thematic Program II of the National Research Program of the Czech Republic.

# References

1. Abonyi, J., Feil, B.: Cluster Analysis for Data Mining and System Identification. Birkhäuser Verlag AG, Basel (2007)
2. Adamic, L.A.: The Small World Web. In: Abiteboul, S., Vercoustre, A.-M. (eds.) ECDL 1999. LNCS, vol. 1696, pp. 443–452. Springer, Heidelberg (1999)
3. Agrawal, R., Gehrke, J., Gunopulos, D., Raghavan, P.: ACM SIGMOD Record 27(2), 94–105 (1998)
4. Al-Sultan, K.: Patttern Recognition 28(9), 1443–1451 (1995)
5. Anonymous, 5384 works that have been based on the self-organizing map (SOM) method developed by Kohonen, Part I (2005), http://www.cis.hut.fi/research/som-bibl/references_a-k.ps
6. Anonymous, 5384 works that have been based on the self-organizing map (SOM) method developed by Kohonen, Part II (2005), http://www.cis.hut.fi/research/som-bibl/references_l-z.ps
7. Barabasi, A.L., Albert, R.: Science 286(5439), 509–512 (1999)
8. Barbará, D., Li, Y., Couto, J.: COOLCAT: An entropy-based algorithm for categorical clustering. In: Proceedings of the 11th international conference on information and knowledge management, pp. 582–589. ACM Press, McLean (2002)
9. Ben-Dor, A., Shamir, R., Yakhini, Z.: Journal of Computational Biology 6(3/4), 281–297 (1999)
10. Berkhin, P.: Survey of Clustering Data Mining Techniques. Accrue Software, Inc., San Jose (2002)
11. Berry, M.W., Browne, M.: Understanding Search Engines: Mathematical Modeling and Text Retrieval: Software, Environments, Tools. Society for Industrial & Applied Mathematics (1999)
12. Boley, D., Gini, M., Gross, R., Han, E.-H., Hastings, K., Karypis, G., Kumar, V., Mobasher, B., Moore, J.: Journal of Decision Support Systems 27(3), 329–341 (1999)
13. Broder, A., Kumar, R., Maghoul, R., Raghavan, P., Rajagopalan, P., Stata, R., Tomkins, A., Wiener, J.: Graph structure in the Web. In: The 9th international WWW Conference, Amsterdam (2000)
14. Cai, D., He, X., Li, Z., Ma, W., Wen, J.: Hierarchical Clustering of WWW Image Search Results Using Visual, Textual and Link Information. In: Proceedings of MM 2004. ACM, New York (2004)
15. Canutescu, A.A., Shelenkov, A.A., Dunbrack, R.L.: Protein Science 12, 2001–2014 (2003)
16. Carpenter, G.A., Grossberg, S.: Applied Optics 26, 4919–4930 (1987)
17. Carpenter, G.A., Grossberg, S.: Proceedings of the IEEE First International Conference on Neural Networks, pp. 737–745 (1987)
18. Carpenter, G.A., Grossberg, S.: Neural Networks 3, 129–152 (1990)
19. Carpenter, G.A., Grossberg, S., Markuzon, N., Reynolds, J.H., Rosen, D.B.: IEEE Transactions on Neural Networks 3(5), 698–713 (1992)
20. Carpenter, G.A., Grossberg, S., Reynolds, J.H.: Neural Networks 4, 565–588 (1991)

21. Carpenter, G.A., Grossberg, S., Reynolds, J.H.: IEEE Transactions on Neural Networks 6(6), 1330–1336 (1995)
22. Carpenter, G.A., Grossberg, S., Rosen, D.B.: Neural Networks 4, 493–504 (1991)
23. Carpenter, G.A., Grossberg, S., Rosen, D.B.: Neural Networks 4, 759–771 (1991)
24. Chatuverdi, A.: Journal of Classification 18, 35–55 (2001)
25. Cheng, C., Fu, A.W., Zhang, Y.: Entropy-Based Subspace Clustering for Mining Numerical Data. In: Proceedings of 5th ACM SIGKDD International Conference on Knowledge Discovery and Data Mining, pp. 84–93. ACM Press, San Diego (1999)
26. Chung, F.R.K.: Spectral Graph Theory. In: CBMS Regional Conference Series in Mathematics, vol. 92. American Mathematical Society, Providence (1997)
27. Dill, S., Kumar, R., McCurley, K., Rajagopalan, S., Sivakumar, D., Tomkins, A.: ACM Trans. Internet Techn. 2(3), 205–223 (2002)
28. Donath, W.E., Hoffman, A.J.: IBM Journal of Research and Development 17, 420–425 (1973)
29. Dopazo, J., Carazo, J.M.: Journal of Molecular Evolution 44, 226–233 (1997)
30. Ebel, H., Mielsch, L.I., Bornholdt, S.: Scale-free topology of e-mail networks. Phys. Rev. E 66 (2002)
31. Ester, M., Kriegel, H., Sander, J., Xu, X.: A density-based algorithm for discovering clusters in large spatial databases with noise. In: Proceedings of the 2nd ACM SIGKDD International Conference on Knowledge Discovery and Data Mining, pp. 226–231. AAAI Press, Portland (1996)
32. Estivill-Castro, V.: ACM SIGKDD Explorations Newsletter 4(1), 65–75 (2002)
33. Ferragin, P., Gulli, A.: A personalized search engine based on Web-snippet hierarchical clustering. In: Proceedings of 14th international conference on World Wide Web 2005, Chiba, Japan, pp. 801–810 (2005)
34. Fiedler, M.: Czech. Math. J. 23, 298–305 (1973)
35. Fiedler, M.: Czech. Math. J. 25(100), 619–633 (1975)
36. Fritzke, B.: Neural Network 7, 1141–1160 (1974)
37. Gan, J., Ma, C., Wu, J.: Data Clustering: Theory, Algorithms, and Applications. ASA-SIAM, Philadelphia (2007)
38. Ganti, V., Gehrke, J., Ramakrishnan, R.: CACTUS – Clustering categorical data using summaries. In: Proceedings of the 5th ACM SIGKDD International Conference on Knowledge Discovery and Data Mining, pp. 73–83 (1999)
39. Gibson, D., Kleinberg, J., Raghavan, P.: Inferring Web Communities from Link Topology. In: Proceedings of the 9th ACM Conference on Hypertext and Hypermedia, pp. 225–234 (1998)
40. Goil, S., Nagesh, H., Choudhary, A.: MAFIA: Efficient and Scalable Subspace Clustering for Very Large Data Sets. Technical Report No. CPDC-TR-9906-010, Northwestern University (1999)
41. Golub, G., Van Loan, C.: Matrix computations. Johns Hopkins University Press (1989)
42. Gordon, A.D.: Classification, 2nd edn. Chapman & Hall/CRC, Boca Raton (1999)
43. Grossberg, S.: Biological Cybernetics 23, 187–202 (1976)
44. Grossberg, S.: Adaptive pattern classification and universal recoding: I. Parallel development and coding of neural feature detectors. In: Anderson, R. (ed.), pp. 121–134 (1988) (Reprinted from Biological Cybernetics 23)
45. Guha, S., Rastogi, R., Shim, K.: ACM SIGMOD Record 28(2), 73–84 (1998)
46. Guha, S., Rastogi, R., Shim, K.: Information Systems 25(5), 345–366 (2000)
47. Guimerà, R., Danon, L., Díaz-Guilera, A., Giralt, F., Arenas, A.: Self-similar community structure in a network of human interactions. Physical Review 68 (2003)

48. Hammouda, K.M., Kamel, M.S.: IEEE Transactions on Knowledge and Fata Engineering 18(10), 1279–1296 (2004)
49. Han, J., Kamber, M.: Data Mining: Concepts and Techniques. Morgan Kaufmann Publishers, San Francisco (2001)
50. Hartigan, J.A.: Clustering Algorithms. John Wiley & Sons, New York (1975)
51. Hartuv, E., Shamir, R.: Information Processing Letters 76(4-6), 175–181 (2000)
52. Hassoun, M.H.: Fundamentals of Artificial Neural Networks. MIT Press, Cambridge (1995)
53. Haveliwala, T., Gionis, A., Indyk, P.: Scalable Techniques for Clustering the Web. In: Proceedings of WebDB (2000)
54. He, X., Ding, C.H.Q., Zha, H., Simon, H.D.: Automatic Topic Identification Using Webpage Clustering. In: Proceedings of the 2001 IEEE International Conference on Data Mining (ICDM 2001), pp. 195–203 (2001)
55. He, Z., Xu, X., Deng, S., Dong, B.: K-Histograms: An Efficient Clustering Algorithm for Categorical Dataset. Technical Report No. Tr-2003-08, Harbin Institute of Technology (2003)
56. Henzinger, M.R.: Improved Data Structures for Fully Dynamic Biconnectivity. Report, Digital Equipment Corporation (1997)
57. Henzinger, M.R.: Internet Mathematics 1(1), 115–126 (2003)
58. Her, J.H., Jun, S.H., Choi, J.H., Lee, J.H.: A Bayesian Neural Network Model for Dynamic Web Document Clustering. In: Proceedings of the IEEE Region 10 Conference (TENCON 1999), vol. 2, pp. 1415–1418 (1999)
59. Herrero, J., Valencia, A., Dopazo, J.: Bioinformatics 17, 126–136 (2001)
60. Hinneburg, A., Keim, D.A.: An efficient approach to clustering in large multimedia databases with noise. In: Proceedings of the 4th ACM SIGKDD International Conference on Knowledge Discovery and Data Mining, pp. 58–65. AAAI Press, Menlo Park (1998)
61. Hinneburg, A., Keim, D.A.: Optimal grid-clustering: Towards breaking the curse of dimensionality in high-dimensional clustering. In: Proceedings of the 25th International Conference on Very Large Data Bases, Edinburgh, pp. 506–517 (1999)
62. Hinton, G.E., Anderson, J.A.: Parallel Models of Associative Memory. Hillsidale, NJ (1989)
63. Hopfield, J.J.: Neural Network and Physical Systems with Emergent Collective Computational Abilities. Proceedings of Acad. Sci. USA 79, 2554–2558 (1982)
64. Höppner, F., Klawon, F., Kruse, R., Runkler, T.: Fuzzy Cluster Analysis. Methods for Classification. In: Data Analysis and Image Recognition. Wiley, New York (2000)
65. Huang, X., Lai, W.: Identification of Clusters in the Web Graph Based on Link Topology. In: Proceedings of the Seventh International Database Engineering and Applications Symposium (IDEAS 2003), pp. 123–130 (2003)
66. Huang, Z.: Data Mining and Knowledge Discovery 2, 283–304 (1998)
67. Húsek, D., Pokorný, J., Řezanková, H., Snášel, V.: Data Clustering: From Documents to the Web. In: Vakali, A., Pallis, G. (eds.) Web Data Management Practices: Emerging Techniques and Technologies, pp. 1–33. Idea Group Publishing, Hershey (2007)
68. Jain, A.K., Dubes, R.C.: Algorithms for Clustering Data. Prentice Hall, New Jersey (1988)
69. Jain, A.K., Murty, M.N., Flynn, P.J.: ACM Computing Surveys 31(3), 264–323 (1999)

70. Joshi, A., Jiang, Z.: Retriever: Improving Web Search Engine Results Using Clustering. Idea Group Publishing (2002)
71. Karypis, G., Han, E., Kumar, V.: IEEE Computer 32(8), 68–75 (1999)
72. Kaski, S., Honkela, T., Lagus, K., Kohonen, T.: Neurocomputing 21, 101–117 (1998)
73. Kasuba, T.: Simplified fuzzy ARTMAP. AI Expert, 18–25 (1993)
74. Kaufman, L., Rousseeuw, P.: Finding Groups in Data: An Introduction to Cluster Analysis. Wiley, New York (1990)
75. Kawamura, M., Okada, M., Hirai, Y.: IEEE Transaction on Neural Networks 10(3), 704–713 (1999)
76. Kleinberg, J.M.: JACM 46(5), 604–632 (1999)
77. Kohonen, T.: Self-Organizing Maps. Proceedings of IEEE 78, 1464–1480 (1991)
78. Kohonen, T.: Self-Organizing Maps, Third Extended Edition. Springer, Heidelberg (2001)
79. Kohonen, T., Kaski, S., Lagus, K., Salogärui, J., Honkela, J., Paatero, V., Saarela, A.: IEEE Transaction on Neural Networks 11, 574–585 (2000)
80. Kosko, B.: Appl. Opt. 26(23), 4947–4960 (1987)
81. Kumar, S.R., Raghavan, P., Rajagopalan, S., Tomkins, A.: Trawling the Web for Emerging Cyber Communities. In: Proceedings of the 8th WWW Conference, pp. 403–416 (1999)
82. Langville, A.N., Meyer, C.D.: SIAM Review 47(1), 135–161 (2005)
83. Lian, W., Cheung, D.W.L., Mamoulis, N., Yiu, S.M.: IEEE Transaction on Knowledge Data Engineering 16(1), 82–96 (2004)
84. Massey, L.: Neural Networks 16(5-6), 771–778 (2003)
85. Matula, D.W.: Cluster analysis via graph theoretic techniques. In: Mullin, R.C., Reid, K.B., Roselle, D. (eds.) Proceedings Louisiana Conference on Combinatorics, Graph Theory and Computing, pp. 199–212 (1970)
86. Matula, D.W.: SIAM Journal of Applied Mathematics 22(3), 459–480 (1972)
87. Matula, D.W.: Graph theoretic techniques for cluster analysis algorithms. In: Van Ryzin, J. (ed.) Classification and Clustering, pp. 95–129 (1987)
88. McCulloch, W.S., Pitts, W.: Bulletin of Mathematical Biophysics 5, 115–133 (1943)
89. Mercer, D.P.: Clustering large datasets. Linacre College (2003)
90. Nagesh, H., Goil, S., Choudhary, A.: Adaptive grids for clustering massive data sets. In: Proceedings of the 1st SIAM ICDM, Chicago, IL, vol. 477 (2001)
91. Newman, M.E.J.: SIAM Review 45, 167–256 (2003)
92. Newman, M.E.J., Balthrop, J., Forrest, S., Williamson, M.M.: Science 304, 527–529 (2004)
93. Ng, R.T., Han, J.: Efficient and effective clustering methods for spatial data mining. In: Proceedings of the 20th International Conference on Very Large Data Bases, pp. 144–155 (1994)
94. Pal, S.K., Talwar, V., Mitra, P.: IEEE Transactions on Neural Networks 13(5), 1163–1177 (2002)
95. Pierrakos, D., Paliouras, G., Papatheodorou, C., Karkaletsis, V., Dikaiakos, M.: Construction of Web Community Directories using Document Clustering and Web Usage Mining. In: Berendt, B., Hotho, A., Mladenic, D., Van Someren, M., Spiliopoulou, M., Stumme, G. (eds.) ECML/PKDD 2003, First European Web Mining Forum, Cavtat, Dubrovnik, Croatia (2003)
96. Řezanková, H., Húsek, D., Snášel, V.: Clustering as a Tool for Data Mining. In: Klíma, M. (ed.) Applications of Mathematics and Statistics in Economy, pp. 203–208. Professional Publishing, Praha (2004)

97. Rice, M.D., Siff, M.: Electronic Notes in Theoretical Computer Science 40, 323–346 (2001)

98. Rumelhart, D.E., McClelland, J.L.: Explorations in the Microstructure of Cognition Vols. 1- 2. MIT Press, Cambridge (1988)

99. Salton, G., Buckley, C.: Information Processing and Management 24(5), 513–523 (1988)

100. Sásik, R., Hwa, T., Iranfar, N., Loomis, W.F.: Percolation Clustering: A Novel Approach to the Clustering of Gene Expression Patterns. Dictyostelium Development PSB Proceedings 6, 335–347 (2001)

101. Schenker, A., Kande, A., Bunke, H., Last, M.: Graph-Theoretic Techniques for Web Content Mining. World Scientific, Singapore (2005)

102. Sharan, R., Shamir, R.: A clustering algorithm for gene expression analysis. In: Miyano, S., Shamir, R., Takagi, T. (eds.) Currents in Computational Molecular Biology, pp. 6–7. Universal Academy Press (2000)

103. Shi, J., Malik, J.: IEEE Transactions on Pattern Analysis and Machine Intelligence 22(8), 888–905 (2000)

104. Tamayo, P., Slonim, D., Mesirov, J., Zhu, Q., Kitareewan, S., Dmitrovsky, E., Lander, E.S., Golub, T.R.: Proceedings of National Acad. Sci. USA 96, 2907–2912 (1999)

105. Tomida, S., Hanai, T., Honda, H., Kobayashi, T.: Genome Informatics 12, 245–246 (2001)

106. Vakali, A., Pokorný, J., Dalamagas, T.: An Overview of Web Clustering Practices. In: Lindner, W., Mesiti, M., Türker, C., Tzitzikas, Y., Vakali, A.I. (eds.) EDBT 2004. LNCS, vol. 3268, pp. 597–606. Springer, Heidelberg (2004)

107. Wang, H.C., Dopazo, J., Carazo, J.M.: Bioinformatics 14(4), 376–377 (1998)

108. Wang, Y., Kitsuregawa, M.: Evaluating Contents-Link Web Page Clustering for Web Search Results. In: CIKM 2002, pp. 499–506. ACM McLean, Virginia (2002)

109. Wang, W., Yang, J., Muntz, R.: STING: A statistical information grid approach to spatial data mining. In: Proceedings of the 23rd International Conference on Very Large Data Bases, pp. 186–195. Morgan Kaufmann Publishers, Athens (1997)

110. White, S., Smyth, P.: A Spectral Clustering Approach to Finding Communities in Graph. SDM (2005)

111. Wu, C., Zhao, S., Chen, H.L., Lo, C.J., McLarty, J.: CABIOS 12(2), 109–118 (1996)

112. Wu, C.H.: Gene Classification Artificial Neural System. In: Doolittle, R.F. (ed.) Methods in Enzymology: Computer Methods for Macromolecular Sequence Analysis. Academic Press, New York (1995)

113. Yao, Y., Chen, L., Chen, Y.Q.: Neural Processing Letters 14, 169–177 (2001)

114. Zamir, O., Etzioni, O.: Web Document Clustering: A Feasibility Demonstration. In: Proceedings of the 21st International ACM SIGIR Conference on Research and Development in Information Retrieval, pp. 46–54 (1998)

115. Zamir, O., Etzioni, O.: The International Journal of Computer and Telecommunications Networking Archive 31(11-16), 1361–1374 (1999)

116. Zamir, O., Etzioni, O., Madanim, O., Karp, R.M.: Fast and Intuitive Clustering of Web Documents. In: Proceedings of the 3rd International Conference on Knowledge Discovery and Data Mining, pp. 287–290 (1997)

117. Zhang, T., Ramakrishnan, R., Livny, M.: ACM SIGMOD Record 25(2), 103–114 (1996)

# Efficient Construction of Image Feature Extraction Programs by Using Linear Genetic Programming with Fitness Retrieval and Intermediate-Result Caching

Ukrit Watchareeruetai, Tetsuya Matsumoto, Yoshinori Takeuchi, Hiroaki Kudo, and Noboru Ohnishi

This chapter describes a bio-inspired approach for automatic construction of feature extraction programs (FEPs) for a given object recognition problem. The goal of the automatic construction of FEPs is to cope with the difficulties in FEP design. Linear genetic programming (LGP) [4]—a variation of evolutionary algorithms—is adopted. A population of FEPs is constructed from a set of basic image processing operations—which are used as primitive operators (POs), and their performances are optimized in the evolutionary process. Here we describe two techniques that improve the efficiency of the LGP-based program construction. One is to use fitness retrieval—to avoid wasteful evaluations of the programs discovered before. The other one is to use intermediate-result caching—to avoid evaluation of the program-parts which were recently executed. The experimental results show that much computation time of the LGP-based FEP construction can be reduced by using these two techniques.

## 1  Introduction

Feature extraction is a significant process in object recognition systems. Recognition accuracies of the systems greatly depend on the quality of extracted features. Usually, FEP is designed by human experts, who have to consider what features would effectively solve the problem at hand. Human

Ukrit Watchareeruetai, Tetsuya Matsumoto, Yoshinori Takeuchi,
Hiroaki Kudo, and Noboru Ohnishi
Department of Media Science, Graduate School of Information Science,
Nagoya University
Furo-cho, Chikusa-ku, Nagoya 464-8603 Japan
e-mail: ukrit@ieee.org,{takeuchi,matumoto,kudo,
ohnishi}@is.nagoya-u.ac.jp

A. Abraham et al. (Eds.): Foundations of Comput. Intel. Vol. 4, SCI 204, pp. 355–375.
springerlink.com                                    © Springer-Verlag Berlin Heidelberg 2009

experts then create an image processing program to extract such potential features from images, and test the created program. If a satisfactory performance is not obtained, they have to choose the other features and revise the program. It is a difficult and time-consuming task. Typically, human experts design FEPs based upon their knowledge and experience. However, in fact, there are a very large number of features (may be infinite) that can be extracted from images. This implies that only a portion of feature space can be explored, and unconventional but potential features may be ignored.

Evolutionary computation (EC) [5]—a powerful optimization/search paradigm inspired by natural evolution—are exploited to cope with these difficulties. In this approach, the program design problem is converted to an optimization problem—to maximize performance of FEP. Various EC techniques have been adopted for automatic construction of FEPs (or even object recognition programs) for a given problem, e.g., genetic algorithm [10], tree-based GPs [2, 13, 18, 19, 29], graph-based GP [20, 22, 23], and LGP [11, 12] based approaches.

In this work, we focus on an approach based on LGP which is similar to the work of Krawiec and Bhanu [11, 12]. In this case, FEPs are constructed from a set of POs, which are basic image processing operations. These operations are commonly used in many image processing programs, and independent of problem domain. Users only input training images and ground truths (reference images) into the evolutionary construction systems, and define an objective function. Then the system randomly generates a population of FEPs and attempts to maximize program performances by using LGP technique. These mean that the FEPs are automatically constructed without domain-specific knowledge.

However, a major problem of this approach is that evolutionary process needs very long computation time because a lot of image processing operation sequences are executed and evaluated. Consequently, we need a method to improve the efficiency of evolutionary construction system. In this work, we describe two efficiency improvement techniques, i.e., fitness retrieval and intermediate-result caching, for avoiding execution of redundant programs and program-parts. The experimental results show that much of computation time can be reduced without any decrease of search performance.

The rest of this chapter is organized as follows. Section 2 describes the details of LGP-based construction of FEPs: program representation, decoding, genetic operators and fitness evaluation. Section 3 explains how fitness retrieval and intermediate-result caching can improve the efficiency of the LGP-based construction of FEPs. In section 4, experimental results show that the use of fitness retrieval and intermediate-result caching can supremely improve the efficiency of the LGP-based construction of FEPs. Section 5 concludes this chapter.

# 2 Construction of Feature Extraction Programs by Using LGP

## 2.1 Feature Extraction Program Representation and Decoding

In LGP-based representation, a FEP is represented as a sequence of POs which are basic image processing operations (e.g., filtering, edge detection, and thresholding). The sequence may be fixed- or variable-lengths. In execution process, each operation in the sequence is sequentially interpreted and executed one-by-one, until all operations in the sequence are finished. Figure 1 shows an example of execution process of an LGP program. The executions are based on a set of shared registers. In particular, an operation fetches inputs from registers, processes them, and stores its output into a register. After that, we go to the next operation and repeat this process until all operations are finished. Three types of shared registers, i.e., image, numerical, and constant registers, are used. Image registers $(R_I)$ store input images and processed images, whereas numerical registers $(R_N)$ store initialized real-values and numerical processed-results. Constant registers $(R_C)$ also store real-value (pre-defined) but they cannot be re-written by any operation, whereas the image and numerical registers can be re-written. Each operation is coded by four components, i.e., one opcode (operation code) and three arguments. The opcode describes what operation will be executed, whereas the three arguments define related input and output registers.

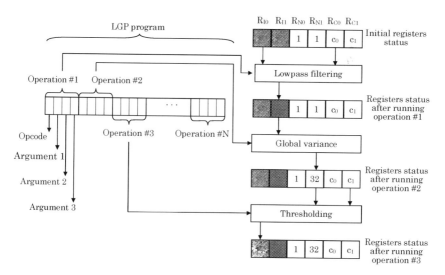

**Fig. 1** Example of LGP program decoding

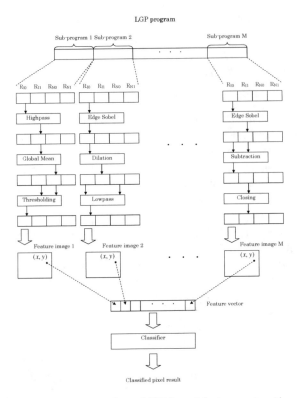

**Fig. 2** Structure of our LGP-based feature extraction program

The LGP used in this work is slightly different from the works of [11, 12]. As shown in Fig. 2, an LGP program consists of multiple sub-programs, which may have different lengths; each is executed independently of the others and produces one feature. In this work, when a sub-program execution is finished, the content stored in image register zero $(R_{I0})$ is used as program output, i.e., a feature image. Each pixel value of the feature images is an element of feature vector corresponding to that pixel. A classifier is adopted to decide to which class that pixel belongs, based upon the feature vector. In this work, we use Bayesian classifier, which is known as an optimal classifier in the sense that it minimizes probability of classification error [24]. Probability density functions of patterns, which are needed in the decision of Bayesian classifier, are estimated by using histogram approximation method [24]. The reason why we use sub-program representation is to reduce the complexity of problems; instead of extracting all features simultaneously by using one program, we independently extract one feature by using one sub-program.

Moreover, before execution of each sub-program, input image is dumped into all image registers, and all numerical registers are reset to one (instead of initializing registers by the output of some pre-defined POs). This is to minimize bias from human decision. The set of POs used here contains 57

**Table 1** List of primitive operators used in this work

| One-input operations | Two-input operations |
|---|---|
| **image → image** | **image, image → image** |
| highpass | image addition |
| Sobel edge detection | image subtraction |
| Laplacian edge detection | image multiplication |
| image negative | image division |
| mean thresholding | **image, real value → image** |
| entropy thresholding | lowpass filter |
| histogram equalization | median filter |
| image scaling | morphological erosion |
| image square root | morphological opening |
| image absolute | morphological closing |
| **image → real value** | local histogram equalization |
| global mean | thresholding |
| global variance | local variance |
| global STD | local skewness |
| global skewness | local kurtosis |
| global kurtosis | local maximum (max filter) |
| global maximum | local minimum (min filter) |
| global minimum | local mode |
| global median | local range |
| global mode | local entropy |
| global range | **branch operations** |
| global entropy | |

basic image processing operations, and no operations working with ROI mask are used (see Table 1). Most of POs in the table can be found in [8].

## 2.2 Fitness Evaluation

In evaluation process, leave-one-out cross-validation [24] is adopted. In particular, we do validation $T$ times for each individual, where $T$ is the number of images in dataset. For each time of validation, we use one image as validation image and use the remaining images for training the classifier, and then we calculate segmentation accuracy

$$Acc = \frac{1}{H \cdot W} \sum_{\substack{0 \leq i < H \\ 0 \leq j < W}} (1 - \frac{|O(i,j) - GT(i,j)|}{255}),$$

where $O(i,j)$ is the classification result of the classifier, whereas $GT(i,j)$ is its ground truth. $H$ and $W$ are image height and width, respectively. Note that, in segmented image, pixel values 0 and 255 mean to object and background pixels, respectively. $Acc$ equals to the number of pixels that are correctly segmented. The average value of $Acc$ over all validations is used as the fitness value of the individual.

## 2.3  Selection and Recombination

### Selection

After all individuals in the current population are evaluated, selection process is done to select some individuals into a mating pool. Here the tournament selection [17] is adopted. In particular, a number of individuals (tournament size) will be randomly selected and the individual with the best fitness (among the selected individual) wins the tournament and is copied into the mating pool. The individuals with better fitness will have higher chances to be selected and have more copies in the mating pool. The individuals in the mating pool will be randomly selected (with equal probability) and be applied by crossover and mutation operators to generate offspring.

### Crossover

In this work, we use two types of crossovers. The first is sub-program-level crossover which randomly selects a sub-program and exchanges the selected sub-program between two parents. This allows parent individuals to exchange their sub-programs without any destruction of potential building-blocks. In other words, a feature image is exchanged between two parents. An example of this type of crossover is shown in Fig. 3.

The second type of crossover is operation-level crossover which randomly exchanges operations of a sub-program between two parents. In particular, a sub-program is randomly selected, then only the operations in the selected sub-program will be randomly exchanged. Here we adopted the parameterized uniform crossover [21] which randomly exchanges the corresponding operation with a pre-defined probability (0.2 here). However, the parameterized uniform crossover is proposed for a fixed-length chromosome. Therefore we have to modify it by using a shift value that is randomly selected from a range of zero to $|L_i^{p1} - L_i^{p2}|$, where $L_i^{p1}$ and $L_i^{p2}$ are the lengths (number of operations) of the selected sub-programs of the first and second parents, respectively. The sub-program with lower length is shifted by this shift value as shown in Fig. 4. Then only the overlapped operations will be exchanged based on the parameterized uniform crossover.

The probabilities that sub-program-level or operation-level crossovers will be adopted are $p_s$ (0.1 here) and $1 - p_s$, respectively.

### Mutation

Mutation operator used in this work is a macro-level mutation, which randomly inserts, deletes, or modifies an operation. The probabilities that

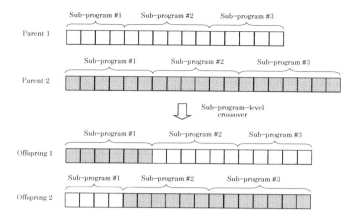

**Fig. 3** Sub-program-level crossover

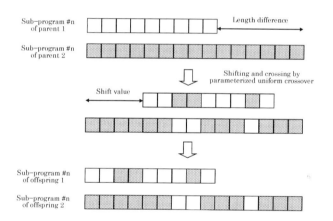

**Fig. 4** Operation-level crossover (parameterized uniform crossover)

insertion, deletion or modification will be selected are equal. Mutation can introduce new genetics materials (operation contents) to the population, whereas the crossover operators just change the structure of the existing genetic materials but cannot produce the new ones. Also the sub-program length can be changed by the mutation but not by the crossover operators used in this work.

In this work, one individual of the next population is generated by copying the best individual of the current population (elitism), $N_c$ individuals by crossover, $N_m = N - N_c - 1$ individuals by mutation, where $N$ is population size. For all experiments in this work, $N = 50$, $N_c = 24$, and $N_m = 25$.

# 3  Improving Efficiency of Evolutionary Program Construction

In [1], Ando and Nagao have attempted to reduce computation time of their evolutionary construction of image processing programs by using a special hardware, i.e., a graphic processing unit (GPU). In this work, we focus on algorithmic approaches of efficiency improvement of the evolutionary construction of FEPs. There are many researches attempt to improve the efficiency of GPs [6, 7, 9, 16, 27, 28], which may be applied with the evolutionary construction of FEPs. However, the two techniques described in this section, i.e., fitness retrieval and intermediate-result caching, concern with the redundancies in the representation and can improve the efficiency of the evolutionary construction without any decreasing of search performance.

## 3.1  *Redundancies and Canonical Transformation*

### Redundancies in LGP

LGP-based representation contains various types of redundancies, e.g., the existence of introns [14] (the operations that have no effect on program output), the difference in operation order, the difference in register usage. The existence of redundancies causes multiple genotypes to represent the (phenotypically) equivalent programs—these programs always give the same output with each other. Figure 5 shows examples of these types of redundancies. All programs (in fact, each is just a sub-program in our sub-structure representation) would have different representations but they represent the equivalent programs. The other types of redundancies are the use of modulo operation in protection mechanism and the difference in sub-program order [26]. Due to these various types of redundancies, it is difficult to verify in the original LGP-based representation whether or not two genotypes represent the equivalent program.

### Canonical Transformation

In [26], we have proposed a method that transforms the original LGP representations which contains many redundancies into canonical forms in which all redundancies are removed. This transformation is called canonical transformation. In the canonical form, it is easy to verify whether two genotypes represent the equivalent program because these redundancies are removed. The canonical transformation consists of five steps. Each step is to remove a type of redundancies as mentioned before. They are briefly explained as follows (the more details are described in [26]):

1. Intron removal: the algorithm proposed in [3] is used to identify introns. These introns are removed from the representation. Note that the remaining

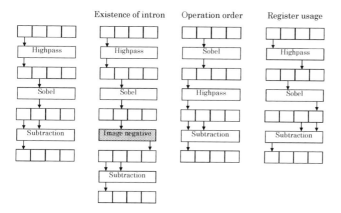

**Fig. 5** Examples of the programs that have different genotypes (representation) but are phenotypically equivalent

operations are called effective operations because they have some effect on the program output.

2. Conversion to modulo form: we replace the original opcode $O$ by $O$ $modulo$ $N_{PO}$ ($N_{PO}$ is the number of POs), and replace the original argument $ARG_i$ by $ARG_i$ $modulo$ $N_{reg}$ ($N_{reg}$ is the number of registers of the corresponding type).

3. Operation reordering: we search backward from the last effective operation by using depth-first-search technique. The order obtained by this backward search is used as the new operation order.

4. Register reassignment: we start register reassignment from the first effective operation. Input registers are assigned based on data dependencies. Output registers are assigned based on the following rules: choose the register with the lowest index among the registers being not write-protected in that time.

5. Sub-program reordering: firstly, sub-programs in an individual are reordered based on their length (the number of effective operations). If there are two or more sub-programs that have the same length, we reorder them based on their genome contents byte-by-byte.

## Phenotypically-Equivalent Individual Identification

In canonical form, operation order and the values of opcode and the three arguments of the individuals that represent the equivalent program are the same. In other words, the canonical forms of the two genotypes that represent the equivalent program will have the same length (the number of operations) and their contents are exactly the same. Consequently, whether two genotypes represent the equivalent program may be verified by comparing the length,

```
 1: population initialization;
 2: call fitness_evaluation();
 3: do{
 4:     selection;
 5:     crossover;
 6:     mutation;
 7:     call fitness_evaluation();
 8: }while(termination criteria are not satisfied);
 9:
10: void fitness_evaluation(){
11:     for all individuals{
12:         transform individual i into its canonical form;
13:         compare it with the canonical forms in memory;
14:         if(individual i is not redundant){
15:             evaluate individual i;
16:             store its canonical form and fitness;
17:         }
18:         else{
19:             retrieve fitness of the matched canonical form;
20:             assign the fitness to individual i;
21:         }
22:     }
23: }
```

**Fig. 6** Pseudo code of the LGP with fitness retrieval

and comparing further their contents byte-by-byte if their lengths are the same. This process is simple and very fast.

## 3.2  Fitness Retrieval

As we described in section 3.1, there are many redundancies in LGP-based representation. Due to these redundancies, there will be a lot of individuals that represent the equivalent programs occurred in the evolutionary search. The use of genetic operators, i.e., crossover and mutation, may alter the representation (genotype) of individuals but may not affect their canonical forms. For example, mutation operator may delete an intron operation from an individual but its canonical form is still unchanged.

We may allocate a space of memory for storing canonical forms and fitness values of the individuals we have discovered in the evolutionary search. These fitness values stored in the memory can be retrieved later, and computation time can be reduced. This is similar to the work of Niehaus et al. [15] who proposed a canonical form for graph-based GP. In particular, before execution of an individual, we find its canonical form and verify whether its canonical form is matched with one of the forms stored in the memory. If it is matched, it means that this individual represent something that we have discovered before in the evolutionary search (we call it redundant individuals.) We need not to execute this individual because its fitness has been known already. We just retrieve its fitness value from the memory instead. However, if it is not matched with any canonical form in the memory, that individual represents something that is new and has not been found yet. In this case, we have to evaluate it and store its canonical form and fitness value into the memory. We call this process fitness retrieval. By using the fitness retrieval, wasteful execution of the individuals representing the program discovered before. Note that the use of fitness retrieval does not alter the original representation of an individual—we just find its canonical forms and compare it with those in the memory, and left their own representation in the population without any change. Consequently, this approach will not decrease search performance of the evolutionary search. Figure 6 describes the pseudo code of the evolutionary search with the fitness retrieval.

## 3.3  *Intermediate-Result Caching*

Another way to reduce computation time of GPs is to save intermediate results, together with parts of programs in canonical forms, and retrieve them later if the same program-parts are found (Fig. 7). We call it intermediate-result caching. In our LGP-based feature synthesis approach, register content, i.e., images and numeric values, will be saved (the content of the constant registers need not to be saved.) Although the idea of the intermediate-result caching is similar to the fitness retrieval, saving intermediate-results of every program-parts discovered in evolutionary search may not practical because of memory limitation. In the case of fitness retrieval, we store only canonical forms (a few to dozens bytes) and their fitness values (4 bytes); consequently, memory limitation is not major concerned. However, in the case of intermediate-result caching, we store canonical forms and register contents for all training images; it needs large memory space. For example, if we have $M$ training images, and LGP are executed on $N_I$ image registers and $N_N$ numerical registers, we have to store $M \times N_I$ images and $M \times N_N$ floating-point values for one cache index. And if $C$ is cache size, we needs to allocate memory to store $C \times M \times N_I$ images and $C \times M \times N_N$ floating-point values. Now, memory limitation becomes big concern.

### Caching Strategy

Due to memory limitation, it is difficult to save all intermediate results we found so far. In our caching system, when memory (cache) space is full, the oldest intermediate-result will be replaced with the latest one. The mechanism of intermediate-result caching used in this work is described as follows:

1. Transform considered individual $i$ into its canonical form $c$. Then compare $c$ with the canonical forms $c_j$ stored in memory, where $j \in \{1, 2, 3, ...N_c\}$ and $N_c$ is the current number of canonical forms stored in memory.
2. If $c$ matches with $c_j$ in memory, set the fitness of $i$ equal to the fitness of $c_j$, stop the algorithm.
3. Defining $n$ = length of $c$ -1.
4. Defining $s_n[c]$ as a string of length $n$ whose string content is created by copying $c$ from the operation #1 to operation #$n$.

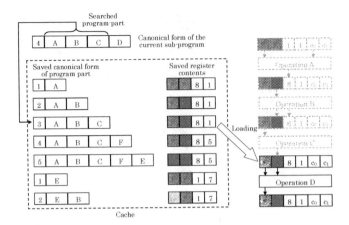

**Fig. 7** Example of caching mechanism: the operations A, B, and C need not to be executed, only the operation D is executed after loading register content from the cache

5. Compare $s_n[c]$ with $s_k$ stored in cache ($k \in \{1, 2, 3, ..., N_s\}$, $N_s$ is cache size)
6. If $s_n[c]$ matches with $s_k$ in cache, load register contents stored in cache number $k$, increase value of $n$ by one, and go to step 10.
7. If $n$ is still greater than zero, decrease value of $n$ by one, and go back to step 5. Otherwise, $n = 1$, initialize register contents and go to the next step.
8. Execute the operation #$n$ of the canonical form $c$, and save the current register contents and string $s_n[c]$ into cache. Increase value of $n$ by one.
9. If $n$ is greater than the length of $c_p$, evaluate the result, save the canonical form $c$ and its fitness into memory, and stop the algorithm. Otherwise, go back to step 8.

## Fitness Evaluation Reordering

Cache size would affect the performance of the caching method. In particular, if cache size is too small, we may rarely find program-part in the cache matched (hit) with the current program-part, and hit-rate becomes very low. Therefore we would have no or little advantage from using caching method. In contrast, as we discussed before, larger cache size may not practical because of memory limitation.

We describe a way to improve cache performance by reordering fitness evaluation based on indexing of canonical form. Before fitness evaluation process, each offspring is assigned one or two numbers, i.e., canonical form indexes of its parents. Therefore an offspring generated by crossover will have two canonical form indexes of its parents, whereas an offspring generated by mutation will have only one index. We then find a histogram of these canonical form indexes from all offspring. The canonical form index that has higher frequency will be assigned a higher rank. Each offspring will be assigned minimum-rank and summation-rank based on canonical form indexes of its parents. We reorder offspring in the population based on these two values. The offspring that has higher minimum-rank or has the same minimum-rank but the higher summation-rank will be assigned lower order. By doing that, the offspring in population are reordered based on canonical form of parents. The offspring that generated by the dominated parents will be executed first. Therefore the probability that the recently executed program-parts stored in cache will be matched with the current program-part would become higher because similar programs are continuously executed.

## 4   Experiments and Results

In this section, two experiments were conducted to measure how much the efficiency of the LGP-based construction of FEPs can be improved by the two techniques. The first experiment is concerned with the fitness retrieval; the computation time needed by the LGP with and without the fitness retrieval was measured. In the second experiment, we measure how much computation time can be reduced further by the addition of the intermediate-result caching. Also the effect of using the fitness evaluation reordering described in section 3.3 was investigated.

### 4.1   Computation Time Reduction by Fitness Retrieval

**Experiment Set-up**

We have conducted an experiment to measure redundancies in LGP based evolutionary search and to measure how much computation time can be reduced by using the fitness retrieval. The evolutionary process with four timers as shown in Fig. 8 was experimented. It was modified from the pseudo code described in Fig. 6. The TIMER1 was used to measure computation time of genetic processes—selection, recombination, and population initialization. The TIMER2 measures the computation time of the canonical transformation and redundant individual identification. The TIMER3 and TIMER4 measure the needed evaluation time and wasteful evaluation time, respectively. In the experiment, we use the LGP representation with sub-program structure (Fig. 2). An LGP program consists of two sub-programs. Each sub-program is variable length; its length is varied between 0 to 20 instructions. The population size is 50. In reproduction process, 24 offspring are generated by crossover operation, 25 offspring by mutation, and the remaining one is the copying of the elitist individual. The maximum number of generation is 100. The evolutionary process was run 30 times independently with different seeds. A Core 2 Duo 1.8 GHz PC with 1024 MB memory was used in the experiment.

```
 1: TIMER1 start;        %computation time of genetic process%
 2: population initialization;
 3: TIMER1 stop;
 4: call fitness_evaluation();
 5: do{
 6:     TIMER1 start;
 7:     selection;
 8:     crossover;
 9:     mutation;
10:     TIMER1 stop;
11:     call fitness_evaluation();
12: }while(termination criteria are not satisfied);
13:
14: void fitness_evaluation(){
15:     for all individuals{
16:         TIMER2 start;        %canonical transformation time%
17:         transform individual i into its canonical form;
18:         compare it with the canonical forms in memory;
19:         TIMER2 stop;
20:         if(individual i is not redundant){
21:             TIMER3 start;        %evaluation time as needed%
22:             evaluate individual i;
23:             TIMER3 stop;
24:             TIMER2 start;
25:             store its canonical form and fitness;
26:             TIMER2 stop;
27:         }
28:         else{
29:             TIMER4 start;        %redundancies evaluation time%
30:             evaluate individual i;
31:             TIMER4 stop;
32:         }
33:     }
34: }
```

**Fig. 8** Pseudo code of the evolutionary search used in the experiment

The test object recognition problem we used is a lawn weed detection problem [25]. The goal of this problem is to detect the area of weeds from lawn fields so that precision spraying, i.e., spraying herbicide only onto the area of detected weeds (instead of spraying into the entire area), can be accomplished. The database we used contains ten lawn weed images. The images were captured under real-light condition by using a top-view camera. The distance from the camera to lawn fields is around 39 cm. Image size is $160 \times 120$ pixels. An image covers lawn area of size $274 \times 205$ mm. Example of lawn weed image and the corresponding ground truth are shown in Fig. 9.

**Results and Discussion**

Table 2 shows the measured computation. The effectiveness of the fitness retrieval technique depends on three things; the degree of redundancies in the evolutionary search, the speed of the canonical transformation, and the complexity of fitness evaluation. Firstly, we found that there are a lot of redundancies occurred in the evolutionary searches; around 76 to 81 %. When the tournament size is increased, redundancies increase too. Figure 10 shows the relation of redundancies with generation number. If the tournament size

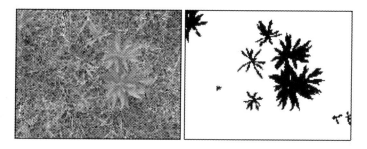

**Fig. 9** An example of lawn weed image and the corresponding ground truth

**Table 2** Performance and measured computation time (sec.) of the LGP with fitness retrieval averaged over 30 independent runs

| Tournament size | two | five | ten |
|---|---|---|---|
| Best-so-far fitness | 0.9446 | 0.9442 | 0.9434 |
| Percentage of redundant individuals | 75.99% | 80.15% | 81.60% |
| Genetic operations | 0.131 | 0.135 | 0.108 |
| Canonical transformation & redundant individual identification | 0.287 | 0.215 | 0.235 |
| Necessary evaluation (non-redundant) | 1399.032 | 1219.620 | 1046.911 |
| Wasteful evaluation (redundant) | 4600.058 | 5009.844 | 4670.332 |
| Conventional LGP | 6000.222 | 6229.600 | 5717.352 |
| LGP with fitness retrieval | 1399.451 | 1219.971 | 1047.255 |
| Computation time reduction | 76.67% | 80.41% | 81.68% |
| Number of redundant individuals | 3799.69 | 4007.30 | 4079.96 |

is too high (five or ten), the number of redundant individuals suddenly increase in the beginning of the evolutionary searches. Within five generations, redundant individuals occupy most space of the population (around 80%) and become steady for all evolutionary search. For the case that tournament size is two, it takes around 30 generations to be steady. The use of fitness retrieval for computation time reduction become fully effective in this period; at each generation, around 80% of wasteful computation time can be ignored.

We found that the computation time needed by the canonical transformation (including the identification of phenotypically-equivalent individuals) is very short; it is slightly longer than the computation time for crossover and mutation processes. Compared with the computation time needed by wasteful evaluation of redundant individuals, it is reasonable to retrieve fitness from memory and avoid this wasteful evaluation. The total computation times of the LGP with and without fitness retrieval technique are also shown in the table. By using this technique, we could save up to 80% of computation time needed by the conventional LGP (without fitness retrieval).

**Fig. 10** Comparison of
redundancies of the three
evolutionary search

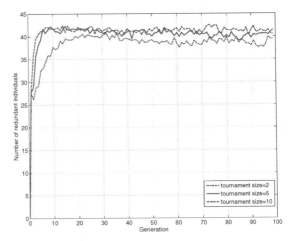

## 4.2   Computation Time Reduction by Intermediate-Result Caching

**Experiment Set-up**

In this experiment, we measure how much computation time can be reduced further by the use of intermediate-result caching, compared with the LGP with only fitness retrieval. Computation time of caching process (to identify program-parts stored in the cache and to store register contents into the cache), necessary execution of program-parts (that did not found in the cache), and wasteful execution of program-parts (that found in the cache) are considered. The tournament size is two. The other parameters are the same as the previous experiment.

**Results and Discussion**

Table 3 shows the measured computation time when the cache size is varied (20, 50 and 100). From the result, the caching process needs much computation time, compared with the canonical transformation. However, the computation time for the caching process does not relate with the change of cache size. As mentioned before, this measured computation time includes the time needed by redundant program-part identification and the time needed for updating cache content. The computation time of the former one would highly depend on the change of cache size. If larger cache size is used, larger number of comparisons of program-parts is needed. However, its computation time is very short (similarly to the computation time needed by the identification of phenotypically-equivalent individuals shown in the previous experiment). In fact, it is the computation time of the latter one that dominates the computation time of the caching process because of the copying register contents (images and numerical numbers) into the cache. The computation time

**Table 3** Computation time reduction by intermediate-result caching averaged over 30 independent runs (sec.)

| Cache size | 20 | 50 | 100 |
|---|---|---|---|
| Genetic operations | 0.026 | 0.034 | 0.041 |
| Canonical transformation | | | |
| & redundant individual identification | 0.212 | 0.196 | 0.199 |
| Caching process | 23.746 | 22.823 | 23.803 |
| Necessary execution of program-parts | 1136.822 | 1085.758 | 1036.881 |
| Wasteful execution of program-parts | 183.772 | 333.857 | 391.819 |
| LGP with fitness retrieval | 1320.832 | 1419.846 | 1428.942 |
| LGP with fitness retrieval and | | | |
| intermediate-result caching | 1160.806 | 1108.812 | 1060.926 |
| Computation time reduction | 12.12% | 21.91% | 25.75% |

of the latter one does not depend on the change in the cache size but depends on the number of cache updating, which would be the same for any cache size.

Although the caching process is not so fast, it needs less computation time than the evaluations of redundant program-parts around eight to 17 times. Consequently, it is reasonable to spend the computation time for the caching process to reduce these wasteful evaluations if memory space is enough to enable the intermediate-result caching. According to the result, around 12 to 25% of the computation time (of the LGP with only fitness retrieval) can be reduced by the use of intermediate-result caching. It implies that there are still a lot of redundancies (redundant program-parts) even after the redundant individuals are identified and avoided.

Table 4 shows the performance of the caching process. Hit rate—a factor that strongly relates with the time reduction rate—is the ratio of the

**Table 4** Performance of cache search averaged over 30 independent runs

| Cache size | 20 | | 50 | | 100 | |
|---|---|---|---|---|---|---|
| | Mean | STD | Mean | STD | Mean | STD |
| Number of cache searches | 7022.40 | 2942.70 | 6288.20 | 3145.04 | 6374.40 | 3874.27 |
| Number of cache hits | 1096.30 | 315.24 | 1418.33 | 466.94 | 1694.37 | 600.25 |
| Hit rate (%) | 17.01 | 4.75 | 24.89 | 6.27 | 29.46 | 5.67 |
| Maximum length of hit | | | | | | |
| program-parts | 9.07 | 1.80 | 9.70 | 2.02 | 10.37 | 2.59 |
| Summation of program-part | | | | | | |
| lengths | 2809.00 | 1134.00 | 3824.00 | 1943.94 | 5021.37 | 2898.96 |
| Lengths per hit | 2.50 | 0.59 | 2.57 | 0.63 | 2.77 | 0.69 |

**Table 5** Computation time reduction by intermediate-result caching without fitness evaluation reordering averaged over 30 independent runs (sec.)

| Cache size | 20 | 50 | 100 |
|---|---|---|---|
| Genetic operations | 0.018 | 0.026 | 0.034 |
| Canonical transformation | | | |
| & redundant individual identification | 0.185 | 0.163 | 0.181 |
| Caching process | 22.334 | 20.073 | 27.471 |
| Necessary execution of program-parts | 1247.004 | 1201.229 | 1373.185 |
| Wasteful execution of program-parts | 82.416 | 98.709 | 207.978 |
| LGP with fitness retrieval | 1329.622 | 1300.128 | 1581.380 |
| LGP with fitness retrieval and | | | |
| intermediate-result caching | 1269.541 | 1222.149 | 1400.873 |
| Computation time reduction | 4.52% | 5.998% | 11.415% |

**Table 6** Performance of cache search without fitness evaluation reordering averaged over 30 independent runs

| Cache size | 20 | | 50 | | 100 | |
|---|---|---|---|---|---|---|
| | Mean | STD | Mean | STD | Mean | STD |
| Number of cache searches | 7010.50 | 3624.47 | 6451.83 | 3884.51 | 8278.83 | 3529.53 |
| Number of cache hits | 612.23 | 266.35 | 610.33 | 243.41 | 1014.53 | 245.45 |
| Hit rate (%) | 9.25 | 2.19 | 10.86 | 3.69 | 13.71 | 4.26 |
| Maximum length of hit | | | | | | |
| program-parts | 8.30 | 2.35 | 7.50 | 2.00 | 9.00 | 2.30 |
| Summation of program-part | | | | | | |
| lengths | 1359.83 | 819.47 | 1367.77 | 741.50 | 2522.83 | 1007.20 |
| Lengths per hit | 2.14 | 0.47 | 2.15 | 0.48 | 2.44 | 0.52 |

number of times that the current program-part is found in the cache (cache hits) and the number of cache search. From the result, the hit rate is varied from 17 to 29% depending on the cache size. Usually, the shorter program-parts would more often be matched because the longer program-parts have more probabilities to be disrupted by crossover and mutation operators. The average value of the length of hit program-part per the number of hits is in range of 2.50 to 2.77. It is quite small but it is more enough to accomplish the goal of intermediate-result caching—to reduce computation time. The program-parts found in the cache sometimes have long length (the number of operations). The longest one we found from all 30 trials has length of 17, and the average of longest length of hit program-parts is around nine to ten. Totally, around 2800 to 5000 wasteful operations can be avoided.

## Effect of The Fitness Evaluation Reordering

Also we compared the effect of using the fitness evaluation reordering (described in section 3.3). Tables 5 and 6 show the performance of the intermediate-result caching techniques without the fitness evaluation reordering. Compared with the results shown in Tables 3 and 4, we found that the performance of intermediate-result caching heavily decreased. Even in the case that the cache size of 100 is provided, the performance can be comparable with only that of the intermediate-result caching with the fitness evaluation reordering with cache size of 20; around 11–12% of computation time can be reduced. In the case of smaller cache size (20 and 50), the computation time reduction rate ranges only from 4.5 to 6 %. The hit rate, length per hit, and the summation of hit program-part length are obviously dropped compared with the case that the fitness evaluation reordering is used. These results suggest that the reordering of evaluation can significantly improve the performance of the intermediate-result caching. The computation time reduction can be accomplished in a certain level even when the small cache size is provided.

## 5 Conclusion

This chapter has described the LGP-based system for constructing FEPs for a given object recognition problem. Then we have explained two techniques for improving the efficiency of the LGP-based construction of FEPs. The first technique is the fitness retrieval, which is to avoid execution of the individuals that represent the programs discovered before. The second technique is the intermediate-result caching, which avoids execution of program-parts that are recently executed. By using the fitness retrieval, we can avoid wasteful evaluation of redundant individuals, and reduce computation time about 80% compared with conventional LGP. By using the intermediate-result caching, we can reduce further computation time around 25% (compared with LGP with fitness retrieval) when cache size of 100 is provided.

**Acknowledgements.** The authors would like to thank the Hori Information Science Promotion Foundation for a research grant.

## References

1. Ando, J., Nagao, T.: Fast tree-structural image processing using GPU. In: Proc. IWAIT 2007, Bangkok, Thailand, pp. 423–428 (2007)
2. Aoki, S., Nagao, T.: Automatic construction of tree-structural image transformations using genetic programming. In: Proc. ICAIP 1999, Venezia, Italy, pp. 136–141 (1999)

3. Brameier, M., Banzhaf, W.: A comparison of linear genetic programming and neural networks in medical data mining. IEEE Transactions on Evolutionary Computation 5(1), 17–26 (2001)
4. Brameier, M., Banzhaf, W.: Linear genetic programming. Springer, Heidelberg (2007)
5. De Jong, K.A.: Evolutionary computation: a unified approach. MIT Press, Cambridge (2006)
6. Fillon, C., Bartoli, A.: A divide & conquer strategy for improving efficiency and probability of success in genetic programming. In: Collet, P., Tomassini, M., Ebner, M., Gustafson, S., Ekárt, A. (eds.) EuroGP 2006. LNCS, vol. 3905, pp. 13–23. Springer, Heidelberg (2006)
7. Gathercole, C., Ross, P.: Dynamic training subset selection for supervised learning in genetic programming. In: Davidor, Y., Schwefel, H.P., Männer, R. (eds.) PPSN 1994. LNCS, vol. 866, pp. 313–321. Springer, Heidelberg (1994)
8. Gonzalez, R.C., Woods, R.E.: Digital Image Processing, 2nd edn. Addison Wesley, Reading (2002)
9. Lasarczyk, C., Dittrich, P., Banzhaf, W.: Dynamic subset selection based on a fitness case topology. Evolutionary Computation 12(2), 223–242 (2004)
10. Nagao, T., Masunaga, S.: Automatic construction of image transformation processes using genetic algorithm. In: Proc. ICIP 1996, Lausanne, Switzerland, vol. 3, pp. 731–734 (1996)
11. Krawiec, K., Bhanu, B.: Visual learning by coevolutionary feature synthesis. IEEE Transactions on Systems, Man, and Cybernetics-Part B 35(3), 409–425 (2005)
12. Krawiec, K., Bhanu, B.: Visual learning by evolutionary and coevolutionary feature synthesis. IEEE Transactions on Evolutionary Computation 11(5), 635–650 (2007)
13. Krawiec, K.: Generative learning of visual concepts using multiobjective genetic programming. Pattern Recognition Letters 28(16), 2385–2400 (2007)
14. Levenick, J.R.: Inserting introns improves genetic algorithm success rate: taking a cue from biology. In: Belew, R.K., Booker, L.B. (eds.) Proc. International Conference on Genetic Algorithm (ICGA 1991), pp. 123–127. Morgan Kaufmann, San Francisco (1991)
15. Niehaus, J., Igel, C., Banzhaf, W.: Reducing the number of fitness evaluations in graph genetic programming using a canonical graph indexed database. Evolutionary Computation 15(2), 199–221 (2007)
16. Poli, R., Langdon, W.B.: Running genetic programming backward. In: Yu, T., Riolo, R., Worzel, B. (eds.) Genetic Programming Theory and Practice III, pp. 125–140. Springer, Heidelberg (2005)
17. Poli, R., Langdon, W.B., McPhee, N.F.: A Filed Guide to Genetic Programming (2008), http://lulu.com, http://www.gp-field-guide.org.uk
18. Roberts, M., Claridge, E.: Co-operative coevolution of image feature construction and object detection. In: Yao, X., Burke, E.K., Lozano, J.A., Smith, J., Merelo-Guervós, J.J., Bullinaria, J.A., Rowe, J.E., Tiňo, P., Kabán, A., Schwefel, H.-P. (eds.) PPSN 2004. LNCS, vol. 3242, pp. 902–911. Springer, Heidelberg (2004)

19. Roberts, M., Claridge, E.: A multi-stage approach to cooperatively coevolving feature construction and object detection. In: Rothlauf, F., Branke, J., Cagnoni, S., Corne, D.W., Drechsler, R., Jin, Y., Machado, P., Marchiori, E., Romero, J., Smith, G.D., Squillero, G. (eds.) EvoWorkshops 2005. LNCS, vol. 3449, pp. 396–406. Springer, Heidelberg (2005)

20. Shirakawa, S., Nagao, T.: Genetic image network (GIN): automatically construction of image processing. In: Proc. IWAIT 2007, Bangkok, Thailand, pp. 643–648 (2007)

21. Spears, W.M., De Jong, K.A.: On the virtues of parameterized uniform crossover. In: Belew, R.K., Booker, L.B. (eds.) Proc. International Conference on Genetic Algorithm (ICGA 1991), pp. 230–236. Morgan Kaufmann, San Francisco (1991)

22. Teller, A., Veloso, M.: A controlled experiment: evolution for learning difficult image classification. In: Pinto-Ferreira, C., Mamede, N.J. (eds.) EPIA 1995. LNCS, vol. 990, pp. 165–185. Springer, Heidelberg (1995)

23. Teller, A., Veloso, M.: PADO: a new learning architecture for object recognition. In: Ikeuchi, K., Veloso, M. (eds.) Symbolic Visual Learning, pp. 77–112. Oxford Univ. Press, Oxford (1997)

24. Theodoridis, S., Koutroumbas, K.: Pattern Recognition, 3rd edn. Academic Press, London (2006)

25. Watchareeruetai, U., Takeuchi, Y., Matsumoto, T., Kudo, H., Ohnishi, N.: Computer vision based methods for detecting weeds in lawns. Machine Vision and Applications 17(5), 287–296 (2006)

26. Watchareeruetai, U., Takeuchi, Y., Matsumoto, T., Kudo, H., Ohnishi, N.: Transformation of redundant representations of linear genetic programming into canonical forms for efficient extraction of image features. In: Proc. IEEE Congress on Evolutionary Computation (CEC 2008), Hong Kong, China, pp. 1996–2003 (2008)

27. Wong, P., Zhang, M.: SCHEME: caching subtrees in genetic programming. In: Proc. IEEE Congress on Evolutionary Computation (CEC 2008), Hong Kong, China, pp. 2683–2690 (2008)

28. Zhang, B.T., Cho, D.Y.: Genetic programming with active data selection. In: McKay, B., Yao, X., Newton, C.S., Kim, J.-H., Furuhashi, T. (eds.) SEAL 1998. LNCS, vol. 1585, pp. 146–153. Springer, Heidelberg (1999)

29. Zhang, M., Ciesielski, V., Andreae, P.: A domain-independent window approach to multiclass object detection using genetic programming. EURASIP Journal on Applied Signal Processing 8, 841–859 (2003)

# Mining Network Traffic Data for Attacks through MOVICAB-IDS

Álvaro Herrero and Emilio Corchado

**Abstract.** This study describes an Intrusion Detection System (IDS) called MOVICAB-IDS (MObile VIsualization Connectionist Agent-Based IDS). This system is based on a dynamic multiagent architecture combining case-base reasoning and an unsupervised neural projection model to visualize and analyze the flow of network traffic data. The formulation of the underlying Intrusion Detection framework is presented in advance. The described IDS enables the most interesting projections of a massive traffic data set to be extracted and depicted through a functional and mobile visualization interface. By its advanced visualization facilities, MOVICAB-IDS allows providing an overview of the network traffic as well as identifying anomalous situations tackled by computer networks, responding to the challenges presented by traffic volume and diversity. To show the performance of the described IDS, it has been tested in different domains containing several interesting attacks and anomalous situations.

**Keywords:** Data Mining, Artificial Neural Networks, Unsupervised Learning, Computer Network Security, Intrusion Detection.

## 1 Introduction

Artificial Neural Networks (ANNs) are connectionist models that can be applied to different problems depending on the neural architecture to be used. They have been probed to successfully perform pattern recognition, information compression, dimensionality reduction, clustering, classification, data visualization, etc. As most of these problems belong to the data mining field, ANNs can then be seen as data mining tools.

ANNs are a set of methods intended to emulate the biological information processors, as the following features of these processors are desired:

• Robustness and fault-tolerance: we (human beings) have our largest number of neurons early on in life. In spite of daily losing many thousands of neurons, we

Álvaro Herrero and Emilio Corchado
Department of Civil Engineering, University of Burgos
C/ Francisco de Vitoria s/n, 09006, Burgos, Spain
Tel.: +34 947 9513; Fax: +34 947 9395
e-mail: {ahcosio,escorchado}@ubu.es

A. Abraham et al. (Eds.): Foundations of Comput. Intel. Vol. 4, SCI 204, pp. 377–394.
springerlink.com &copy; Springer-Verlag Berlin Heidelberg 2009

continue to function for many years without an associated deterioration in our
capabilities.

- Flexibility: we do not require to be reprogrammed when facing a new environment because we adapt to it (learning).
- Fuzzyness: we can handle fuzzy, probabilistic, noisy and inconsistent data.
- Efficiency: the above listed features are supported by a highly parallel, small, compact and little-power dissipating mechanism.

As with other machine learning paradigms, the interesting facet of ANNs leaning is not just that the input patterns may be learned/classified/identified precisely but that this learning has the capacity to generalise. That is, while learning will take place on a set of training patterns, an important property of the learning is that the network can generalise its results on a set of test patterns which it has not seen during learning. As an important consequence, there is a danger of overlearning a set of training patterns so that new patterns (not included in the training set) are not properly classified.

Intrusion Detection Systems (IDSs) have become a required element in addition to the computer security infrastructure of most organizations. In the context of computer networks, an IDS can roughly be defined as a tool designed to detect suspicious patterns that may be related to a network or system attack. Intrusion Detection (ID) is then a field focused on the identification of attempted or ongoing attacks in a computer system (Host IDS - HIDS) or network (Network IDS - NIDS). The accurate detection of computer and network system intrusions in real-time has always been an interesting and challenging problem for system administrators and information security researchers. It could mainly be attributed to the dynamic nature of systems and networks, the creativity of attackers, the wide range of computer hardware and operating systems and so on. Such complexity rises when dealing with distributed network-based systems and insecure networks such as the Internet.

Since initial works on the field, ID has been approached from several different standpoints; many different forms of Artificial Intelligence (AI) have been applied mainly to classify traffic data as "normal" or "attack", for example, Genetic Programming [1], [2], Machine Learning [3], [4], [5], [6], [7], [8], Expert Systems [9], [10], [11], [12] or ANNs [13], [14], [15], [16], [17], [18] among others, statistical [19] and signature verification [20] techniques. Most of these systems can generate different alarms when anomalous situations occur, but they can not provide a general overview of what is happening inside a network.

From an opposite point of view, a great variety of visualization-based approaches have been proposed as well [21], [22], [23], [24], [25]. In this case, the ID issue is enabled by providing a visual depiction of the network topology or the traffic. These visualization tools rely on the human ability to recognize different features and detect anomalies through graphical devices [26].

This work describes MOVICAB-IDS, an NIDS characterized by the use of several AI techniques, such as Artificial Neural Networks (ANNs), Multiagent Systems (MAS) [27] and Case-Based Reasoning (CBR) [28]. MOBICAB-IDS is based on a general framework for NIDS (see Section 3).

Embedding ANNs in the deliberative agents of a dynamic Multiagent System (MAS) let us take advantage of some of the properties of the neural paradigm (generalization and pattern recognition) and agents (reactivity, proactivity and sociability) making the ID task possible. Additionally, MOVICAB-IDS provides the network administrator with a mobile visualization.

This research proposes MOVICAB-IDS as a complementary tool to other network security ones. That is, MOVICAB-IDS can work in unison with other defence mechanisms (even if they are IDSs), to provide an intuitive depiction of both normal and anomalous traffic.

In contrast to other security tools, IDSs need to be monitored to realize most of its benefits [29]. So, an IDS would be useless if nobody looks at its outputs. In keeping with this idea, MOVICAB-IDS goes one step further than previously mentioned visualization tools, combining all the features extracted from the packet headers to depict each simple packet. Thus, it provides the network administrator with a snapshot of network traffic, protocol interactions and traffic volume in order to identify anomalous situations. To do so, an unsupervised neural projection architecture (see next section) is applied.

# 2 Unsupervised Projection Models

The identification of patterns that exist across dimensional boundaries in high dimensional data sets is a challenging task. Such patterns may become visible if changes are made to the spatial coordinates. Projection methods project high-dimensional data points onto a lower dimensional space in order to identify "interesting" patterns. In some cases, these interesting patterns are selected in terms of any specific index or projection, such as the data variance –as is the case of Principal Component Analysis (PCA) [30], [31], [32]- or higher order statistics (the skew or kurtosis index in the case of Exploratory Projection Pursuit (EPP) [33]). Having identified the interesting projections, the data is then projected onto a lower dimensional subspace plotted in two or three dimensions, which makes it possible to examine its structure at a glance. The remaining dimensions are discarded as they mainly relate to a very small percentage of the information or the data set structure. In that way, the patterns identified through a multivariable data set may be visually analysed with greater ease.

## 2.1 Statistical Projection Models

PCA is a standard statistical technique for compressing data; it can be shown to give the best linear compression of the data in terms of least mean square error. This statistical technique describes the variation in a set of multivariate data in terms of a set of uncorrelated variables each of which is a linear combination of the original variables. Its goal is to derive new variables, in decreasing order of importance, which are linear combinations of the original variables and are uncorrelated with each other. It should be noted that even if we are able to characterize the data with a few variables, it does not follow that an interpretation will ensue.

EPP is a more recent statistical method aimed at solving the difficult problem of identifying structure in high dimensional data. It does this by projecting the data onto a low dimensional subspace in which we search for its structure by eye.

However not all projections will reveal the data's structure equally well. It therefore defines an index that measures how "interesting" a given projection is, and then represents the data in terms of projections that maximise that index.

Then, the first step for EPP is to define which indices represent interesting directions. Concerning projections, "interestingness" is usually defined with respect to the fact that most projections of high-dimensional data give almost Gaussian distributions [34]. If we wish to identify "interesting" features in data, we should therefore look for those directions onto which the data-projections are as far from the Gaussian as possible.

Two simple measures of deviation from a Gaussian distribution are based on the higher order moments of the distribution. Skewness is based on the normalised third moment and measures the deviation of the distribution from bilateral symmetry. Kurtosis is based on the normalised fourth moment and measures the heaviness of the tails of a distribution. A bimodal distribution will often have a negative kurtosis and therefore negative kurtosis can signal that a particular distribution shows evidence of clustering.

## 2.2 *Cooperative Maximum Likelihood Hebbian Learning*

Cooperative Maximum Likelihood Hebbian Learning (CMLHL) is an EPP model based on Maximum Likelihood Hebbian Learning (MLHL) [35], [36] and including lateral connections [37], [38] which have been derived from the Rectified Gaussian Distribution [39]. The Rectified Gaussian Distribution is a modification of the standard Gaussian distribution in which the variables are constrained to be non-negative, enabling the use of non-convex energy functions. In a more precise way, CMLHL includes lateral connections based on the mode of the cooperative distribution that is closely spaced along a non-linear continuous manifold. By including these lateral connections, the resultant net can find the independent factors of a data set but does so in a way that captures some type of global ordering in the data set.

Considering an N-dimensional input vector ($x$), an M-dimensional output vector ($y$) and with $W_{ij}$ being the weight (linking input $j$ to output $i$), CMLHL can be expressed as:

Feed-forward step:

$$y_i = \sum_{j=1}^{N} W_{ij} x_j, \forall i .$$
(1)

Lateral activation passing:

$$y_i(t+1) = [y_i(t) + \tau(b - Ay)]^+ .$$
(2)

Feedback step:

$$e_j = x_j - \sum_{i=1}^{M} W_{ij} y_i, \forall j .$$
(3)

Weight change:

$$\Delta W_{ij} = \eta . y_i . sign(e_j) | e_j |^{p-1} . \tag{4}$$

Where: $\eta$ is the learning rate, $\tau$ is the "strength" of the lateral connections, $b$ the bias parameter and $p$ a parameter related to the energy function [35], [36], [37].

$A$ is a symmetric matrix used to modify the response to the data whose effect is based on the relation between the distances among the output neurons. It is based on the Cooperative Distribution, but to speed learning up, it can be simplified to:

$$A(i, j) = \delta_{ij} - \cos(2\pi(i - j)/M). \tag{5}$$

Where $\delta_{ij}$ is the Kronecker delta.

## 2.3 Self-Organizing Map

The Self-Organizing Map (SOM) [40] was developed as a visualization tool for representing high dimensional data on a low dimensional display using unsupervised learning but. This model is not a projection architecture but a topology preserving mapping model instead. A SOM, composed of a discrete array of L nodes arranged on an N-dimensional lattice, maps these nodes into a D-dimensional data space while preserving their ordering. The dimensionality of the lattice (N) is normally smaller than that of the data, in order to perform the dimensionality reduction. The SOM can be viewed as a non-linear extension of PCA, where the map manifold is a globally non-linear representation of the training data [41].

Typically, the array of nodes is one or two-dimensional, with all nodes connected to the N inputs by an N-dimensional weight vector. The self-organization process is commonly implemented as an iterative on-line algorithm, although a batch version also exists. An input vector is presented to the network and a winning node, whose weight vector is the closest (in terms of Euclidean distance) to the input, is chosen.

So the SOM is a vector quantiser, and data vectors are quantised to the reference vector in the map that is closest to the input vector. The weights of the winning node and the nodes close to it are then updated to move closer to the input vector. When this algorithm is iterated sufficiently, the map self-organizes to produce a topology-preserving mapping of the lattice of weight vectors to the input space based on the statistics of the training data. This neural model is applied here for comparative purposes as it is one of the most widely used unsupervised neural models for visualizing structure in high-dimensional data sets.

## 3   A General Framework for NIDS

To detect anomalous situations in a computer network, a 5-step framework is proposed:

- 1st step.- Network Traffic Capture.
- 2nd step.- Data Pre-processing.
- 3rd step.- Segmentation.

- 4th step.- Data Analysis.
- 5th step.- Visualization.

The ID process starts when packets travelling over the network are captured by using sniffing techniques. That is, one of the network interfaces is set up in promiscuous mode, gathering all the information travelling along the network. Every single packet is captured and the information contained in it is stored.

The captured data is selected and pre-processed. Traffic is selected by taking into account the protocol at transportation level and a set of features contained in the headers of the captured packets is selected from the raw network traffic.

IDSs have to deal with the practical problem of high volumes of quite diverse data. To deal with this problem, we propose to split the traffic into different groups, taking into account the protocol (either UDP, TCP, ICMP...) over IP. In the data pre-processing step, the system performs a data selection from the captured information. As a result, the selected data finally contains the following 5 variables extracted from the packet headers: source port (port number from where the source host sent the packet), destination port (destination host port number to which the packet is sent), size (total packet size in bytes), timestamp (the time when the packet was sent) and protocol (each packet is assigned the code of the protocol it belongs to).

The pre-processed data stream is divided into simple segments and accumulated ones, consisting of the addition of several consecutive simple segments. This allows the network administrator to perform a more local and detailed analysis on some suspicious situations, while preserving a general overview of the network traffic.

Once simple and accumulated segments have been built, a projection model is applied to analyse them. The data analysis task is based on the use of a dimensionality reduction technique to drive a compact two- or three-dimensional visualization of the 5-dimensional packet data. This technique must be able to provide a projection showing the internal structure of a data set.

Finally, the projection of each segment is presented to the network administrator for scrutiny and monitoring.

## 4 MOVICAB-IDS

In keeping with the above proposed framework, MOVICAB-IDS (MObile VIsualization Connectionist Agent-Based IDS) is an IDS for distributed computer networks, incorporating several AI techniques, such as ANNs, MAS and CBR. To process the continuous data flow coming from the network traffic, MOVICAB-IDS splits massive traffic data into segments and analyze them, thereby providing administrators with an intuitive snapshot to analyse the kinds of events taking place in a computer network.

MOVICAB-IDS contains different types of agents; some of them are reactive agents while others are deliberative (CBR-BDI) agents.

CBR-BDI agents [42], [43], [44] provide planning based on previous experiences as CBR systems use memories (past experiences) to solve new problems. The main idea when working with CBR systems is the concept of case, that can be seen as a past experience described by the 3-tuple <Problem, Solution, Results>. A case is composed of a problem description (initial state), the solution applied to solve the problem (the sequences of actions executed in order to achieve the objectives) and the result obtained after applying the solution (the final state and the evaluation of the plan executed). The following sections describe the six agents making up MOVICAB-IDS: Sniffer, Preprocessor, Analyzer, ConfigurationManager, Coordinator and Visualizer

## 4.1 Sniffer

This reactive agent is in charge of capturing traffic data. The continuous traffic flow is captured and split into segments (of preconfigured size) in order to send it through the network for further process. Finally, the readiness of the segmented data is communicated.

## 4.2 Preprocessor

After splitting traffic data, the generated segments must be preprocessed to apply subsequent analysis. Once the data has been preprocessed, an analysis for this new piece of data is requested.

## 4.3 Analyzer

This is a hybrid CBR-BDI agent. It has got embedded the CMLHL (See Section 2.2) model within the adaptation stage of its CBR system that helps to analyze preprocessed traffic data. This agent generates a solution (or achieve its goals) by retrieving a previously analyzed case and analyzing the new one using the CMLHL architecture. As it is known, the CBR life cycle consists of four steps [28]:

- **Retrieval:** when a new analysis is requested, the Analyzer agent finds the most similar case to the new one in the database.
- **Reuse:** the solution of the most similar case is reused. This solution consists of the values of the parameters used to train the CMLHL model. Then, a set of trainings (for the CMLHL model with a combination of parameter values varying in a specified range) is proposed by tacking into account the distance between the new case and the most similar one. That is, if they are very similar, a reduced set of trainings are going to be performed. On the contrary, if the most similar case is far away from the new one, a great number of trainings with very different parameter values are going to be generated.
- **Revision:** the CMLHL model is trained with the new dataset and the combination of parameter values generated in the reuse stage. When the new projections of the dataset are ready, they are shown to the network administrator through the Visualizer agent. One of these projections must be chosen as the best one (the one that provides the clearest snapshot of the traffic evolution).

- **Retention:** the Analyzer agent stores the new case containing the problem-descriptor and the solution (parameter values used to generate the chosen projection) in the case base for future reuse (See Table 1).

## 4.4  ConfigurationManager

The processes of data capture, split, preprocess and analysis depends on the values of several parameters, as for example: packets to capture, segment length, features to extract, etc. All this information is managed by the ConfigurationManager reactive agent, which is in charge of providing this information to the Sniffer, Preprocessor and Analyzer agents.

## 4.5  Coordinator

In order to improve the system efficiency and perform an almost real-time processing, the preprocessed data must be dynamically and optimally assigned. The Coordinator Agent is in charge of allocating the pending analyses to the available Analyzer Agents.

## 4.6  Visualizer

The analyzed data is presented to the network administrator by means of a functional and mobile visualization through this interface agent. To improve the accessibility of the system, the administrator may visualize the results on a mobile device enabling informed decisions to be taken anywhere and at any time. Depending on the platform where the information will be shown, the offered visualization facilities will be different.

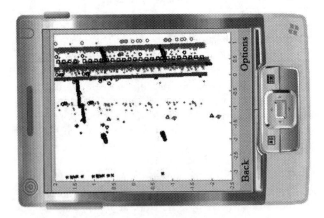

**Fig. 1** MOVICAB-IDS sample mobile visualization

MOVICAB-IDS is accessible from any wireless device, such as a PDA, palmtop, laptop or mobile phone to give more accessibility to network administrators, enabling permanent mobile visualization, monitoring and supervision of their networks.

Fig. 1 shows an example of the visualization provided by MOVICAB-IDS on a mobile device. An emulator was used to test the visualization on a mobile platform.

# 5 Attacks to Be Detected

Among all the implemented network protocols, there are several of them that can be considered quite more dangerous (in terms of the network security), such as the Simple Network Management Protocol (SNMP) [45], ICMP (Internet Control Message Protocol), TFTP (Trivial File Transfer Protocol) and so on. SNMP was identified as one of the top five most vulnerable services by CISCO [46], specially the two first versions of this protocol that are the most widely used at present time. An attack based on this protocol may severely compromise the security of the whole network [47]. Most of the security tools focus their attention on attacks coming from the internet but attacks are just as likely to come from inside the network as from the outside, however. Due to these reasons, the experimental setting of this work is focused on the identification of anomalous situations concerning SNMP.

SNMP was oriented to manage nodes in the Internet community [45]. That is, it is used to control routers, bridges and some other network elements, reading and writing a wide variety of information (such as operating system, version, routing tables, default TTL and so on) about these devices. All this information is stored in the Management Information Base (MIB), so it can be defined in broad terms as the database used by SNMP to store information about the elements that it controls. The IAB (Internet Activities Board) recommended that all IP and TCP implementations were network manageable [48]. The implementation of the MIB and at least one of the management protocols like SNMP is the consequence of this suggestion.

Three main anomalous situations related to the SNMP are distributed throughout the different datasets in this study, namely: scans, SNMP community searches and MIB (Management Information Base) information transfers.

## 5.1 Scans

A port scan may be defined as series of messages sent to different port numbers of a host to gain information on its activity status. These messages can be sent by an external agent attempting to access a host to find out more about the network services this host is providing. A port scan provides information on where to probe for weaknesses, for which reason scanning generally precedes any further intrusive activity [49]. On the contrary, in a network scan the same port is the target for a number of hosts (usually all the available hosts in a network IP range). A network scan is one of the most common techniques used to identify services that might then be accessed without permission [50].

In this experimental study, the data sets contain network scans aimed at port numbers 1434 (registered port assigned to Microsoft-SQL-Monitor, the target of the W32.SQLExp.Worm) and 65788 (as an example of dynamic or private port).

## 5.2  SNMP Community Search

The community string can be seen as the SNMP password for versions 1 and 2. An SNMP community search is characterized by the intruder sending SNMP queries to the same port number of different hosts trying to guess the SNMP community string. The community string for each one of these searches is selected according to a strategy (brute force, dictionary, etc.). Once the community string has been obtained, all the sensitive information stored in the MIB is available for the intruder.

## 5.3  MIB Information Transfer

This situation is a transfer of some (or all the) information contained in the SNMP MIB, generally through the *get* (or *get-bulk*) command. This kind of transfer is potentially quite a dangerous situation, however, the "normal" behaviour of a network may include queries to the MIB. This is a situation in which visualization based IDSs are quite useful; this situations are visualized in a "special" way but it is the network administrator responsibility to decide whether it is a "normal" MIB transfer (known by himself) or it is not.

## 6  MOVICAB-IDS Results

In this section, the experimental results of applying MOVICAB-IDS are presented by different snapshots. Each one of them depicts all the packets contained in the data set whose projection is shown. MOVICAB-IDS plots the packets in different colours and shapes taking into account the original protocol information, what leads to a more intuitive visualization for the administrator.

The traffic contained in the simple (S) and accumulated (A) segments whose results are shown in this section can be roughly described as:

- $S_1$: It only contains normal traffic. That is, no anomalous situations are included in this segment.
- $S_2$: It contains normal traffic and an MIB information transfer generated by the *get-bulk* SNMP command.
- $A_1$: The examples of anomalous situations that this segment contains are two network scans aimed at port numbers 1434 and 65788.
- $A_2$: In addition to the anomalous situations contained in $A_1$ (network scans aimed at port numbers 1434 and 65788) it contains normal traffic and SNMP community searches aimed at port numbers 161, 1161 y 2161 of all the machines in an IP address range. Three different community names were used for each one of these community searches.

Figs. 2, 3, 4 and 5 show some examples of how CMLHL performs when applied to simple and accumulated segments: Fig. 2 shows the projection of a simple segment ($S_1$) containing only normal traffic, with no anomalous situations. This is then, the way in which CMLHL depicts normal traffic, by means of packets evolving in parallel straight directions. Any sign of non-parallel evolution or high concentration of packets is viewed as an anomaly. It can be seen in Fig. 2 how all the packets (related to "normal" traffic) evolve in "normal" parallel directions over time.

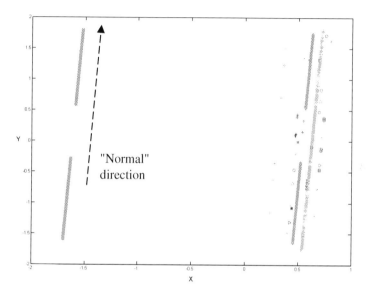

**Fig. 2** MOVICAB-IDS projection of $S_1$

On the other hand, Fig. 3 (projection of data set $S_2$) shows how the system identifies an anomalous situation related to an MIB information transfer. This situation (Groups 1 and 2) is identified as anomalous due to its high concentration of packets (in comparison to the "normal" traffic) and its evolution does not fit straight lines as the normal traffic does.

The following figure (Fig. 4) shows the projection of an 18 minute-long accumulated segment ($A_1$). As can be seen, the network scans contained in the data sets are identified (Groups 1 and 2).

Finally, the projection of $A_2$ is depicted in Fig. 5. In this projection, network scans are labeled as Group 1 and 2, while community searches are labeled as Groups 3, 4, 5 and 6.

All these figures (Fig. 2-5) show in a clear way how a network administrator is able to identify some attacks when the packet traffic evolves in non-parallel directions to the "normal" one and also when the density of packets is much higher than normal situations.

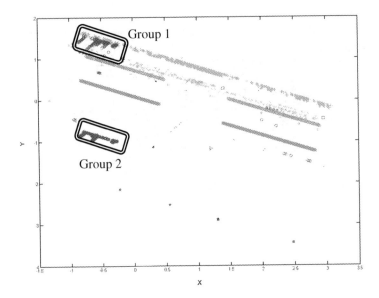

**Fig. 3** MOVICAB-IDS projection of $S_2$

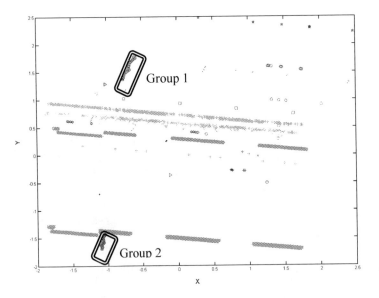

**Fig. 4** MOVICAB-IDS projection of $A_1$

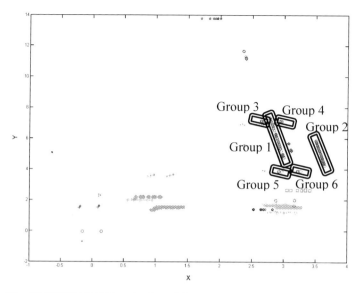

**Fig. 5** MOVICAB-IDS projection of $A_2$

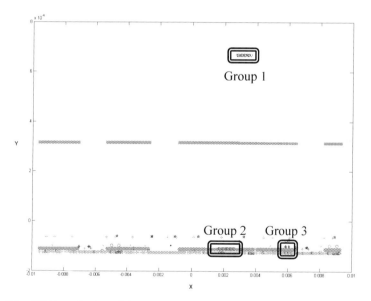

**Fig. 6** PCA projection of $A_2$

## 7 Comparative Study

To contrast the validity of the model, the CMLHL outcome has been compared with that obtained by two other well-known unsupervised models such as PCA and SOM. For the sake of simplicity, only projections concerning one dataset ($A_2$) are provided in this section. For this dataset, only the best results (from a projection point of view) obtained after tuning the models are shown in Figs. 6 and 7.

PCA was applied to the $A_2$ segment (Fig. 6). This technique, failed to detect the anomalous situations (network scans and SNMP community search), although the two principal components amount to 99.9% of the data's variance. Both the network scans (Groups 1 and 2) and the SNMP community searches (Group 3) contained in $A_2$ are not identified as anomalous traffic because in this projection all the packets evolve in parallel lines.

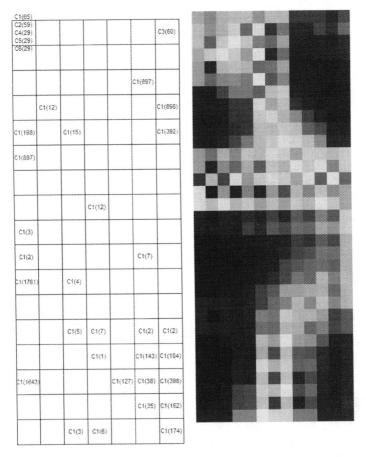

a) SOM labelled map.                              b) Associated U-matrix.

**Fig. 7** SOM results for $A_2$

Finally, the SOM mapping of the $A_2$ segment is depicted in Fig. 7. For visualization purposes, all the packets in the analysed segment were labelled according to the following classes: C1 for normal traffic, C2 for the network scan aimed at the port number 1434, C3 for the network scan aimed at the port number 65788, C4 for the community search aimed at the port number 161, C5 for the community search aimed at the port number 1161, and C6 for the community search aimed at the port number 2161. Fig 7.a depicts the labelled generated map in which the classes identified by each neuron are shown. The number of instances (packets) belonging to each one of the classes identified by the neurons are shown in parentheses. It can be clearly seen that the SOM is able to cluster the presented segment. The neuron in the upper-right corner of the rectangular lattice gathers all the packets labelled as C3, that is, all the packets belonging to the network scan aimed at port number 65788. No other class of traffic is identified by the neighbouring neurons. The neuron in the upper-left corner of the lattice gathers all the packets associated to the other anomalous situations (C2, C4, C5 and C6). Additionally, 65 packets associated to normal traffic were also identified by this neuron. The rest of the normal traffic is identified by some other different neurons. Figure 7.b shows the associated U-matrix. The quality measures associated to this SOM results are: quantization error $= 0.122$ and topographic error $= 0.11$.

# 8  Conclusions

This work describes a neural projection NIDS which offers network administrators greater accessibility using any mobile device due to its visualization features. The system can deal with a high-volume network traffic data stream by pre-processing and splitting it into simple and accumulated segments. The presented IDS is capable of identifying anomalous situations by means of temporal visualization of the system response. Using time information allows us to identify some anomalous situations that would be unidentifiable otherwise.

This research constitutes one of the first attempts to identify anomalous situations working at packet-level. That is, the analysis does not rely on summarized information (such as TCP connections). On the contrary, MOVICAB-IDS analyses the data extracted from each single packet header. Thus, it is not able to identify attacks concerning the packet payload.

Usually, time information is not used in ANN-based IDSs. On the contrary, it can be employed as one of the inputs to the neural model embedded in MOVICAB-IDS. In the case of this model, time information provides an idea of how the traffic data evolves. It helps to identify anomalous situations by taking into account such aspects as high packet density and temporal evolution in non parallel directions.

Some existing IDSs need a "clean" (free of attacks) training dataset. This is not the case of MOVICAB-IDS, which can be trained with a dataset containing known or even new anomalous situations, due to its generalization capabilities.

Finally, we propose the use of MOVICAB-IDS in combination with other security tools (specially other IDS) to overcome their limitations (e.g: identification of 0-day attacks).

Further work will focus on the upgrading of the described MAS.

**Acknowledgments.** This research has been partially supported by the project BU006A08 of the JCyL.

# References

1. Case, J., Fedor, M.S., Schoffstall, M.L., Davin, C.: Simple Network Management Protocol (SNMP). RFC-1157. (1990)
2. Lu, W., Traore, I.: Detecting New Forms of Network Intrusion Using Genetic Programming. Computational Intelligence 20(3), 475–494 (2004)
3. Julisch, K.: Chapter 1 - Data Mining for Intrusion Detection: A Critical Review. In: Applications of Data Mining in Computer Security. Advances in Information Security. Springer, Heidelberg (2002)
4. Lee, W., Stolfo, S.J.: A framework for constructing features and models for intrusion detection systems. In: ACM Transactions on Information and System Security (TIS-SEC), vol. 3(4), pp. 227–261. ACM Press, New York (2000)
5. Liao, Y.H., Vemuri, V.R.: Use of K-Nearest Neighbor Classifier for Intrusion Detection. Computers & Security 21(5), 439–448 (2002)
6. Lee, W., Stolfo, S.J., Mok, K.W.: Adaptive Intrusion Detection: A Data Mining Approach. Artificial Intelligence Review 14(6), 533–567 (2000)
7. Giacinto, G., Roli, F., Didaci, L.: Fusion of Multiple Classifiers for Intrusion Detection in Computer Networks. Pattern Recognition Letters 24(12), 1795–1803 (2003)
8. Chebrolu, S., Abraham, A., Thomas, J.P.: Feature Deduction and Ensemble Design of Intrusion Detection Systems. Computers & Security 24(4), 295–307 (2005)
9. Denning, D.E.: An Intrusion-Detection Model. IEEE Transactions on Software Engineering 13(2), 222–232 (1987)
10. Lunt, T.F.: IDES: An Intelligent System for Detecting Intruders. In: Proceedings of the Symposium: Computer Security, Threat and Countermeasures (1990)
11. Vaccaro, H.S., Liepins, G.E.: Detection of Anomalous Computer Session Activity. In: Liepins, G.E. (ed.) Proceedings of the 1989 IEEE Symposium on Security and Privacy, pp. 280–289 (1989)
12. Sebring, M., Shellhouse, E., Hanna, M., Whitehurst, R.: Expert Systems in Intrusion Detection: A Case Study. In: Proceedings of the 11th National Computer Security Conference, pp. 74–81 (1988)
13. Zanero, S., Savaresi, S.: Unsupervised Learning Techniques for an Intrusion Detection System. In: Proc. of the ACM Symposium on Applied Computing, pp. 412–419 (2004)
14. Corchado, E., Herrero, A., Sáiz, J.M.: Detecting Compounded Anomalous SNMP Situations Using Cooperative Unsupervised Pattern Recognition. In: Duch, W., Kacprzyk, J., Oja, E., Zadro ny, S. (eds.) ICANN 2005. LNCS, vol. 3697, pp. 905–910. Springer, Heidelberg (2005)
15. Herrero, A., Corchado, E., Sáiz, J.M.: An Unsupervised Cooperative Pattern Recognition Model to Identify Anomalous Massive SNMP Data Sending. In: Wang, L., Chen, K., S. Ong, Y. (eds.) ICNC 2005. LNCS, vol. 3610, pp. 778–782. Springer, Heidelberg (2005)
16. Sarasamma, S.T., Zhu, Q.M.A., Huff, J.: Hierarchical Kohonen Net for Anomaly Detection in Network Security. IEEE Transactions on Systems Man and Cybernetics, Part B 35(2), 302–312 (2005)

17. Mukkamala, S., Sung, A.H.: Feature Selection for Intrusion Detection Using Neural Networks and Support Vector Machines. Transportation Security and Infrastructure Protection, 33–39 (2003)
18. Zhang, C.L., Jiang, J., Kamel, M.: Intrusion Detection Using Hierarchical Neural Networks. Pattern Recognition Letters 26(6), 779–791 (2005)
19. Marchette, D.J.: Computer Intrusion Detection and Network Monitoring: A Statistical Viewpoint. Information Science and Statistics. Springer, New York (2001)
20. Roesch, M.: Snort–Lightweight Intrusion Detection for Networks. In: Proc. of the 13th Systems Administration Conf (LISA 1999), pp. 229–238 (1999)
21. Muelder, C., Ma, K.L., Bartoletti, T.: Interactive Visualization for Network and Port Scan Detection. In: Zamboni, D., Kruegel, C. (eds.) RAID 2005. LNCS, vol. 3858, pp. 265–283. Springer, Heidelberg (2006)
22. Nyarko, K., Capers, T., Scott, C., Ladeji-Osias, K.A.: Network Intrusion Visualization with NIVA, an Intrusion Detection Visual Analyzer with Haptic Integration. In: Capers, T. (ed.) Proceedings of the 10th Symposium on Haptic Interfaces for Virtual Environment and Teleoperator Systems, 2002 (HAPTICS 2002), pp. 277–284 (2002)
23. Labib, K., Vemuri, V.R.: An Application of Principal Component Analysis to the Detection and Visualization of Computer Network Attacks. Annals of Telecommunications 61(1-2), 218–234 (2006)
24. Becker, R.A., Eick, S.G., Wilks, A.R.: Visualizing Network Data. IEEE Transactions on Visualization and Computer Graphics 1(1), 16–28 (1995)
25. Ren, P., Gao, Y., Li, Z.C., Chen, Y., Watson, B.: IDGraphs: Intrusion Detection and Analysis Using Stream Compositing. IEEE Computer Graphics and Applications 26(2), 28–39 (2006)
26. Ahlberg, C., Shneiderman, B.: Visual Information Seeking: Tight Coupling of Dynamic Query Filters with Starfield Displays. In: Readings in information visualization: using vision to think, pp. 244–250. Morgan Kaufmann Publishers Inc., San Francisco (1999)
27. Wooldridge, M., Jennings, N.R.: Agent theories, architectures, and languages: A survey. Intelligent Agents (1995)
28. Aamodt, A., Plaza, E.: Case-Based Reasoning - Foundational Issues, Methodological Variations, and System Approaches. AI Communications 7(1), 39–59 (1994)
29. Chuvakin, A.: Monitoring IDS. Information Security Journal: A Global Perspective 12(6), 12–16 (2004)
30. Hotelling, H.: Analysis of a Complex of Statistical Variables Into Principal Components. Journal of Education Psychology 24, 417–444 (1933)
31. Pearson, K.: On Lines and Planes of Closest Fit to Systems of Points in Space. Philosophical Magazine 2(6), 559–572 (1901)
32. Oja, E.: Neural networks, principal components, and subspaces. Int. Journal of Neural Systems 1, 61–68 (1989)
33. Friedman, J.H., Tukey, J.W.: A Projection Pursuit Algorithm for Exploratory Data-Analysis. IEEE Transactions on Computers 23(9), 881–890 (1974)
34. Diaconis, P., Freedman, D.: Asymptotics of Graphical Projection Pursuit. The Annals of Statistics 12(3), 793–815 (1984)
35. Corchado, E., MacDonald, D., Fyfe, C.: Maximum and Minimum Likelihood Hebbian Learning for Exploratory Projection Pursuit. Data Mining and Knowledge Discovery 8(3), 203–225 (2004)

36. Fyfe, C., Corchado, E.: Maximum Likelihood Hebbian Rules. In: Proc. of the 10th European Symposium on Artificial Neural Networks (ESANN 2002), pp. 143–148 (2002)

37. Corchado, E., Fyfe, C.: Connectionist Techniques for the Identification and Suppression of Interfering Underlying Factors. Int. Journal of Pattern Recognition and Artificial Intelligence 17(8), 1447–1466 (2003)

38. Corchado, E., Han, Y., Fyfe, C.: Structuring Global Responses of Local Filters Using Lateral Connections. Journal of Experimental & Theoretical Artificial Intelligence 15(4), 473–487 (2003)

39. Seung, H.S., Socci, N.D., Lee, D.: The Rectified Gaussian Distribution. Advances in Neural Information Processing Systems 10, 350–356 (1998)

40. Kohonen, T.: The Self-Organizing Map. Proceedings of the IEEE 78(9), 1464–1480 (1990)

41. Ritter, H., Martinetz, T., Schulten, K.: Neural Computation and Self-Organizing Maps; An Introduction. Addison-Wesley Longman Publishing Co., Inc. (1992)

42. Carrascosa, C., Bajo, J., Julián, V., Corchado, J.M., Botti, V.: Hybrid Multi-agent Architecture as a Real-Time Problem-Solving Model. Expert Systems with Applications: An International Journal 34(1), 2–17 (2008)

43. Corchado, J.M., Laza, R.: Constructing Deliberative Agents with Case-Based Reasoning Technology. International Journal of Intelligent Systems 18(12), 1227–1241 (2003)

44. Pellicer, M.A., Corchado, J.M.: Development of CBR-BDI Agents. International Journal of Computer Science and Applications 2(1), 25–32 (2005)

45. Case, J., Fedor, M.S., Schoffstall, M.L., Davin, C.: Simple Network Management Protocol (SNMP). RFC-1157 (1990)

46. Cisco Secure Consulting. Vulnerability Statistics Report (2000)

47. Myerson, J.M.: Identifying Enterprise Network Vulnerabilities. Int. Journal of Network Management 12(3), 135–144 (2002)

48. Postel, J.: IAB Official Protocol Standards. RFC-1100 (1989)

49. Stephen, L.: The Spinning Cube of Potential Doom. Commun. ACM 47(6), 25–26 (2004)

50. Kulsoom, A., Lee, C., Conti, G., Copeland, J.A.: Visualizing Network Data for Intrusion Detection. In: Proc. of the Sixth Annual IEEE Information Assurance Workshop - Systems, Man and Cybernetics (SMC), 2005, pp. 100–108 (2005)

# Author Index